Discrete-time signal processing

Discrete-time signal processing

An introduction

A. W. M. van den Enden

N. A. M. Verhoeckx

Philips Research Laboratories,
Eindhoven, The Netherlands

Translated by
D. A. E. Roberts, MITI

Prentice Hall

New York   London   Toronto   Sydney   Tokyo

Part I of this book is a translation of the Dutch book
Digitale Signaalbewerking (ISBN 90 6674 722 6)
which was published in 1987 by Delta Press BV,
Overberg (gem. Amerongen), The Netherlands.

First published in Great Britain in 1989 by
Prentice Hall International (UK) Ltd,
66 Wood Lane End, Hemel Hempstead,
Hertfordshire, HP2 4RG
A division of
Simon & Schuster International Group

Printed and bound in Great Britain at the
University Press, Cambridge

Library of Congress Cataloging-in-Publication Data

Enden, A. W. M. van den.
Discrete-time signal processing: an introduction/
A. W. M. van den Enden and N. A. M. Verhoeckx.
p.   cm.
Bibliography: p.
Includes index.
ISBN 0-13-216763-8 : $60.00
1. Signal processing—Digital techniques.   2. Discrete-time
systems. I. Verhoeckx, N. A. M. II. Title.
TK5102.5.E53   1989
621.38′043–dc19   89-3480

British Library Cataloguing in Publication Data

Enden, A. W. M. van den
Discrete-time signal processing.
1. Discrete time systems. Signals. Processing
I. Title   II. Verhoeckx, N. A. M.
621.38′ 043
ISBN 0-13-216763-8
ISBN 0-13-216755-7 pbk

---

1 2 3 4 5 93 92 91 90 89

Contents

PART II

11 *An integrated switched-capacitor filter for viewdata*

12 *Digital signal processing in television receivers*

Foreword

The processing of signals is an activity that has long formed the basis of the work of many an electronic engineer. Intuitively perhaps we therefore tend to assume that the theoretical basis of this activity is a solid edifice. And indeed for the processing of signals that vary continuously with time this is so. In the mid-sixties, however, when the digital computer had reached such a state of technical competence that the discrete-time processing of signals became a practical possibility – as digital processing – cracks in the building began to appear. Although it was well known that digitization of signals was possible without the loss of relevant information, no adequate description of the process now known as 'digital signal processing' was as yet available. The earliest attempts to fill this gap, often by rather contrived extensions of the existing theory, were not found to be very effective, since they did not take full account of the essential differences that distinguish digital signals from the analog signals then normally in use: a discrete character in time and a discrete character in amplitude.

The late sixties and the seventies saw the growth of a very lively activity centred on the development of an appropriate system of methods and theories for the analysis and description of various forms of discrete-time signal processing. Some of the useful contributions to the work came from Philips Research Laboratories. As an active participant in that process I am delighted to see how quickly these new theories have been accepted by the technical community. This is of course mainly due to the revolutionary developments in IC technology that have sprung up in the last ten years or so. Signal-processing methods that at first were only suitable for use on large computers – and *not* in real time then, either – are now performed by a chip no bigger than a fingernail so rapidly that sometimes more than one signal can be processed at the same time.

At first discrete-time signal processing was limited to a few specialized fields like data transmission. However, with the invention of digital audio systems – the compact disc is the most obvious and widely encountered example – and of digital video a little later, the new discipline spread its wings. This created a substantial problem, however, since almost every system designer now had to have some proficiency in the underlying theory.

The need for a training course on the subject was keenly felt. The existing system designers needed 're-educating', and the newcomers needed help too, since most of the educational institutions (the universities and polytechnics) found it hard to keep in step with the rapid evolution in the field of discrete-time signal processing as they set up their curriculums. Based on the expertise existing at the research laboratories, a course was defined that provided the essentials of the theory and at the same time imparted a sound appreciation of the problems encountered in practical application. This course has elicited a very encouraging response, and is found to meet a real need. Many Philips staff have since followed the course, and it has enabled them to formulate and analyze their work in modern signal processing.

In the last ten years a wide variety of textbooks on the subject have appeared, some dealing with the theory, some concentrating on application. I am nevertheless very glad that the originators of the Philips course have been persuaded to publish their lecture notes in book form. I am convinced that it will be of great help both in introducing the novice to this field and in providing the existing system designers with the background for thorough design and analysis of their systems.

Having spent many hours with the authors during the creation of the course, in lively discussion of both the technical and the educational aspects of the material treated in this book, I know how thoroughly it has been put together, with emphasis on detail yet with practical relevance in mind. Much effort has been spent in incorporating the feedback that came from teaching the course to hundreds of systems designers working with a wide variety of applications. The addition of a number of 'reprints' from Philips Technical Review elucidating some of the specific applications will also prove very worthwhile. I should like to compliment the authors on the result of their efforts, which I hope will enable many more to master the fundamental principles of this fascinating subject and so to promulgate the use of discrete-time signal processing in many and varied fields of application.

Eindhoven, 1989   T. A. C. M. Claasen
                                                         Deputy Director,
                                              Philips Research Laboratories

Preface

While we were writing this book, the person we tried to keep in mind was the undergraduate student. We wanted to provide him (or her) with an accessible and thorough explanation of the fundamentals of a relatively new part of the mental equipment of today's student of electronic engineering: *discrete-time signal processing*. Once this was considered an advanced professional technology, yet now we find it used in the home – in the specialized form of *digital signal processing* – in the Compact Disc. Since complex integrated circuits ('chips') are now within the reach of all, discrete-time signal processing has become a serious candidate for a multitude of applications. This means that many designers of electronic circuits and systems are suddenly confronted with new prospects – not just digital versions of existing circuits, but other discrete-time alternatives as well, such as switched-capacitor filters and charge-coupled devices. And since today's students will have to tackle similar questions tomorrow they must be properly prepared.

Although there are a great many parallels between the old familiar analog world and the new discrete-time approach, there are also significant differences; without the necessary new theoretical resources, the detailed consequences of working in discrete time or with discrete amplitudes are not easy to assess. This explains the torrent of courses, publications and books that attempt to satisfy this need.

At Philips this situation led the Internal Training Department some years ago to ask the authors of this book to set up a course on 'discrete-time signal processing'. This course was originally intended for undergraduates and those who had qualified at institutes of technology, but was soon also found to be extremely useful to others such as practicing engineers and postgraduates. The lecture notes also managed to find their way to other interested readers, who included a large number of lecturers at institutes of technology. The reactions were generally positive to very positive, and eventually encouraged us to rewrite our notes as the first part (Chapters 1 to 10) of the book you are now reading. It is intended for the same target group as the original course. We therefore start with the premise that the reader will have some familiarity with continuous-time signals and systems, with particular reference to the Fourier integral, convolution and poles and zeros.

To demonstrate the practicability of many of the theoretical concepts introduced in the preceding chapters, we have included the second part (Chapters 11 to 15) of this book. This part consists of five reprints of recent articles from Philips Technical Review that describe a number of up-to-date applications and integrated processors.

In more detail, the book is arranged as follows. The introductory first chapter (largely taken from 'Digital signal processing – growth of a technology' by J. B. H. Peek[†]) gives the how and why of digital signal processing and a description of its advantages, disadvantages and fields of application. The second chapter, 'Continuous-time signals and systems', gives a concise summary of the relevant theory. We have included this chapter to revive latent knowledge of the subject and so that we can usefully compare similarities and differences with respect to the discrete-time theory given later in the book. The third chapter treats the conversion of continuous-time signals into discrete-time signals and vice versa. In Chapter 4 the foundations of discrete-time signals and systems are treated, with 'Fourier transform for discrete signals' (FTD) and 'z-transform' (ZT) as the keywords. In the next chapter we consider the widely used 'Discrete Fourier transform' (DFT) and 'Fast Fourier transform' (FFT).

Chapter 6, 'Overview of signal transforms', occupies a special place. It attempts to give an answer to students' ever-recurring questions about the interrelationships between the most important (continuous and discrete) signal transforms. This chapter can perhaps be omitted on a first reading of the book, but on the other hand it is indispensable for a full understanding of the subject.

The next two chapters are clearly more practical in orientation and deal entirely with discrete-time filters (filter structures and design methods). In Chapters 9 and 10 subjects finally appear that have no true counterpart in continuous-time theory; first, multirate systems and next, finite-word-length effects in digital signals and systems. Exercises are provided for Chapters 2 to 10.

Chapters 11 to 15 that constitute Part II of this book give a good impression of the 'state of the art' in discrete-time signal processing. Chapter 11 describes the fundamentals of switched-capacitor filters by considering the realization of an integrated SC filter for one particular application (Viewdata). In Chapters 12 to 15 three fields of application of digital signal processing of ever increasing importance are described (digital television receivers, digital audio and digital filtering in analog-to-digital conversion), and two integrated digital signal processors.

An extensive list of references, a subject index and four appendixes complete the book. The appendixes deal in turn with the Fourier integral (important properties and examples), expansion in partial fractions, the inverse z-transform by means of the complex contour integral, and the answers to the exercises. The detailed workings of the answers to the exercises can be obtained from the publishers as a 'teacher's manual'.

A condensed version of Part I of this book – some 35 pages – has already appeared as a journal article.[‡] Those who have a special interest in terminology may like to consult a list of terms that we have found very useful; it deserves to be as widely known as possible.[§]

Here we should like to thank the Philips Internal Training Department for permission to use the lecture notes as the starting point for this book, and Philips Technical Review for permission to reprint their articles. We are also very grateful to the management of Philips Research Laboratories for the encouragement, cooperation and facilities that they have always given us.

[†]Peek, J. B. H., 1985. Digital signal processing – growth of a technology. *Philips Tech. Rev.*, **42**: 103–9.
[‡]Van den Enden, A. W. M., Verhoeckx, N. A. M., 1985. Digital signal processing: theoretical background. *Philips Tech. Rev.*, **42**: 111–44.
[§]Rabiner, L. R. *et al.*, 1972. Terminology in digital signal processing. *IEEE Trans.*, **AU-20**: 322–37.

We hope, as authors, that this book fulfils a need. Our main aim has been to give a clear description of the basic essentials in the field of discrete-time signal processing, quite independently of the exact way in which this knowledge is applied, and thus also independently of the rapid developments in fields such as semiconductor technology. Consequently the main part of our book has a certain timelessness, and we hope it will therefore serve its reader long and well. But most of all, we should like it to be a pleasant introduction to a fairly young subject – one that in our firm opinion has a great future marked out for it.

Eindhoven, 1989   A. W. M. van den Enden
                                                                                     N. A. M. Verhoeckx

                                                                           Philips Research Laboratories
                                                                                     P.O Box 80000
                                                                                     5600 JA Eindhoven
                                                                                     The Netherlands

Symbols and abbreviations

A/D	analog/digital
$A(e^{j\theta})$, $A(e^{j\omega T})$	amplitude characteristic, e.g. of $X(e^{j\theta})$, $X(e^{j\omega T})$
$A[k]$	amplitude characteristic, e.g. of $X[k]$
$A(\omega)$	amplitude characteristic, e.g. of $X(\omega)$
a_i, b_i	filter coefficient
a_k, b_k	real Fourier-series coefficient
B	length of a binary word (in bits)
CT/DT	continuous-time/discrete-time
D/A	digital/analog
dB	decibel
$D(e^{j\theta})$	denominator, e.g. of $H(e^{j\theta})$
DFT	discrete Fourier transform
DT/CT	discrete-time/continuous-time
e	base of the natural logarithms ($= 2.718282 \ldots$)
$e[n]$	discrete(-time) noise signal
f	frequency (in Hz)
FFT	fast Fourier transform
FIR	finite impulse response
FS	Fourier series
FTC	Fourier transform for continuous(-time) signals
FTD	Fourier transform for discrete(-time) signals
$h[n]$	impulse response of a discrete(-time) system
$h(t)$	impulse response of a continuous(-time) system
$H(e^{j\theta})$, $H(e^{j\omega T})$	frequency response of a discrete(-time) system
$H[k]$	DFT of $h[n]$
$H(p)$	system function of a continuous(-time) system
$H(z)$	system function of a discrete(-time) system
$H(\omega)$	frequency response of a continuous(-time) system
$H^*(\ldots)$, $H^*[\ldots]$	complex conjugate function of $H(\ldots)$, $H[\ldots]$
i	discrete variable
$I(e^{j\theta})$, $I(e^{j\omega T})$	imaginary part, e.g. of $X(e^{j\theta})$, $X(e^{j\omega T})$
$I[k]$	imaginary part, e.g. of $X[k]$
$I(\omega)$	imaginary part, e.g. of $X(\omega)$
$I_0(x)$	Bessel function of order zero

IDFT	inverse discrete Fourier transform
IFTC	inverse Fourier transform for continuous(-time) signals
IFTD	inverse Fourier transform for discrete(-time) signals
IIR	infinite impulse response
ILT	inverse Laplace transform
Im z	imaginary part of z
IZT	inverse z-transform
j	discrete variable
j	imaginary unit ($= \sqrt{-1}$)
k	discrete-frequency variable
L	length of an impulse response
LSB	least-significant bit
LT	Laplace transform
LTC	linear time-invariant continuous(-time)
LTD	linear time-invariant discrete(-time)
M	number of poles of a system function
n	discrete-time variable
N	number of points in a DFT; number of zeros in a system function
$N(e^{j\theta})$	numerator, e.g. of $H(e^{j\theta})$
NRDF	non-recursive discrete(-time) filter
p	complex frequency for continuous(-time) signals
P	overflow operation
P_x	power of signal x
q	quantization step
Q	quantization operation
R	interpolation or decimation factor
$R(e^{j\theta}), R(e^{j\omega T})$	real part, e.g. of $X(e^{j\theta})$, $X(e^{j\omega T})$
$R[k]$	real part, e.g. of $X[k]$
$R(\omega)$	real part, e.g. of $X(\omega)$
RDF	recursive discrete(-time) filter
Re z	real part of z
S	scaling factor
SRD	sampling rate decreaser
SRI	sampling rate increaser
t	continuous-time variable
T	period of a continuous signal; sampling interval
$u[n]$	discrete(-time) unit step function
$u(t)$	continuous(-time) unit step function
V	spectral period
$w(t), w[n]$	(usually:) window function
$W(\omega), W(e^{j\theta})$	Fourier transform of $w(t)$, $w[n]$
W_N	'twiddle factor' ($W_N = e^{-j2\pi/N}$)
$x[n], x_d[n]$	arbitrary discrete(-time) signal (normalized form)
$x[nT], x_d[nT]$	arbitrary discrete(-time) signal (non-normalized form)
$x_p[n]$	periodic discrete(-time) signal

$x_Q[n]$	quantized version of $x[n]$
$x(t), x_c(t)$	arbitrary continuous(-time) signal
$x_h(t)$	periodic continuous(-time) signal
$X(e^{j\theta}), X_d(e^{j\theta})$	FTD of $x[n], x_d[n]$ (normalized form)
$X(e^{j\omega T}), X_d(e^{j\omega T})$	FTD of $x[nT], x_d[nT]$ (non-normalized form)
$X[k], X_p[k]$	DFT of $x[n], x_p[n]$
$X(p), X_c(p)$	LT of $x(t), x_c(t)$
$X(z), X_d(z)$	ZT of $x[n], x_d[n]$
$X(\omega), X_c(\omega)$	FTC of $x(t), x_c(t)$
$x[n] \circ\!\!-\!\!\circ X[k]$	DFT pair
$x[n] \circ\!\!-\!\!\circ X(e^{j\theta})$	FTD pair (normalized form)
$x[n] \circ\!\!-\!\!\circ X(z)$	ZT pair
$x[nT] \circ\!\!-\!\!\circ X(e^{j\omega T})$	FTD pair (non-normalized form)
$x(t) \circ\!\!-\!\!\circ X(\omega)$	FTC pair
$x(t) \circ\!\!-\!\!\circ X(p)$	LT pair
$x_h(t) \circ\!\!-\!\!\circ \alpha_k$	FS pair
$x[n] * h[n]$	*linear* convolution of $x[n]$ and $h[n]$
$x[n] \circledast h[n]$	*circular* convolution of $x[n]$ and $h[n]$
$x(t) * h(t)$	*linear* convolution of $x(t)$ and $h(t)$
$y(\ \ldots\ ), y[\ \ldots\ ]$	see $x(\ \ldots\ ), x[\ \ldots\ ]$
$Y(\ \ldots\ ), Y[\ \ldots\ ]$	see $X(\ \ldots\ ), X[\ \ldots\ ]$
z	complex frequency for discrete(-time) signals
ZT	z-transform
α_k	complex Fourier-series coefficient
δ_1, δ_2	ripple (in filter attenuation)
$\delta[n]$	discrete(-time) unit pulse
$\delta(t)$	delta-function or Dirac pulse
θ	relative angular frequency ($= \omega T$)
τ	continuous-time variable
τ_g	group delay
$\varphi(e^{j\theta}), \varphi(e^{j\omega T})$	phase characteristic, e.g. of $X(e^{j\theta}), X(e^{j\omega T})$
$\varphi[k]$	phase characteristic, e.g. of $X[k]$
$\varphi(\omega)$	phase characteristic, e.g. of $X(\omega)$
ξ, ψ	phase angle
ω, ν	angular frequency (in rad/s)
ω_a	angular frequency for continuous(-time) signals
ω_d	angular frequency for discrete(-time) signals
ω_0	fundamental angular frequency
Σ	summation sign
π	$= 3.141593 \ldots$
Π	product sign
$\oint_C$	contour integral (counterclockwise) in the complex plane, over a closed path C

PART I

1
Introduction

1.1 BACKGROUND

It is hard to imagine a life without numbers. All unawares, we have gradually become used to using numbers in dealing with all the various commonplaces of everyday living. For many people the day begins with a look at the clock, and much depends on whether it says 'ten to six' or 'five past eleven'. During the rest of the day we are bombarded with numerical data, from the prices in the shops to ages of traffic accident victims in the newspapers, from the latest quotations of stock-market prices to the quantities in a recipe.

In about 500 BC the Greek philosopher and mathematician Pythagoras and his followers based a complete philosophy of life on the conviction that 'number is the essence of all'. Although this is a rather sweeping generalization, it is an aphorism that might well apply to some modern technological developments. It seems very relevant to microelectronics: electronic circuits with components of extremely small dimensions. One of the keywords in this subject today is 'digitization', derived by way of 'digit' (figure), from the Latin 'digitus' (finger). The complicated circuits – called integrated circuits, ICs or chips – that can be made by microelectronics technology are by their nature eminently suitable for manipulating numbers (Table 1.1). This has had two important consequences.

TABLE 1.1 Some of the techniques that have facilitated the manipulation of numbers, and their dates. The last column gives a figure for the number of elementary arithmetic operations (additions) per second. The term VLSI (very-large-scale integration) indicates that a large number (more than 50 000 to 100 000) of components are integrated on a single chip.

Technique	Characteristic components	Dates	Number of additions per second
Mechanical	Gearwheel	1650–1900	1
Electromechanical	Relay, switch	1900–1945	10
Electronics	Thermionic valve	1945–1955	10^4
Semiconductor electronics	Transistor	1955–1965	10^5
Microelectronics	Integrated circuit	1965–1975	10^6
VLSI microelectronics	Microprocessor	1975–	10^7

In the first place this has led to the almost exponential growth in the use of the computer in all kinds of applications where numbers have traditionally been used, as in bookkeeping, stock control, numerical computation and statistics. These applications come under the general heading of *data processing*.

In the second place there is a tendency to use numbers in situations where they would not have been used before. Here too there are often advantages to be gained from microelectronics. One of the best-known examples is the digital wristwatch, which has largely taken the place of the mechanical spring-driven watch with hands and a dial. An even more recent example is the Compact Disc player, replacing the conventional record player. In this example the music information is recorded on the disc in digital form, that is to say as a sequence of numbers. On playback the numbers are processed and converted back to music. This is referred to as *digital signal processing*.

The electronic equipment used for digital signal processing has many features in common with the electronic equipment used for data processing, because both are essentially concerned with the manipulation of numbers. The individual elementary components are the same 'logic circuits' – such as gates, adders and storage or memory cells. The aim pursued, however, and the manner of attaining it (the algorithm) are completely different. Sometimes it is the similarities between the two fields that are striking, sometimes the dissimilarities. It is quite possible that a computer in a scientific institution might at one moment be occupied in processing the payroll, and at another moment it might be processing signals received from interplanetary space probes. This can be done on the same hardware by using different programs. In most cases, however, we see that the hardware is also tailored to specific applications. Although built up from the same elementary components, there are quite different digital chips for watches, pocket calculators, video games, Compact Disc players and digital TVs. To a large extent these differences stem from the totally different algorithms required for signal processing as opposed to data processing.

All indications suggest that both data processing and digital signal processing have yet to develop to their full extent. Much is still to be gained (improved quality, or lower costs, or both – and much more) through this continuing digitization of technology (Figs 1.1 and 1.2).

1.2 SIGNALS AND SIGNAL PROCESSING

Signals exist in many and various forms, ranging from drum signals in the bush to the stop signal from the traffic policeman or the combined radio/television signal received through a central antenna system, as in cable television. The typical common characteristic of all these signals is that they represent a message, or information. The message does not depend greatly on the nature of the signal. The acoustic drum signal can be converted by a microphone into an electrical signal for transmission to the other end of the world. There it can be converted into an optical signal for recording on a Compact Disc. Then the opposite route can be followed, from optical signal via electrical signal to acoustic signal, without any degradation of the original drum message.

In the present state of the technology, electrical signals offer the widest scope for transmission, storage and manipulation ('signal processing'). In signal transmission,

FIGURE 1.1 With thermionic valves, numbers could be manipulated by electronics for the first time. This photograph shows a special counter tube from the late forties; it counted from 0 to 9. The count could be read from a light spot on the front of the tube.

however, the arrival of the fiber-optic cable has considerably increased the significance of optical signals, and perhaps one day there will be 'optical computers'.

In theoretical considerations the physical nature of a signal is irrelevant. Every signal is 'abstracted' to a function of one or more independent variables, representing for example time or position in space.

From now on we shall confine ourselves to signals that can be represented as a function of a single variable, which we generally take to be 'time', while the value of the function itself is denoted as 'the (instantaneous) amplitude'. The signals fall into different categories. Time can be either a continuous variable t, which may assume any arbitrary real value, or a discrete variable n, which can only represent integers. We can also distinguish between signals that have a continuous amplitude, which may assume any value (between two extremes), and signals that have a discrete amplitude, which only has a limited number of different values. In this way we have signals of four kinds, which are shown schematically in Fig. 1.3 and which will be denoted as $x(t)$, $x_Q(t)$, $x[n]$ and $x_Q[n]$. The most familiar type of signal is undoubtedly $x(t)$, usually referred to as an analog signal. In the past this type of signal has received the most attention, since the components and circuits available were most suitable for processing such signals. But the situation has changed greatly in the last twenty years or so. Technological developments have given much greater significance to both types of discrete-time signals. This applies both to the digital signal $x_Q[n]$ and to the sampled-data signal $x[n]$.

Digital signals can be represented as sequences of numbers with a limited range of possible different values, and it is therefore easy to process them with the same logic circuits as those from which digital computers are formed. The binary system of numbers is nearly always used here; each number takes the form of a string of 'ones' and 'zeros' (bits) and is therefore also called a 'binary word'.

(a)

(b)

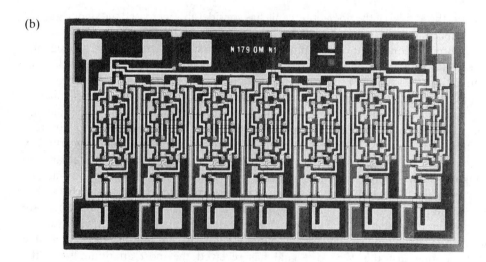

FIGURE 1.2 (a) An application of the counter tube of Fig. 1.1: a decimal counter that could count from 0 to $10^7 - 1$ at a maximum rate of 30 000 units per second. (b) Integrated circuit (binary divider) from the late sixties. The dimensions of this chip are only 2×1.1 mm^2. Although this circuit does not fulfil exactly the same function as the one in (a), a general comparison demonstrates the radical changes that have taken place during the first twenty years in the development of electronic digital techniques. And this was only the beginning.

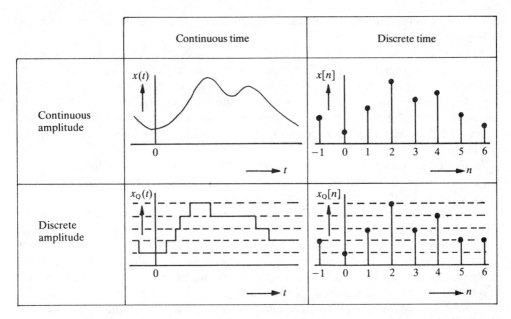

FIGURE 1.3 Signals can be either continuous or discrete in both time and amplitude. They can therefore be subdivided into four types, as illustrated here. The signal $x(t)$ is analog and $x_Q[n]$ digital. The discrete-time signal of continuous amplitude $x[n]$ is generally referred to as a sampled-data signal. The signals of type $x[n]$ and $x_Q[n]$ are both frequently referred to simply as discrete signals. The continuous-time signal of discrete amplitude $x_Q(t)$ is not commonly known by any alternative name.

Sampled-data signals have become so important mainly because of the considerable progress in electrical circuits in which signals are represented as electric charges that are transferred from one point to another at regular intervals and thus processed. Typical devices of this kind are charged-coupled devices (CCDs), charge-transfer devices (CTDs)[1] and switched-capacitor filters (SCFs).[2]

Signals of the type $x_Q(t)$, where the transition from one discrete amplitude to another can take place at random times, are the least commonly used. If they are used, the number of possible amplitudes is often limited to two. Examples of such signals are to be found in certain forms of pulse modulation, e.g. in the LaserVision video-disc system.[3]

Formerly, continuous-time and discrete-time signals were always treated as two separate worlds. All this changed in 1949, when C. E. Shannon introduced his sampling theorem.[4] This showed that any continuous-time signal $x(t)$ whose frequency spectrum is limited in bandwidth can be represented without loss of information as a sequence of samples $x[n]$ of the original signal, or, in other words, as a discrete-time signal.

In the derivation of the sampling theorem it is assumed that the samples $x[n]$ can take all real values, so that the instantaneous values of $x(t)$ can be reproduced exactly. Here we come to the fundamental difference between a general discrete-time signal $x[n]$ and a digital signal $x_Q[n]$: although both consist of a sequence of samples or numbers, with a digital signal the number of possible different values is essentially limited. The transition from a discrete-time signal of the general type to a digital signal always implies a form of approximation ('quantization'), and although the quantization may be arbitrarily fine, it is irreversible.

The required information cannot always be read directly from the signals and it may be that the signals are not free from disturbances such as noise. The usefulness of the signals can then be increased by means of signal processing. Familiar concepts encountered here are interpolation, extrapolation, smoothing, filtering and prediction.[5,6] A historic example of signal processing is to be found in the work of Sir Arthur Schuster in the 1890s; he investigated the occurrence of periodicities in certain meteorological phenomena.[7]

In general a signal-processing system is referred to by the same name as the signal being processed: an analog system processes analog signals; a discrete-time system processes discrete-time signals.

In this book we shall examine the theoretical fundamentals of digital signal processing. In doing so we must however bear in mind that the particular part of the theory that deals with the discrete-amplitude nature of the signals is in fact the least suitable for a systematic description and is therefore generally the least accessible. To some extent this is because the resultant 'finite-word-length effects' appear as nonlinear effects, which can be responsible for undesirable behavior of the system, such as oscillations. Also the discrete-amplitude nature of the signals and the operations often results in an apparent addition of noise-like disturbances, called 'quantization noise', whose consequences can only be properly described in statistical terms.

In the design or analysis of digital signal-processing systems the discrete amplitudes are really a serious complication. The usual practice is therefore deliberately to ignore this aspect at first and only consider the discrete-time aspect. The consequences of the discrete amplitude are then considered separately and taken into account if necessary. This is why much of the literature on digital signal processing[8,9] really belongs to the wider category of general discrete-time signal processing. This is also true for large parts of this book.

Some modern textbooks give a combined treatment of discrete-time and continuous-time signals; see for example refs [10, 11]. From now on we shall limit the term 'digital' as far as possible to those situations in which we do in fact take the finite word length into account. We shall also use the terms 'continuous' and 'discrete' without further qualification, tacitly referring to the time aspect and not to the amplitude aspect.

Finally, it should be emphasized that not all digital signals are derived by sampling and quantization from an analog signal; some electrical signals are digital right from the outset. A simple example is a digital tone generator that supplies directly a sequence of numbers of sinusoidally varying value. Here again the successive signal values are referred to as 'samples'.

1.3 ADVANTAGES AND DISADVANTAGES OF DIGITAL SIGNAL PROCESSING

The advantages of digital signal processing fall into three categories: those that are fundamental in nature, those that come about through the use of microelectronics, and those that are most evident from a comparison with conventional analog signal processing.

The first category consists mainly of advantageous effects that are a direct consequence of using a limited number of discrete sample values. If the electrical representation of a sample value is unambiguously recognizable (i.e. each bit is uniquely recognizable as a 'zero' or a 'one'), small relative errors between the bits are unimportant. This has the following positive consequences:

- The tolerances on the value of the components used to build the circuits do not need to be very close.
- The sensitivity to external effects (temperature and interfering signals) and internal effects (ageing and drift) is low.
- The accuracy of operation can be precisely controlled by selecting the number of different possible sample values (directly related to the word length).
- The circuits are completely reproducible (i.e. identical circuits behave identically), so that no trimming is required during manufacture, for example.
- The number of successive operations that can be performed on a signal is theoretically unlimited, since undesired accumulation of interfering effects such as noise can be avoided.

Another advantage that we would like to put in this category is the flexibility that can be achieved by making a circuit or system programmable. This makes it possible to modify a particular processing function without having to make radical changes in the hardware.

Making digital signal processing equipment in the form of chips gives the following advantages in the second category:

- Small dimensions
- High reliability
- Capability of complex processing
- Low price.

In a comparison with analog signal processing the advantages of the third category are very obvious. Some processing functions are too complex for practical analog execution, or may for fundamental reasons be difficult or impossible in an analog system, or only possible by approximation. On the other hand, digital implementation of the same functions may often be quite straightforward. Some examples are:

- The 'ideal memory', whose contents can in principle remain completely uncorrupted for an indefinite period; with such a memory a digital 'ideal integrator' can be made and very-low-frequency signals can be processed digitally.
- Phase-linear filters, which are important in data transmission and television applications.
- Circuits in which certain processing operations have to be exactly equivalent, e.g. for the compensation of two effects.
- Self-regulating ('adaptive') systems.
- Signal transforms, e.g. from the time domain to the frequency domain and vice versa, as in the discrete Fourier transform (DFT). Working with numbers allows various special mathematical techniques to be used (the most familiar is the fast Fourier transform or FFT technique, which considerably reduces the number of calculations required).

- The processing of two-dimensional signals, such as images.
- New ways of suppressing the effects of interference by error correction, and of improving security.

This is a fairly long list of advantages; however, there are some disadvantages. The principal ones are:

- In the present state of the technology, digital signal processing always requires a certain amount of electrical power: passive digital circuits do not yet exist. (But note that in modern analog systems as well active circuits are definitely going to outnumber passive circuits.)
- Digital signal processing cannot be applied to signals at the higher frequencies, although the upper limit is steadily rising.
- When digital signal processing is used in an analog environment analog-to-digital and digital-to-analog converters are necessary, and these may be fairly complex.
- In the present state of the technology there can be difficulties in the analog-to-digital or digital-to-analog conversion of very weak or very strong signals. The digital processing of very weak signals (e.g. antenna signals) and strong signals (e.g. for driving a loudspeaker or a television picture tube) must therefore still be accompanied by analog preamplification or post-treatment, as appropriate.
- The same information (e.g. music) requires a larger bandwidth as a digital signal than it does as an analog signal.
- The design and manufacture of digital chips is a highly specialized technology, often requiring large quantities of manpower, money, specialized knowledge and special equipment, which increase as chips become more complex (Fig. 1.4).

In each practical case it will be necessary to consider the pros and cons very carefully to decide whether digital signal processing will provide the best results. Originally it was mainly in the professional applications (see later) that the balance first moved towards the digital side. Recently, however, we have seen the same development in much more ordinary consumer equipment. An example already mentioned is the Compact Disc player. Now wide-ranging applications in the television field also seem to be becoming an economic proposition.

1.4 HARDWARE MODULES FOR DIGITAL SIGNAL PROCESSING

The theoretical principles of digital signal processing apply just as much to the execution of a particular algorithm in the form of software for a general-purpose computer as to an implementation in the form of a specific piece of hardware. The great practical and economic significance of digital signal processing is especially apparent in the second of these two forms. We shall therefore take a look now at the modules used here.

The most elementary components are the electrical circuits (gates) that perform logic functions such as AND, OR, NAND, NOR and NOT. Each only requires a few electronic components, such as transistors. An assembly of gates results in somewhat more complex components, such as parts of counters (half-adders) and memory cells (bistables or flip-flops). From these in turn larger modules can be formed, such as adders, multipliers

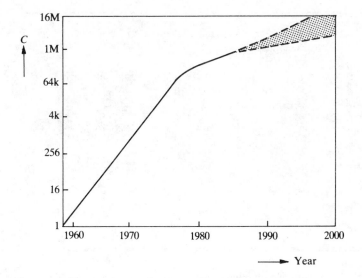

FIGURE 1.4 The highest degree of complexity C of integrated circuits, expressed in terms of the number of components (e.g. transistors) per chip, has increased approximately exponentially. C has doubled every year from 1959 to the late seventies; since then the rate of increase has decreased slightly, and now doubles every two years. In the near future chips with a million components will probably not be exceptional. The trend follows 'Moore's law'; in 1964 and in 1975 G. E. Moore produced the forecasts confirmed by the development shown here. On the vertical axis k represents 2^{10} ($= 1024$) and M represents 2^{20} ($= 1\,048\,576$).

and serial memories (shift registers). In principle we then have all the elements we need for making any of the circuits belonging to the very important class of linear time-invariant digital systems.† The only functions required for such systems are addition, multiplication and delay (or temporary storage in a memory). Systems of this class include most digital filters in common use.

As well as shift registers, some other types of electronic memory have been developed that have characteristics that in some respects are significantly different. Two such memories are the 'random access memory', or RAM, whose name emphasizes the difference from shift registers with their serial access, and the permanent 'read only memory', or ROM, whose contents can be read as often as required but not rewritten.

Complete digital systems can be built with the modules mentioned above used as separate components. Sometimes a ready-made microprocessor chip – really a small computer in itself – is used for controlling the interaction between the different components. From this point two further steps can be taken in the development toward components of higher complexity. These are the 'signal processor' and the 'custom IC'. The signal processor is a more elaborate type of microprocessor with provision for more efficient implementation of typical signal-processing functions, such as large numbers of

---

†Because the word lengths are finite, leading to quantization and overflow effects (see Chapter 10), digital systems are never *strictly speaking* linear. The combination here of the two concepts 'linear' and 'digital' indicates that *in practice* these effects are negligible and that there are no other nonlinear operations.

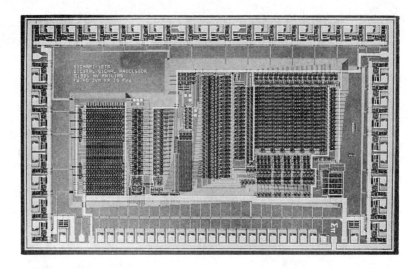

FIGURE 1.5 Photograph of a 'number cruncher' for use in a digital signal processor. This circuit has been optimized in certain respects. For example, the minimum of time is lost on internal transfer of signal samples, and little time is taken up by the most common signal operations, such as addition of multiplication results. The circuit is primarily intended for processing 16-bit signals, but its modular design readily permits extension to larger word lengths. When memories and input–output circuits are added on the same chip, the result is a complete programmable integrated digital signal processor. The IC shown here is a product of CMOS technology. It contains about 19 000 transistors on an area of 15.5 mm^2. The smallest dimensions are 2 μm and the cycle time for the calculation is only 0.1 μs.

multiplications (Fig. 1.5). Since the signal processor is programmable, it can be used for a variety of applications.[12,13,59] In mass production it may be cheaper to use digital signal processing based on custom ICs (Fig. 1.6). As the name implies, these are 'custom-made' for one specific application.

Here again it is important to emphasize the importance of digital memories in the growing application of digital signal processing. In audio applications, for example, developments were greatly stimulated by the appearance of the Compact Disc.[14] In essence, every individual optical disc is simply a huge permanent digital memory with an effective storage capacity of the order of 5 gigabits (5 000 000 000 bits), with 1.4 megabits representing one second of stereo sound. Digital signal processing in color television receivers will only really become interesting when an inexpensive digital memory is available that can record and play back the information in a complete television field (about 2 megabits) 50 or 60 times a second. It now seems that this is economically feasible (see Fig. 1.7).

One category of components that should not be left out here is that of analog-to-digital and digital-to-analog converters (A/D and D/A converters for short). In considering digital signal processing in an analog environment these are often assumed to be ideal components. But since they are – literally – the link between the analog world and the digital world, any errors either in amplitude or in time that are made in the conversions are usually directly reflected in the ultimate quality of the processed signal. When such

FIGURE 1.6 A custom-made chip for digital signal processing in Compact Disc players. The circuits on this chip are designed specifically for certain tasks. One important task is the replacement of fragments of 'mutilated' sound (linear interpolation for 'error concealment'; this should not be confused with error correction, already performed on the digital signal). The other main task is the calculation of three audio samples between each two successive values stored on the Compact Disc (interpolation for increasing the sampling rate by a factor of four). Digital filtering is of vital significance in both tasks. Because of the stereo nature of the system, all major circuits are duplicated on the chip: one half of the chip processes the left sound signal, the other the right sound signal. *1* clock and control circuits. *2a*, *2b* input registers. *3a*, *3b* random-access memories (RAMs) for storing the filter input signals. *4a*, *4b* read-only memories (ROMs) for storing filter coefficients. *5a*, *5b* multipliers/accumulators for performing the filter calculations. *6* output circuits. This chip is a product of 2.6-μm NMOS technology. It contains about 17 000 transistors on an area of about 16 mm^2.

FIGURE 1.7 The quality of television receivers can be improved by using digital signal processing. This requires image-storage devices that can be produced economically. An important step in that direction is the charge-coupled device (CCD) shown here, which can store 308 kbit of picture information with sufficiently high input and output rates. The photograph shows a 28-pin package complete with cooling fins, alongside a partly unpackaged unit. The substrate is a 'slice' or 'wafer' from an earlier phase in the production process. A large number of unmounted chips of the same type can be seen on the substrate.

converters[15, 16] are included in a digital system the designer should therefore give full attention to their performance.

1.5 APPLICATIONS OF DIGITAL SIGNAL PROCESSING

As far as is known, the first applications of digital signal processing for civil purposes were in geophysics (soil research for oil and gas exploration), and then in radio astronomy and radar astronomy. Digital correlation techniques[17] have been used in these fields since the end of the fifties. The calculations were usually made on general-purpose computers. Since then the possible uses have increased very considerably in number and versatility, largely because of the introduction of special signal-processing hardware. Significant examples are to be found in telecommunications, for instance in modems[18, 19] for data transmission to and from computers – one of the first commercial applications of adaptive digital signal processing (automatic equalization). Digital techniques are also used in telephony, for example in vocoders[20] and transmultiplexers,[21] mostly for long-distance communications. It looks as if in the not too distant future telephone signals will be digitized at source (i.e. at the subscriber's location) and transmitted all the way to their destination in digital form.

Another important field is that of medical applications; this applies not only to investigations of the functioning of heart and brain, but also to examinations of other organs and even of the unborn child. Image processing is becoming increasingly important here, and can be used in combination with X-ray and NMR techniques.[22] Professional digital image processing is also applied to photographs received from weather satellites or from space vehicles at an even greater distance from the Earth. This began in about 1964 with space flights to the moon (Ranger VII) and to the planet Mars (Mariner 4).[23] Television studios are also making increasing use of digital techniques for manipulating picture signals; see also ref. [24] for the special CAROT application.

There are interesting applications in instruments based on digital signal transforms (DFT, FFT) for spectrum analysis and other analytical purposes, and also for industrial process control, especially in the chemical industry.

Most of the samples of applications mentioned here are of a professional nature. Digital signal processing has been brought within the reach of the ordinary consumer with the introduction of the Compact Disc. In the near future this will be followed by applications in the television receiver,[25] the radio receiver and the telephone set. It seems to be a natural and inevitable development. Perhaps the day will come when people will ask: '*Analog* signal processing, now what's that?'

2
Continuous-time signals and systems

2.1 INTRODUCTION

Although this book deals primarily with *discrete* signals and systems, we thought it appropriate to include a short chapter on *continuous* signals and systems here. There are a number of reasons for this.

In the first place, we have assumed that the reader has some familiarity with the 'continuous' theory. This chapter should therefore strengthen the acquaintance in places where it might be a little hazy. This will help later in identifying a number of clear parallels between continuous theory and discrete theory, which make the subject easier to grasp. In other cases, however, it will be apparent that there are differences that must be fully appreciated to avoid errors in interpretation.

Another reason for the inclusion of this chapter is that in practice we often start from an analog (and hence continuous) signal, which we want to convert into a digital (and hence discrete) version. Such conversions are exactly where continuous theory and discrete theory meet – and it is necessary to master both. This will appear from the next chapter. Later in the book (in Chapter 6, 'Overview of signal transforms') we shall try to give a brief account of a number of discrete and continuous methods of calculation – in particular signal transforms – and their interrelationships. Later on (in Chapter 8, 'Design methods for discrete filters') we shall see that we sometimes wish to replace an existing analog system by a digital system. There again we must be at home in both camps.

2.2 THE FOURIER INTEGRAL

As electronic engineers, we very often have to solve problems of one of the following two types:

- System analysis: what output signal will a given system supply for a given input signal?
- System synthesis: what system will supply a given output signal as the result of a given input signal?

For continuous systems these questions can often be answered by making use of a differential equation (with time as the independent variable) that gives the relation between the input signal and the output signal. The solution of this equation can be difficult, however, especially if it is of high order.

A much simpler solution is based on the use of signal transforms, which offer a method of description with the real frequency (ω or f) or the 'complex frequency' (p) as the independent variable. The original differential equation in t can then usually be replaced by a much more manageable algebraic equation in ω, f or p. After solving this new equation we find as the result a *frequency function* and, after a subsequent inverse transformation, the corresponding *time function*.

One of the best-known transforms for a continuous function $x(t)$ is without doubt the *Fourier integral* which we shall also call the Fourier transform for continuous signals (FTC). This is defined as:

$$X(\omega) = \int_{-\infty}^{\infty} x(t)e^{-j\omega t}\,dt \qquad \text{(FTC)} \qquad (2.1)$$

When we calculate this integral the function $x(t)$ is uniquely replaced by (or 'transformed into') the function $X(\omega)$. In this book $x(t)$ will in general be a real function of time. $X(\omega)$ is the corresponding complex frequency spectrum. (We follow the usual convention of indicating a time function by a lower-case letter and its Fourier transform by the corresponding upper-case letter.)

Since $X(\omega)$ is a complex quantity, it can also be written as:

$$X(\omega) = R(\omega) + jI(\omega) = A(\omega)e^{j\varphi(\omega)} \qquad (2.2)$$

where $R(\omega)$ = real part of $X(\omega)$

$\quad I(\omega)$ = imaginary part of $X(\omega)$

$\quad A(\omega)$ = modulus or amplitude of $X(\omega)$

$\qquad = \sqrt{(R(\omega)^2 + I(\omega)^2)}$

$\quad \varphi(\omega)$ = argument or phase of $X(\omega)$

$\qquad = \begin{cases} \arctan\{I(\omega)/R(\omega)\} & \text{for } R(\omega) > 0 \\ \arctan\{I(\omega)/R(\omega)\} + \pi & \text{for } R(\omega) < 0 \end{cases}$

If we can calculate the integral of (2.1), we can therefore 'transform' $x(t)$ into $X(\omega)$. With the 'inverse transformation' – the inverse Fourier transform for continuous signals (IFTC) – we can recover $x(t)$ uniquely from $X(\omega)$:

$$x(t) = \frac{1}{2\pi} \int_{-\infty}^{\infty} X(\omega)e^{j\omega t}\,d\omega \qquad \text{(IFTC)} \qquad (2.3)$$

The functions $x(t)$ and $X(\omega)$ form a 'Fourier pair'; this is indicated as follows:

$$x(t) \; \circ\!\!-\!\!-\!\!\circ \; X(\omega) \qquad (2.4)$$

We shall now illustrate the above with the aid of two examples. (A survey of important Fourier pairs and the most widely used properties of the Fourier integral is given in Appendix I.)

Example 1

The time function $x(t)$ of Fig. 2.1(a) is given by ($\alpha > 0$):

$$x(t) = \begin{cases} e^{-\alpha t} & \text{for} \quad t > 0 \\ 0 & \text{for} \quad t < 0 \end{cases} \tag{2.5}$$

From (2.1) we then find:

$$X(\omega) = \frac{1}{\alpha + j\omega} = \frac{\alpha}{\alpha^2 + \omega^2} - \frac{j\omega}{\alpha^2 + \omega^2} = \frac{1}{\sqrt{(\alpha^2 + \omega^2)}} e^{-j \arctan(\omega/\alpha)} \tag{2.6}$$

Therefore:

$$R(\omega) = \frac{\alpha}{\alpha^2 + \omega^2}, \qquad I(\omega) = \frac{-\omega}{\alpha^2 + \omega^2} \tag{2.7a}$$

and

$$A(\omega) = \frac{1}{\sqrt{(\alpha^2 + \omega^2)}}, \qquad \varphi(\omega) = -\arctan(\omega/\alpha) \tag{2.7b}$$

These last four frequency functions are shown in Fig. 2.1(b) and (c).

Example 2

Figure 2.2(a) shows a rectangular function $x(t)$, for which:

$$x(t) = \begin{cases} A & \text{for} \quad |t| < \tau \\ 0 & \text{for} \quad |t| > \tau \end{cases} \tag{2.8}$$

We find that its Fourier transform is:

$$X(\omega) = \frac{2A \sin(\omega \tau)}{\omega} \tag{2.9}$$

This function is shown in Fig. 2.2(b).

2.3 DISCONTINUITIES IN $x(t)$ OR $X(\omega)$

In Examples 1 and 2 just given we are dealing with time functions that contain discontinuities or jumps (at $t = 0$ in Example 1 and at $t = -\tau$ and $t = \tau$ in Example 2). We have deliberately not specified the value of the functions *at* the discontinuities more closely. And as long as we take an arbitrary (but finite) value there, the Fourier transforms (2.6) and (2.9) remain unchanged. If we transform (2.6) back again by the IFTC, however, we find a very specific value at $t = 0$: exactly $\frac{1}{2}$. Similarly, the inverse transformation of (2.9) for $t = \pm\tau$ gives an $x(t)$ with the value $A/2$. In general, inverse

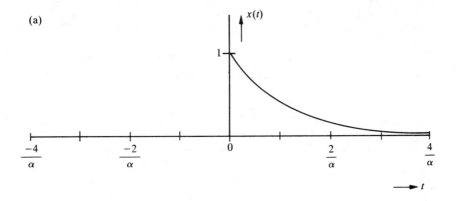

(a)

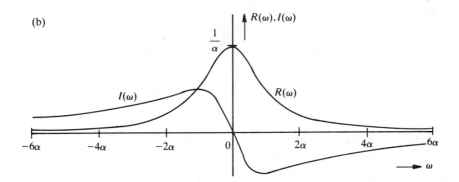

(b)

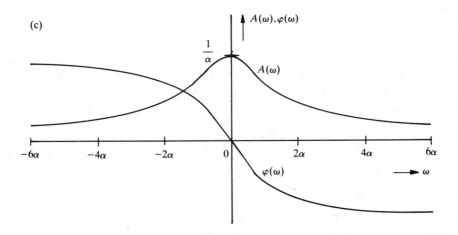

(c)

FIGURE 2.1 Time function $x(t)$ and the associated Fourier transform $X(\omega) = A(\omega)e^{j\varphi(\omega)} = R(\omega) + jI(\omega)$.

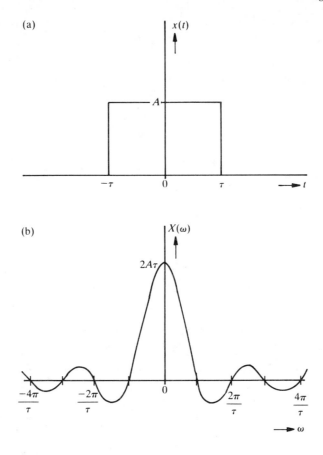

FIGURE 2.2 A rectangular function $x(t)$ and its Fourier transform $X(\omega)$.

transformation at the original discontinuity gives a well-defined value, which is exactly equal to the mean of the two values on either side of the discontinuity. So as to come back to exactly the same $x(t)$ after forward and inverse transformation, we must therefore from the outset assign the appropriate mean value to each discontinuity.

The same thing is also true for discontinuities in $X(\omega)$. We shall merely state, without further proof, that in general (for both functions of time and functions of frequency) one-to-one correspondence is obtained in this way after forward and inverse transformation, and that we then recover the original function exactly.

2.4 THE DELTA FUNCTION OR DIRAC PULSE

A very special signal, which does not actually exist in practice, but is widely used in the theory of continuous signals and systems – especially in connection with signal transforms – is the delta function or Dirac pulse. This function is what is known as a 'generalized function'. A well-founded theoretical description of the delta function can only be

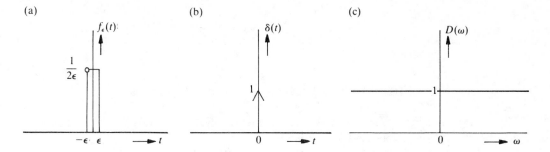

FIGURE 2.3 Possible definition of a delta function $\delta(t) = \lim_{\epsilon \to 0} f_\epsilon(t)$ and its Fourier transform $D(\omega) = \lim_{\epsilon \to 0} F_\epsilon(\omega) = 1$.

obtained with the aid of 'distribution theory', which is well beyond the scope of this book (see for example Appendix A in ref. [26] or Appendix I in ref. [27]).

For the moment we want to summarize the delta function and its most important properties without worrying too much about the theoretical problems that can arise here. One of the many possible ways of defining the delta function is as follows. Let us start with the rectangular pulse $f_\epsilon(t)$ given in Fig. 2.3(a) and defined as:

$$f_\epsilon(t) = \begin{cases} \dfrac{1}{2\epsilon} & \text{for} \quad |t| < \epsilon \\[2mm] 0 & \text{for} \quad |t| > \epsilon \end{cases} \tag{2.10}$$

From (2.9) the spectrum of this pulse is:

$$F_\epsilon(\omega) = \frac{\sin(\omega\epsilon)}{\omega\epsilon} \tag{2.11}$$

If we make the parameter ϵ continually smaller, the pulse $f_\epsilon(t)$ becomes narrower and higher, while the area remains constant at 1. In the limiting case for $\epsilon \to 0$ we have the delta function $\delta(t)$:

$$\delta(t) = \lim_{\epsilon \to 0} f_\epsilon(t) \tag{2.12}$$

$\delta(t)$ is represented symbolically by a vertical arrow of length 1 (see Fig. 2.3(b)).

Besides definition (2.12) there are also other ways of defining exactly the same delta function. Two alternatives are:

$$\begin{cases} \delta(t) = 0 & \text{for} \quad t \neq 0 \\[4mm] \displaystyle\int_{-\infty}^{\infty} \delta(t)\,dt = 1 \end{cases} \tag{2.13}$$

or

$$\left\{ \begin{array}{l} \displaystyle\int_{-\infty}^{\infty} x(t)\delta(t)\,\mathrm{d}t = x(0) \\[4mm] \text{where } x(t) \text{ is an arbitrary function with no discontinuity at } t = 0. \end{array} \right. \tag{2.14}$$

These last two definitions are not easy to represent visually, but they do give a more direct insight into certain widely used properties of $\delta(t)$.

The Fourier transform of $\delta(t)$ follows from the limiting value of (2.11). For very small ϵ we have $\sin(\omega\epsilon) = \omega\epsilon$, so that the spectrum $D(\omega) = \lim\limits_{\epsilon\to 0} F_\epsilon(\omega)$ of $\delta(t)$ has a constant value of 1 (Fig. 2.3(c)) or:

$$\delta(t) \;\circ\!\!-\!\!-\!\!\circ\; 1 \tag{2.15}$$

We shall now state a few important properties of $\delta(t)$ without proof:

$\delta(t - t_0)$ and $\delta(t_0 - t)$ represent a delta function at $t = t_0$ $\hfill$ (2.16a)

$$\int_{-\infty}^{\infty} \delta(t - t_0)x(t)\,\mathrm{d}t = \int_{-\infty}^{\infty} \delta(t_0 - t)x(t)\,\mathrm{d}t = x(t_0) \tag{2.16b}$$

$$\delta(at) = \frac{1}{|a|}\delta(t) \tag{2.16c}$$

$$x(t)\delta(t - t_0) = x(t_0)\delta(t - t_0) \tag{2.16d}$$

$\delta^2(t)$ is undefined, and $\hfill$ (2.16e)

$$x(t) * \delta(t - t_0) = x(t - t_0) \tag{2.16f}$$

(The asterisk in the last equation indicates the convolution operation; see also section 2.9.)

2.5 PERIODIC CONTINUOUS SIGNALS

If $x(t)$ is a periodic function, there are considerable problems with the calculation of the FTC integral in (2.1). However, by using delta functions in the frequency domain [with ω as the independent variable, e.g. $\delta(\omega)$ and $\delta(\omega - \omega_0)$] it is possible to assign a Fourier transform or frequency spectrum $X(\omega)$ to all kinds of periodic signals $x(t)$. It is of course well known that periodic continuous signals only contain very particular frequency components, or 'harmonics'. The spectrum $X(\omega)$ therefore consists of delta functions of different areas at the various harmonics. (Here again in diagrams of these spectra we represent a delta function by an arrow whose *length* is a measure of the area of the delta function.) Some examples of these can be found in the Table of Appendix I, which shows a number of widely used Fourier pairs.

For periodic continuous signals we *can* use the IFTC to recover $x(t)$ from $X(\omega)$, but as just indicated, it is *not* practical to use the FTC for the other direction. So how do we find $X(\omega)$ if we know $x(t)$? We do this by making use of the Fourier series (FS); from now on in this book we shall also consider this to be a signal transform. In the 'forward transform' we now make use of only one period (of length T) of the periodic signal $x(t)$, and from this we derive two sequences of (real) numbers $a_k(k = 0, 1, 2, \ldots)$ and b_k ($k = 1, 2, 3, \ldots$) given by:

$$a_k = \frac{2}{T} \int_{-T/2}^{T/2} x(t)\cos(k\omega_0 t)\,\mathbf{dt}$$

and

$$b_k = \frac{2}{T} \int_{-T/2}^{T/2} x(t)\sin(k\omega_0 t)\,dt$$

with $\omega_0 = 2\pi/T$

(2.17a)

(2.17b)

The 'inverse transform' is:

$$x(t) = \frac{a_0}{2} + \sum_{k=1}^{\infty} [a_k \cos(k\omega_0 t) + b_k \sin(k\omega_0 t)] \quad \text{with } \omega_0 = 2\pi/T$$

(2.18)

From this last expression we see that we are dealing with a series expansion of $x(t)$; it shows clearly that the function $x(t)$ can be expressed as the sum of a constant term $a_0/2$ that represents the mean value (e.g. a direct voltage or current) and an infinite sequence of sine and cosine terms (the 'harmonics'), each multiplied by a coefficient a_k or b_k. In the above equations ω_0 is the fundamental angular frequency.

By making use of the general relations:

$$\cos(k\omega_0 t) = \frac{1}{2} e^{jk\omega_0 t} + \frac{1}{2} e^{-jk\omega_0 t}$$

(2.19a)

and

$$\sin(k\omega_0 t) = \frac{1}{2j} e^{jk\omega_0 t} - \frac{1}{2j} e^{-jk\omega_0 t}$$

(2.19b)

we can rewrite (2.18) much more concisely as:

$$x(t) = \sum_{k=-\infty}^{\infty} \alpha_k e^{jk\omega_0 t}$$

(2.20a)

with:

$$\alpha_k = \frac{1}{T} \int_{-T/2}^{T/2} x(t) e^{-jk\omega_0 t}\,dt$$

(2.20b)

where $\omega_0 = 2\pi/T$ and $\alpha_k = (a_k - jb_k)/2$. In general the coefficients α_k will therefore have a complex value. We are now at the point we were actually concerned with, since it can be shown (see Exercise 2.4 of section 2.11) that the spectrum $X(\omega)$ of the original periodic signal $x(t)$ of period T is expressed exactly by:

$$X(\omega) = 2\pi \sum_{k=-\infty}^{\infty} \alpha_k \delta(\omega - k\omega_0) \text{ with } \omega_0 = \frac{2\pi}{T} \qquad (2.21)$$

Example 3
Figure 2.4(a) shows a periodic continuous signal $x(t)$. From the Fourier series we find the Fourier coefficients α_k of Fig. 2.4(b) for $x(t)$. These are:

$$\alpha_k = \begin{cases} \dfrac{2\sin(k\pi/2)}{k\pi} & \text{for } k \text{ odd} \\[4mm] 0 & \text{for } k \text{ even} \end{cases} \qquad (2.22a)$$

α_k is a discrete function (k can only be an integer).
With the aid of (2.21) we can derive from this the spectrum $X(\omega)$ of Fig. 2.4(c). This is:

$$X(\omega) = \sum_{\substack{\text{odd} \\ k}} \frac{4\sin(k\pi/2)}{k} \delta\left(\omega - k\frac{2\pi}{T}\right) \qquad (2.22b)$$

$X(\omega)$ is a continuous function, because $X(\omega)$ is defined for every value of ω, even though $X(\omega)$ differs from zero only if ω is an odd multiple of $2\pi/T$.

2.6 CONTINUOUS SYSTEMS

In the preceding sections we have seen how we can describe continuous *signals* both in the form of a time function and in the form of a frequency function. In this section we turn our attention to continuous *systems*. We shall confine ourselves here to the important class of linear time-invariant continuous systems (LTC systems).

Such a system is shown in Fig. 2.5. It has one input, with an input signal $x(t)$ applied to it, and one output, where the output signal $y(t)$ appears. At this moment we are completely uninterested in the actual components from which the system is formed or the precise details of the operations performed in it. We define the properties *linearity* and *time-invariance* on the basis of $x(t)$ and $y(t)$.

Linearity. A continuous system is linear if the input signal $ax_1(t) + bx_2(t)$ produces an output signal $ay_1(t) + by_2(t)$, where a and b are arbitrary constants. Here $x_1(t)$ and $x_2(t)$ are arbitrary input signals and $y_1(t)$ and $y_2(t)$ the corresponding output signals.

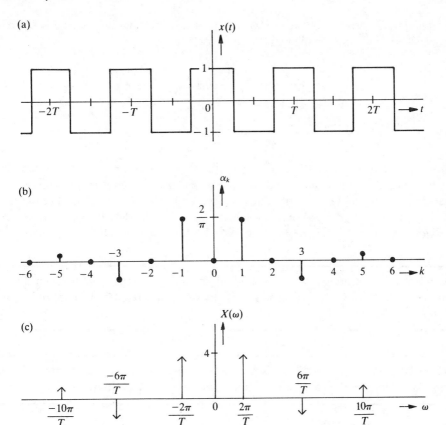

FIGURE 2.4 Periodic square wave $x(t)$ and two of its frequency spectra. The quantities α_k are the coefficients given by the FS of $x(t)$; $X(\omega)$ consists of delta functions and can be converted back into $x(t)$ via the IFTC.

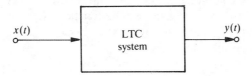

FIGURE 2.5 Linear time-invariant continuous system.

Time-invariance. A continuous system is time-invariant if the input signal $x(t-\tau)$ produces an output signal $y(t-\tau)$, where τ is an arbitrary delay, $x(t)$ an arbitrary input signal and $y(t)$ the corresponding output signal.

In general, the relation between $x(t)$ and $y(t)$ in any LTC system can be described by a linear differential equation in t with constant coefficients, e.g.:

$$y(t) = b_0 x(t) + b_1 \frac{dx(t)}{dt} + \ldots + b_N \frac{d^N x(t)}{dt^N}$$

$$+ a_1 \frac{dy(t)}{dt} + \ldots + a_M \frac{d^M y(t)}{dt^M} \tag{2.23}$$

where the coefficients $a_1, \ldots, a_M$ and $b_0, \ldots, b_N$ characterize the system. The solution of (2.23) is by no means always simple, and we are therefore pleased to turn to other more manageable methods of description. This will form the subject of the rest of this chapter. However, we should first like to introduce the definitions of two concepts (*stability* and *causality*) that are important for the practical realizability of systems.

Stability. A continuous system is stable if any arbitrary input signal $x(t)$ of finite amplitude (i.e. $|x(t)|_{max} \le A$) produces an output signal $y(t)$ of finite amplitude (i.e. $|y(t)|_{max} \le B$).

Causality. A continuous system is causal if the output signal at any instant $t = t_0$ is not dependent on the values of the input signal at times *later* than t_0.

Causality could also be loosely defined by 'there is no output signal as long as there is no input signal'.

2.7 THE IMPULSE RESPONSE

An LTC system can be completely described by its impulse response, i.e. by the output signal $h(t)$ obtained if the input signal is a delta function $\delta(t)$ (see Fig. 2.6). If we know $h(t)$, we can calculate the output signal $y(t)$ that corresponds to any input signal $x(t)$. We can explain this with the aid of Fig. 2.7.

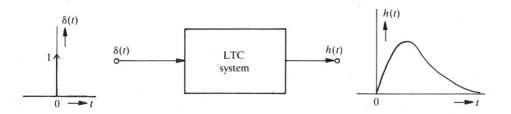

FIGURE 2.6 Impulse response $h(t)$ of an LTC system.

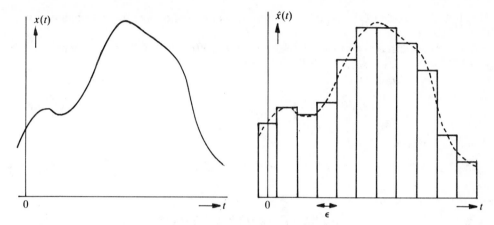

FIGURE 2.7 A continuous signal $x(t)$ can be approximated by a signal $\hat{x}(t)$ composed of an infinite train of narrow pulses of width ϵ.

In the figure we have approximated a function $x(t)$ by a function $\hat{x}(t)$ that consists of adjoining narrow rectangles of width ϵ and height equal to the original value of the function at the centre-line of the rectangle.

Because of the linearity, we can find the output signal $\hat{y}(t)$ that corresponds to $\hat{x}(t)$ by calculating the output signals resulting from each of the little rectangles separately and then adding them together. In an LTC system we therefore only need to know the output signal of one rectangle to be able to calculate, for a given $\hat{x}(t)$, the output signal $\hat{y}(t)$. In addition we see that as ϵ becomes smaller, $\hat{x}(t)$ becomes a better approximation to $x(t)$ and $\hat{y}(t)$ becomes a better approximation to $y(t)$. In the limiting case where $\epsilon \rightarrow 0$ we can replace the rectangle at $t = \tau$ by a delta function $x(\tau)\delta(t-\tau)$ and we can write:

$$x(t) = \int_{-\infty}^{\infty} x(\tau)\delta(t-\tau)\,\mathrm{d}\tau \qquad (2.24)$$

(This equation also follows directly from the second integral of (2.16b), on substituting $t = \tau$ and $t_0 = t$ in that integral.) We can therefore determine $y(t)$ for a given $x(t)$ if we know the output signal that corresponds to $\delta(t)$, and by definition this is the impulse response $h(t)$. More formally we can write:

- The input signal $\delta(t)$ produces the output signal $h(t)$.
- Because of time invariance the input signal $\delta(t-\tau)$ then produces the output signal $h(t-\tau)$.
- Because of linearity the input signal $x(\tau)\delta(t-\tau)$ then produces the output signal $x(\tau)h(t-\tau)$.
- Therefore the input signal $\int_{-\infty}^{\infty} x(\tau)\delta(t-\tau)\,\mathrm{d}\tau$ produces the output signal $\int_{-\infty}^{\infty} x(\tau)h(t-\tau)\,\mathrm{d}\tau$.

Summarizing, we thus find that an LTC system with impulse response $h(t)$ and input signal $x(t)$ provides an output signal $y(t)$ given by:

$$y(t) = \int_{-\infty}^{\infty} x(\tau)h(t-\tau)\,\mathrm{d}\tau \qquad (2.25)$$

The right-hand side of this equation represents a convolution integral; we shall return to this in section 2.9.

We can see directly from the impulse response $h(t)$ of an LTC system whether the system is stable or causal or both.

For a *stable* system:

$$\int_{-\infty}^{\infty} |h(t)|\,dt < \infty \qquad (2.26)$$

For a *causal* system:

$$h(t) = 0 \quad \text{for} \quad t < 0 \qquad (2.27)$$

2.8 THE FREQUENCY RESPONSE

The Fourier transform of the impulse response $h(t)$ of an LTC system is called the *frequency response $H(\omega)$*. Just like $h(t)$, $H(\omega)$ gives a complete description of this system. The importance of $H(\omega)$ is mainly that it makes it possible to determine the output signal $y(t)$ for a given input signal $x(t)$ without having to calculate the convolution integral (2.25). For, if we again designate the Fourier transforms of $x(t)$ and $y(t)$ by $X(\omega)$ and $Y(\omega)$, then:

$$Y(\omega) = X(\omega) \cdot H(\omega) \qquad (2.28)$$

Inverse transformation of $Y(\omega)$ gives $y(t)$.

The two methods of calculating $y(t)$ from $x(t)$ and $h(t)$ (convolution in the time domain and multiplication in the frequency domain) are set out schematically in Fig. 2.8.

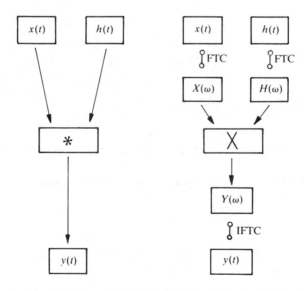

FIGURE 2.8 Two methods of determining the output signal $y(t)$ of an LTC system: convolution in the time domain or multiplication in the frequency domain.

2.9 THE CONVOLUTION INTEGRAL

As we noted earlier, (2.25) is called the convolution integral, or just the convolution, of two continuous time functions $x(t)$ and $h(t)$ and is often written concisely as:

$$y(t) = x(t) * h(t) \qquad (2.29)$$

By substituting $\tau = t - \tau'$ in (2.25) we find after some manipulation that we can also write:

$$y(t) = \int_{-\infty}^{\infty} x(t - \tau')h(\tau')d\tau' \qquad (2.30)$$

The roles of $x(t)$ and $h(t)$ are now reversed as compared with (2.25). We therefore have:

$$y(t) = x(t) * h(t) = h(t) * x(t) \qquad (2.31)$$

To illustrate the idea of convolution, the calculation of (2.25) for $t = t_1$ is shown graphically in Fig. 2.9. We start with the $h(\tau)$ and $x(\tau)$ shown and perform the following operations:

I. 'Fold $h(\tau)$ over' in time[†]: this gives $h(-\tau)$.
II. Shift $h(-\tau)$ through a time t_1: this gives $h(t_1 - \tau)$.
III. Multiply $x(\tau)$ by $h(t_1 - \tau)$: this gives $x(\tau)h(t_1 - \tau)$.
IV. Integrate the function $x(\tau)h(t_1 - \tau)$ over all τ: this gives the value of the hatched area in Fig. 2.9.

$$y(t_1) = \int_{-\infty}^{\infty} x(\tau)h(t_1 - \tau)d\tau$$

V. By varying t_1 from $t_1 = -\infty$ to $t_1 = \infty$ we find all the remaining values of $y(t)$.

The property that convolution in the time domain corresponds to multiplication in the frequency domain (see Fig. 2.8) can be represented in the form of an equation as follows:

$$y(t) = x(t) * h(t) \; \circ\!\!-\!\!-\!\!\circ \; Y(\omega) = X(\omega) \cdot H(\omega) \qquad (2.32)$$

In exactly the same way as for continuous time functions the convolution of two continuous frequency functions $X(\omega)$ and $H(\omega)$ is defined as:

$$X(\omega) * H(\omega) = \int_{-\infty}^{\infty} X(v)H(\omega - v)dv \qquad (2.33)$$

Again we have:

$$X(\omega) * H(\omega) = H(\omega) * X(\omega) \qquad (2.34)$$

[†]Convolution comes from the Latin verb *convolvere*, which means 'to turn round', 'to roll together'.

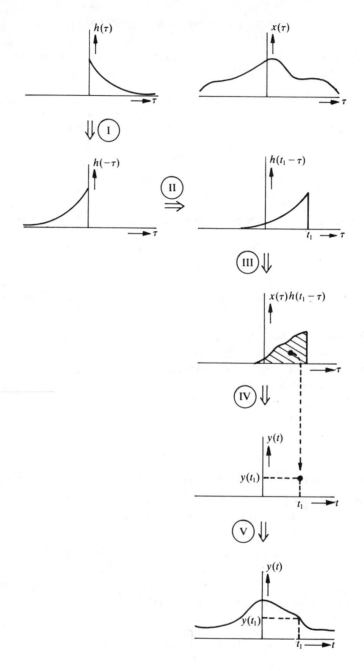

FIGURE 2.9 Graphical representation of the convolution $y(t) = x(t) * h(t) = \int_{-\infty}^{\infty} x(\tau) h(t - \tau) d\tau$.

Later we shall also need the counterpart of property (2.32): multiplication in the time domain corresponds to convolution in the frequency domain ($\times$ a constant factor $1/2\pi$):

$$y(t) = x(t) \cdot h(t) \quad \circ\!\!-\!\!-\!\!-\!\!\circ \quad Y(\omega) = \frac{1}{2\pi} X(\omega) * H(\omega) \tag{2.35}$$

2.10 POLES AND ZEROS

Yet another method for describing LTC systems depends upon poles and zeros and can be used to give a very direct derivation of certain properties of a system. The theory on which it is founded is based on the introduction of the complex frequency variable p (where $p = \sigma + j\omega$) and the Laplace transform (see for example ref. [28]). Here however, without any claim to mathematical rigor, we should like to turn to poles and zeros by way of the Fourier transform. We start from the time description of an LTC system by a differential equation, as given in (2.23). If once again we have:

$$x(t) \quad \circ\!\!-\!\!-\!\!\circ \quad X(\omega) \quad \text{and} \quad y(t) \quad \circ\!\!-\!\!-\!\!\circ \quad Y(\omega) \tag{2.36}$$

then it follows from (2.23) and the property, so far not mentioned:

$$\frac{d^n x(t)}{dt^n} \quad \circ\!\!-\!\!-\!\!\circ \quad (j\omega)^n \cdot X(\omega) \tag{2.37}$$

that also:

$$Y(\omega) = b_0 X(\omega) + b_1 j\omega X(\omega) + \ldots + b_N (j\omega)^N X(\omega)$$
$$+ a_1 j\omega Y(\omega) + \ldots + a_M (j\omega)^M Y(\omega) \tag{2.38}$$

From this equation and (2.28) the frequency response is given by:

$$H(\omega) = \frac{Y(\omega)}{X(\omega)} = \frac{b_0 + b_1 j\omega + \ldots + b_N (j\omega)^N}{1 - a_1 j\omega - \ldots - a_M (j\omega)^M} \tag{2.39}$$

In this equation ω is a *real* variable. We replace $j\omega$ by the *complex* variable $p = \sigma + j\omega$ and then find[†] what is called the *system function*.

$$H(p) = \frac{b_0 + b_1 p + \ldots + b_N p^N}{1 - a_1 p - \ldots - a_M p^M} \tag{2.40}$$

Both p and $H(p)$ are complex quantities and we cannot represent the relation between p and $H(p)$ in simple graphic form here (as we could the relation between $H(\omega)$ and ω).

[†]Strictly speaking, it is mathematically not quite correct to use the same function H on the left-hand sides of both (2.39) and (2.40). For the right-hand side of (2.39) becomes the right-hand side of (2.40) when we substitute p for $j\omega$, while on the left-hand side we substitute p for ω, not $j\omega$. To avoid this problem we ought to have consistently used $j\omega$ as function argument instead of ω, with say $H(j\omega)$, $X(j\omega)$ and $Y(j\omega)$ instead of $H(\omega)$, $X(\omega)$ and $Y(\omega)$. We prefer to tolerate this minor imperfection and keep the notation simple.

Meanwhile, however, we do know that we shall find the original frequency response for $p = j\omega$.

We can rewrite (2.40) as:

$$H(p) = \frac{b_N(p - z_1)(p - z_2) \, \ldots \, (p - z_N)}{-a_M(p - p_1)(p - p_2) \, \ldots \, (p - p_M)} = \frac{b_N \prod\limits_{i=1}^{N} (p - z_i)}{-a_M \prod\limits_{j=1}^{M} (p - p_j)} \tag{2.41}$$

The (complex) quantities z_i are called the zeros of $H(p)$ and the (complex) quantities p_j are called the poles of $H(p)$; for if $p = z_i$, then $H(p) = 0$ and if $p = p_j$ then $H(p) = \infty$. We see from equation (2.41) that $H(p)$, except for the constant b_N/a_M, is completely determined by the values of z_i and p_j. We can easily represent these poles and zeros graphically in the 'p-plane'. This is done for a completely arbitrary system in Fig. 2.10.

Since for a physically realizable system it is always true that the coefficients a_i and b_j from which we started (2.23) are real, the zeros can only be real or occur as complex-conjugate pairs; the same is true for the poles. This means that Fig. 2.10 has mirror symmetry with respect to the horizontal axis.

We can draw a number of direct conclusions from the poles-and-zeros description of an LTC system:

1. In a stable system *all* the poles are to the left of the vertical axis (so the system of Fig. 2.10 is not stable!). Zeros can be on either the left or the right of the vertical axis.
2. For $p_0 = j\omega_0$, i.e. for all values of p that lie on the vertical axis, $H(p_0)$ has exactly the same value as the frequency response $H(\omega_0) = A(\omega_0)e^{j\varphi(\omega_0)}$. For a particular frequency ω_0 we find the amplitude $A(\omega_0)$ by dividing the product of all the distances between $j\omega_0$ and the zeros by the product of all the distances between $j\omega_0$ and the poles. We find the phase $\varphi(\omega_0)$ by adding all the phase contributions of the zeros and subtracting all the phase contributions of the poles from this sum. This can be derived from (2.41). Figure 2.11 shows the situation for a second-order system (i.e. $M = 2$) with two zeros ($N = 2$):

$$A(\omega_0) = \frac{R_1 R_2}{R_3 R_4} \text{ and } \varphi(\omega_0) = \beta_1 + \beta_2 - \beta_3 - \beta_4 \tag{2.42}$$

(Here we have put $b_N/a_M = -1$ for simplicity.) We thus see that each pole or each zero has the most effect in a frequency range that corresponds to the part of the vertical axis nearest to that pole or zero. Furthermore, the relative effect of a pole or zero increases the closer it is to the vertical axis. In the extreme case of a zero actually on the vertical axis, this gives rise to a frequency response with zero amplitude and a jump in phase of π radians at the corresponding frequency ($\omega = \omega_1$ in Fig. 2.10). On the other hand, a pole on the vertical axis gives a frequency response with an infinitely large amplitude and a phase jump of π radians at the corresponding frequency ($\omega = \omega_2$ in Fig. 2.10).
3. If there are no zeros in the right-hand half-plane, we have a 'minimum-phase network'.[†] By reflecting (some of) the zeros in the vertical axis, we change the phase

---

† A minimum-phase network is a circuit in which for a given amplitude characteristic the group delay $\tau_g(\omega) = -d\varphi/d\omega$ has the lowest possible value for *all* frequencies.

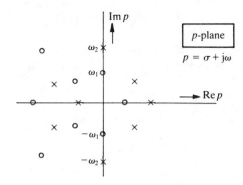

FIGURE 2.10 Poles (x) and zeros (o) of a system function $H(p)$.

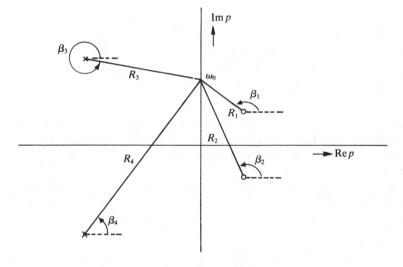

FIGURE 2.11 Graphical determinaton of $H(\omega)$ for $\omega = \omega_0$.

characteristic of the system (which becomes non-minimum-phase), but not the amplitude characteristic (verify this with the aid of (2.42) for Fig. 2.11).

4. A stable 'all-pass' network or phase-shifter (i.e. $|H(\omega)| = 1$ for all ω) has only poles in the left-hand half-plane and only zeros in the right-hand half-plane. Poles and zeros occur only in pairs that have mirror symmetry with respect to the vertical axis (verify this).

5. Poles-and-zeros plots of systems connected in cascade can be added together; if a pole and a zero coincide, they cancel one another.

6. In a physically realizable system M is always larger than or equal to N; the number of poles is thus larger than or equal to the number of zeros. (Why is this? Hint: look at (2.39) for $\omega \to \infty$.)

2.11 EXERCISES

Exercises to section 2.2

2.1 Use the FTC to determine the Fourier transform $X(\omega)$ of the following functions of time:

(a) $x(t) = \begin{cases} 1 - |t|/\tau & \text{for } |t| < \tau \\ 0 & \text{elsewhere} \end{cases}$

(b) $x(t) = e^{-\alpha|t|}$ with $\alpha > 0$

2.2 Use the IFTC to determine the time function that corresponds to:

$$X(\omega) = \begin{cases} 1 & \text{for } |\omega| < \omega_0 \\ 0 & \text{for } |\omega| > \omega_0 \end{cases}$$

Exercises to section 2.5

2.3 Determine both the real Fourier-series coefficients a_k and b_k and also the complex coefficients α_k of the periodic functions (of period T) given in Fig. 2.12.

2.4 Verify (2.21) by showing that applying the IFTC to this equation gives $x(t)$ exactly as in (2.20a).

2.5 Determine the complex Fourier-series coefficients α_k and the Fourier transform $X(\omega)$ of the functions:

(a) $x(t) = 1$
(b) $x(t) = \cos(\omega_0 t)$
(c) $x(t) = \sin(\omega_0 t)$
(d) $x(t) = e^{j\omega_0 t}$

2.6 The function $x(t)$ is defined as follows:

$$x(t) = \sum_{n = -\infty}^{\infty} \delta(t - nT)$$

Draw this periodic function. Then determine the Fourier-series coefficients α_k and the Fourier transform $X(\omega)$.

Exercise to section 2.8

2.7 For the network shown in Fig. 2.13, determine:
 (a) The differential equations that give the relation between $x(t)$ and $y(t)$;

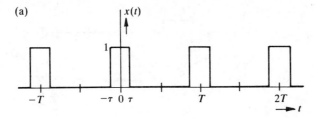

(a)

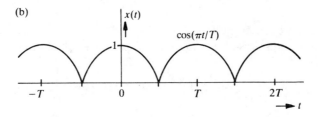

(b)

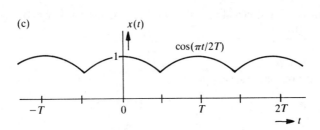

(c)

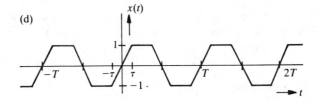

(d)

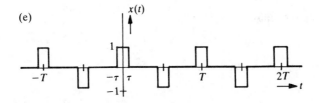

(e)

FIGURE 2.12 Exercise 2.3.

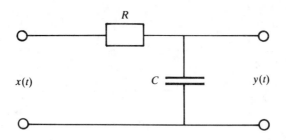

FIGURE 2.13 Exercise 2.7.

(b) The frequency response $H(\omega)$;
(c) The impulse response $h(t)$; start with $H(\omega)$ and use the Fourier pair:

$$x(t) = e^{-\alpha t} u(t) \; \circ\!\!-\!\!-\!\!\circ \; X(\omega) = \frac{1}{\alpha + j\omega}$$

[$u(t)$ is the unit-step function with $u(t) = 0$ for $t<0$ and $u(t) = 1$ for $t>0$. What value ought we to take for $u(t)$ at $t = 0$?]
(d) The output signal $y(t)$ if $x(t) = u(t)$ (the Fourier transform of $u(t)$ can be found in Appendix I).

Exercises to section 2.9
2.8 A system has impulse response $h(t)$:

$$h(t) = \begin{cases} 1 & \text{for } |t|<1 \\ 0 & \text{for } |t|>1 \end{cases}$$

Determine the output signal $y(t)$, both by using convolution in the time domain (as shown graphically in Fig. 2.9), and also by using multiplication in the frequency domain, for the following three input signals (you can use the Fourier pairs from Appendix I if you wish):

(a) The unit-step function: $x(t) = \begin{cases} 0 & \text{for } t<0 \\ 1 & \text{for } t>0 \end{cases}$

(b) A rectangular pulse: $x(t) = \begin{cases} 1 & \text{for } |t|<1 \\ 0 & \text{for } |t|>1 \end{cases}$

(c) A sine function that starts at $t = 0$:

$$x(t) = \begin{cases} 0 & \text{for } t<0 \\ \sin(\pi t) & \text{for } t>0 \end{cases}$$

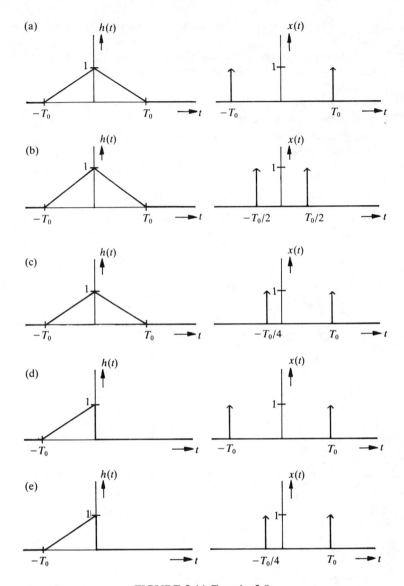

FIGURE 2.14 Exercise 2.9.

2.9 Draw for Figs 2.14(a) to (e) the result of the convolution of $h(t)$ and $x(t)$.

Exercise to section 2.10

2.10 Two systems H_1 and H_2 have the system functions:

$$H_1(p) = \frac{p^2 + 25}{(p-2)(p^2 + 6p + 25)} \quad \text{and} \quad H_2(p) = \frac{(p-2)(p^2 - 6p + 25)}{p^2 + 25}$$

(a) Calculate and draw the poles and zeros of H_1, H_2 and the cascade connection of H_1 and H_2.

(b) Comment (without using formulae) on the networks H_1, H_2 and the cascade connection of H_1 and H_2 with regard to:
   - Stability
   - Frequencies for which the frequency response is zero or infinity
   - Minimum-phase property
   - 'All-pass' character
   - Physical realizability.

3

Conversion from continuous-time signals to discrete-time signals and vice versa

3.1 INTRODUCTION

If we wish to process a continuous (and generally analog) signal $x_c(t)$ in a discrete system, we must first convert $x_c(t)$ into a discrete signal $x_d[n]$ without loss of information. But is this always possible? The answer is to be found in the *sampling theorem*:[†]

The sampling theorem. If a continuous signal $x_c(t)$ contains no frequency components at frequencies above $\omega_{max} = 2\pi f_{max}$ [rad/s], all the information about $x_c(t)$ is entirely contained in the values $x_c(nT)$, provided that $T < 1/(2f_{max})$ or $1/T > 2f_{max}$.

The values $x_c(nT)$ can be obtained from $x_c(t)$ by 'sampling' at a sampling rate $f_s = 1/T$. If we now define a discrete signal $x_d[n]$ such that:

$$x_d[n] = x_c(nT) \qquad (3.1)$$

we have completed the entire transition from continuous time (CT) to discrete time (DT) (see Fig. 3.1). Equation (3.1) means that the values of the successive samples of $x_d[n]$ correspond with the values $x_c(nT)$ exactly. It is entirely irrelevant *how* the samples of $x_d[n]$ are represented. They can be represented equally well by a 'packet' of electrical charge (as in a switched-capacitor filter) or by a binary word (as in a computer or a digital filter). The only condition is that the charge packets or binary words represent the values of $x_c(nT)$ accurately.

---

†The sampling theorem is usually attributed to Shannon, who introduced it in information theory in the late forties. At about the same time Kotelnikov was doing much the same thing in Russia. However, the actual theoretical foundation had already been laid a few decades earlier, by E. T. and J. A. Whittaker. It is probably better to follow the example of A. J. Jerri[4] and to refer to 'the WKS sampling theorem', using the first three letters of all three surnames.

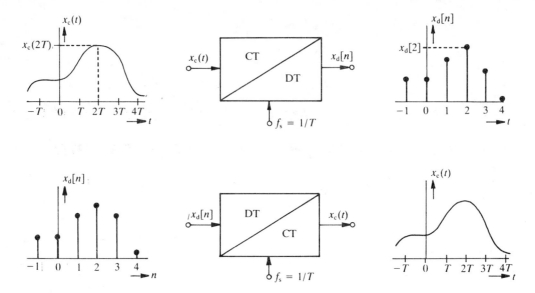

FIGURE 3.1   To convert a continuous-time signal into a discrete-time signal a CT/DT converter is used; the inverse operation is performed in a DT/CT converter.

In drawings we represent the successive samples of a discrete signal by 'pins with round heads'; in this way extra emphasis is placed on the difference from, say, delta functions, which are after all continuous-time functions. We also use square brackets for discrete signals, e.g. $x_d[n]$, where n represents an integer. In this notation there is thus no clear reference to the quantity 'absolute time' (expressed in seconds or other units of time). This can be a disadvantage, for example if the discrete sample has been obtained by sampling a continuous signal at a sampling interval of T seconds and we want to give a frequency description of the discrete signal in terms of absolute frequency (i.e. in Hz or rad/s). This simple notation may also have its disadvantages in situations where signals with different sampling intervals T_1 and T_2 (in other words, with different sampling rates) are to be treated simultaneously. In all these cases we shall use the completely equivalent notation $x[nT]$, $x[nT_1]$, $x[nT_2]$, etc. for $x[n]$.[†]

The relation in the time domain between a continuous signal $x_c(t)$ and the discrete signal $x_d[n]$ obtained from it by CD/DT conversion is given by (3.1). It is also easily recognized in the example of Fig. 3.1. But what is the relation between the spectrum $X_c(\omega)$ of $x_c(t)$ and the spectrum of $x_d[n]$? And how can we get back from discrete time to continuous time, or in other words, how can we recover $x_c(t)$ from $x_d[n]$? In answering these questions in the next two sections, we begin by taking a few apparently sideways steps.

[†]Sometimes the opposite is true: we then make little use of the fact that $x[n]$ is a function of the variable n, and only wish to make a clear distinction between discrete signals and continuous signals. The more compact notation x_n can then be used (see for example Chapter 12).

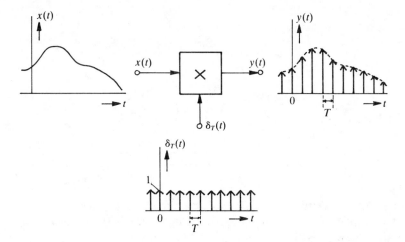

FIGURE 3.2 Multiplying a continuous signal $x(t)$ by a periodic train of Dirac pulses gives a sampled signal $y(t)$.

3.2 SAMPLING WITH DIRAC PULSES

Let us look more closely at the diagram of Fig. 3.2. Here a continuous signal $x(t)$ is multiplied by a periodic train or Dirac pulses of period T. This can be considered as a form of sampling. The output signal $y(t)$ is a continuous function given by:

$$y(t) = x(t)\delta_T(t) = x(t) \sum_{n=-\infty}^{\infty} \delta(t-nT) \tag{3.2a}$$

$$= \sum_{n=-\infty}^{\infty} x(t)\delta(t-nT) \tag{3.2b}$$

$$= \sum_{n=-\infty}^{\infty} x(nT)\delta(t-nT) \tag{3.2c}$$

The signal $y(t)$ thus consists of a train of equidistant (i.e. equally spaced) Dirac pulses. The area of the Dirac pulse that occurs at $t = nT$ is equal to the value of $x(t)$ at that instant. We find the spectrum $Y(\omega)$ of $y(t)$ by determining the Fourier transform of (3.2). In (3.2a), $y(t)$ is expressed as the product of the two time functions that are multiplied together in Fig. 3.2. From (2.35) we can find $Y(\omega)$ as a constant factor $1/(2\pi)$ times the *convolution* of two corresponding frequency functions, which are given by:

$$x(t) \; \circ\!\!-\!\!\circ \; X(\omega) \tag{3.3}$$

and (see also Appendix I)

$$\sum_{n=-\infty}^{\infty} \delta(t-nT) \; \circ\!\!-\!\!\circ \; \frac{2\pi}{T} \sum_{k=-\infty}^{\infty} \delta\left(\omega - \frac{2\pi k}{T}\right) \tag{3.4}$$

Therefore:

$$Y(\omega) = \frac{1}{2\pi} X(\omega) * \frac{2\pi}{T} \sum_{k=-\infty}^{\infty} \delta\left(\omega - \frac{2\pi k}{T}\right) \qquad (3.5a)$$

$$= \frac{1}{T} \sum_{k=-\infty}^{\infty} X(\omega) * \delta\left(\omega - \frac{2\pi k}{T}\right) \qquad (3.5b)$$

$$= \frac{1}{T} \sum_{k=-\infty}^{\infty} X\left(\omega - \frac{2\pi k}{T}\right) \qquad (3.5c)$$

Equation (3.5c) shows that $Y(\omega)$ consists of an infinite sequence of versions of $X(\omega)$ shifted by multiples of $2\pi/T$ and multiplied by the factor $1/T$. This can be seen in Fig. 3.3(a) for the case where $X(\omega) = 0$ for $|\omega| > \pi/T$ and in Fig. 3.3(b) for the case where this does *not* hold.

In this second case the condition of the sampling theorem is *not* satisfied. The result is that the shifted versions of $X(\omega)$ overlap one another and the frequency content of $Y(\omega)$ for certain values of ω is determined by more than one of the shifted versions of $X(\omega)$ (two in this example). This is known as *aliasing*. If this effect occurs, we can never find out from $Y(\omega)$ exactly what $X(\omega)$ and $x(t)$ were like before the sampling.

3.3 SIGNAL RECONSTRUCTION

From Fig. 3.3(a) we can also see how we can reconstruct $X(\omega)$ from $Y(\omega)$ and hence $x(t)$ from $y(t)$, provided that $x(t)$ satisfies the condition of the sampling theorem. All we have to do is to pass $Y(\omega)$ through an ideal low-pass filter of frequency response $L(\omega)$ where:

$$L(\omega) = \begin{cases} T & \text{for } |\omega| < \pi/T \\ 0 & \text{for } |\omega| > \pi/T \end{cases} \qquad (3.6)$$

The impulse response of this filter (see also Appendix I) is:

$$l(t) = \frac{\sin(\pi t/T)}{\pi t/T} \qquad (3.7)$$

The curves for $L(\omega)$ and $l(t)$ are shown in Fig. 3.4. (The characteristic shape of the curve for $l(t)$ is called the '$(\sin x)/x$ pulse'.) If we apply the signal $y(t)$ of (3.2c) to the input of this filter, we obtain an output signal that must be exactly the same as $x(t)$. Therefore:

$$x(t) = l(t) * y(t) = \frac{\sin(\pi t/T)}{\pi t/T} * \sum_{n=-\infty}^{\infty} x(nT)\delta(t-nT)$$

$$= \sum_{n=-\infty}^{\infty} x(nT) \left\{ \delta(t-nT) * \frac{\sin(\pi t/T)}{\pi t/T} \right\} \qquad (3.8)$$

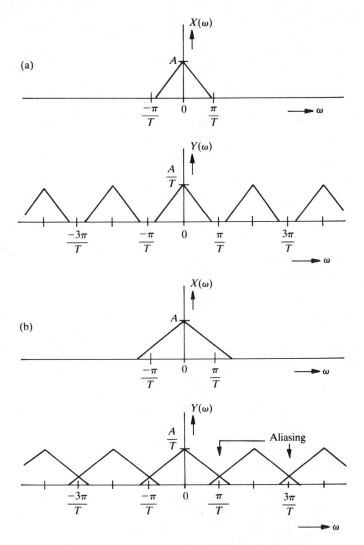

FIGURE 3.3 Spectra of $X(\omega)$ and $Y(\omega)$ from Fig. 3.2 for two different situations: in (a) aliasing does not occur and in (b) it does.

Using (2.16f) we have:

$$x(t) = \sum_{n=-\infty}^{\infty} x(nT) \left(\frac{\sin(\pi(t-nT)/T)}{\pi(t-nT)/T} \right) \tag{3.9}$$

The original function $x(t)$ can thus be obtained by adding together an infinite number of $(\sin x)/x$ pulses. The nth $(\sin x)/x$ pulse here is shifted through a distance nT with respect to the origin and multiplied ('weighted') by a factor $x(nT)$. This recovery process is called *interpolation*. It is illustrated in Fig. 3.5. Each separate $(\sin x)/x$ pulse is indicated by a different kind of line.

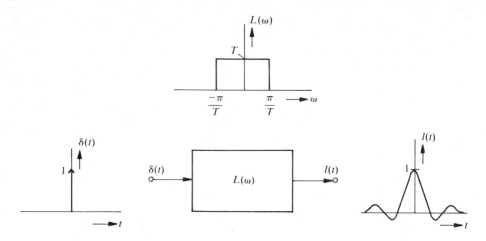

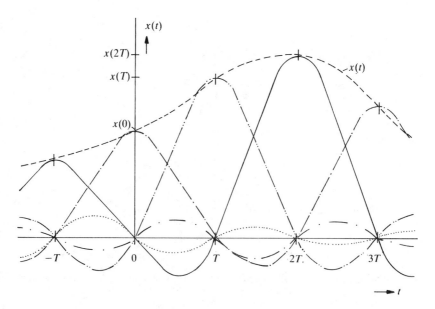

FIGURE 3.4 Ideal low-pass filter with frequency response $L(\omega)$ and impulse response $l(t)$.

FIGURE 3.5 Interpolation: the signal $x(t)$ can be reconstructed from the samples $x(nT)$ by summation of weighted and shifted $(\sin x)/x$ pulses.

3.4 PRACTICAL CONSIDERATIONS

In using Dirac pulses and an ideal low-pass filter in the previous two sections we were working at a rather abstract level, because:

- the Dirac pulse is a signal that does not exist in practice, since it is infinitely high and infinitely narrow;

- an ideal low-pass filter has an impulse response that runs from $t = -\infty$ to $t = \infty$; which means that such a filter is not causal, and therefore not realizable.

Strictly speaking, in sections 3.2 and 3.3 we went right to the boundary between the continuous-time domain and the discrete-time domain, but did not actually cross it. The final step, which is described by (3.1), has not yet been taken. However, we can now make use of the key position occupied by the Dirac pulse at the boundary between continuous and discrete time.

Le us calculate $Y(\omega)$ once again, but this time starting from (3.2c). We now use the Fourier pair (see Appendix I):

$$\delta(t - nT) \quad \circ\!\!-\!\!-\!\!\circ \quad e^{-j\omega nT} \tag{3.10}$$

Since $x(nT)$ is not a function of t, this remains unchanged as a kind of constant on Fourier transformation, and from (3.2c) we find:

$$Y(\omega) = \sum_{n=-\infty}^{\infty} x(nT)e^{-j\omega nT} \tag{3.11}$$

In Chapter 4 we shall see that a special Fourier transform, called the FTD, can be defined for discrete signals. For a discrete signal for which the successive samples correspond with $x(nT)$ exactly, the FTD gives a Fourier transform that we designate as $X(e^{j\omega T})$, which is exactly equal to the right-hand side of (3.11). We can summarize the above again in a number of diagrams; this is done in Fig. 3.6. So we have now answered the first question of section 3.1 about the relation between spectra before and after CT/DT conversion.

For the conversion in the inverse direction (from DT to CT) we still have the problem of the ideal low-pass filter from section 3.3. In practice this is usually solved by using a filter with a rectangular impulse response $k(t)$:

$$k(t) = \begin{cases} 1 & \text{for } 0 < t < T \\ 0 & \text{for } t < 0 \text{ and } t > T \end{cases} \tag{3.12}$$

At the output of this filter (also known as a zero-order hold circuit) we do not recover $x(t)$, but a stepwise approximation $\hat{x}(t)$ with Fourier transform $\hat{X}(\omega)$, for which (see also Fig. 3.7):

$$\hat{X}(\omega) = K(\omega) \cdot Y(\omega)$$

$$= \frac{2\sin(\omega T/2)}{\omega} e^{-j\omega T/2} Y(\omega) \tag{3.13}$$

Figure 3.8 shows $X(\omega)$, $Y(\omega)$ and $\hat{X}(\omega)$ (the term $e^{-j\omega T/2}$, which corresponds to a delay of $T/2$, is not considered here). We should see two important differences between the two functions $X(\omega)$ and $\hat{X}(\omega)$, which should be identical in the ideal case:

- $\hat{X}(\omega)$ has a $(\sin \omega T/2)/\omega$ distortion in the range $|\omega| < \pi/T$.
- $\hat{X}(\omega)$ contains frequency contributions around multiples of $2\pi/T$.

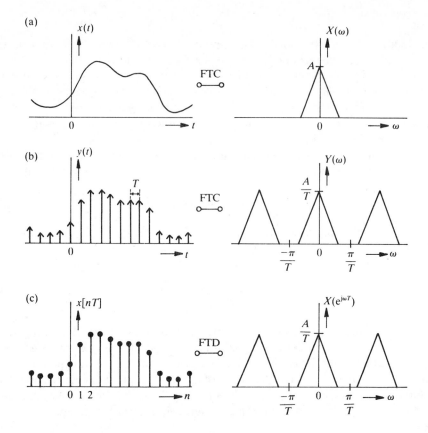

FIGURE 3.6 (a) The spectrum $X(\omega)$ of a continuous signal $x(t)$ can be determined via the FTC. (b) After sampling with a periodic train of Dirac pulses a continuous signal $y(t)$ is obtained whose spectrum $Y(\omega)$ can also be determined via the FTC. (c) The discrete signal $x[n]$ or $x[nT]$ can also be derived from $x(t)$. The spectrum $X(e^{j\omega T})$ of this discrete signal can be determined via the FTD. It turns out that $X(e^{j\omega T}) = Y(\omega)$.

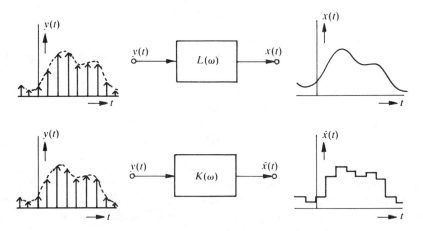

FIGURE 3.7 If a zero-order hold circuit $K(\omega)$ is used in the reconstruction of $x(t)$ from $y(t)$ instead of an ideal low-pass filter $L(\omega)$, we obtain a stepwise approximation $\hat{x}(t)$.

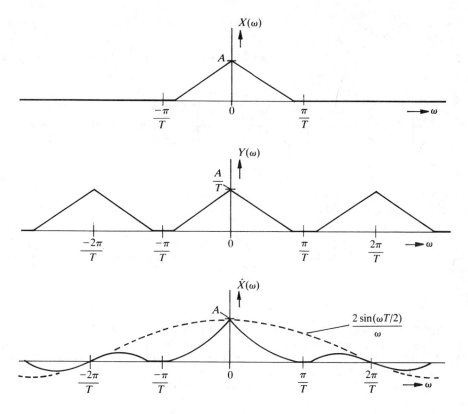

FIGURE 3.8 Spectra $X(\omega)$, $Y(\omega)$ and $\hat{X}(\omega)$ of the signals $x(t)$, $y(t)$ and $\hat{x}(t)$ from Fig. 3.7.

3.5 PRACTICAL SYSTEMS

A practical system for conversion from CT to DT and vice versa is shown in Fig. 3.9. We have tried to include all the aspects dealt with so far in this chapter in the diagram. The actual CT/DT converter is preceded by a low-pass prefilter with a pass band extending to half the sampling rate (π/T). This prevents aliasing that might be caused by unwanted high frequencies (such as noise!). The periodic repetition in the spectrum $X_d(e^{j\omega T})$ of $x_d[n]$ that we have encountered earlier is produced in the CT/DT converter. The DT/CT converter gives a stepwise output signal $\hat{x}_c(t)$ that is an approximation to $\bar{x}_c(t)$. Including a low-pass postfilter that limits the bandwidth again to half the sampling rate gives a smooth approximation $\tilde{x}_c(t)$ to $\bar{x}_c(t)$. The spectra corresponding to all the signals in Fig. 3.9 are also shown schematically.

In the ideal case $\tilde{X}_c(\omega)$ would be equal to $\bar{X}_c(\omega)$. However, the consequences of $(\sin x)/x$ distortion are still clearly visible in the range $|\omega| < \pi/T$. Something can be done about this, nevertheless; in the discrete-time operations usually performed on $x_d[n]$ between point 2 and point 3, it is generally not too difficult to include the complementary $x/(\sin x)$ distortion.

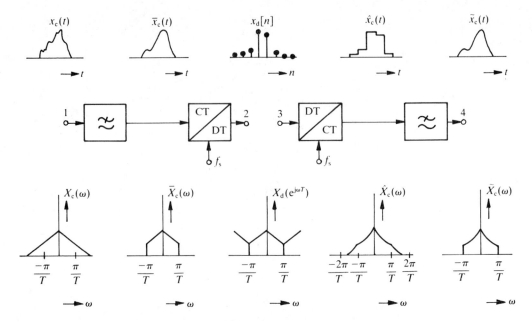

FIGURE 3.9 In a practical situation the signal $x_c(t)$ is first processed by a low-pass filter before CT/DT conversion to ensure that $\bar{x}_c(t)$ will satisfy the condition of the sampling theorem. After DT/CT conversion the initial result is usually a stepwise approximation $\hat{x}_c(t)$ to $\bar{x}_c(t)$. The steep transitions can be removed with an extra postfilter. This gives $\tilde{x}_c(t)$. The spectrum corresponding to each signal is also shown.

3.6 CONCLUDING REMARKS

To conclude this chapter we should like to emphasize once again that not every discrete signal is derived by sampling a continuous signal; there are also signals that are discrete in origin. A simple example is a digital tone generator that directly provides a sequence of numbers whose value varies sinusoidally with time. However, the successive signal values are called 'samples' in this case too.

The widely used term 'analog/digital converter' (often abbreviated to A/D converter) refers to the combination of the CT/DT converter extensively discussed in this chapter with a converter from continuous to discrete amplitude (a 'quantization circuit'; see also Chapter 10 'Finite word length in digital signals and systems').

We can now define the term digital/analog converter: a DT/CT converter that is only suitable for discrete-time signals with a discrete amplitude variation (and therefore only suitable for digital signals).

3.7 EXERCISES

3.1 Given $x(t) = \sin(2\pi t)$,
   (a) Determine $x(nT_1)$ for $-6 \leq n \leq 8$ and T_1 corresponding to a sampling rate of 4 Hz.

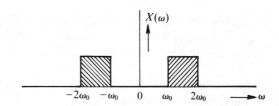

FIGURE 3.10 Exercise 3.2.

(a)

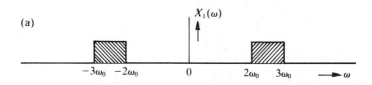

(b)

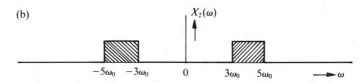

FIGURE 3.11 Exercise 3.3.

(b) Determine $x(t)$ for $t = 1/8$ from (3.9), using $x(nT_1)$ as just determined, with $-6 \leq n \leq 8$, and compare this with the exact value of $\sin(2\pi/8)$. What do you notice?

(c) Determine $x(nT_2)$ for $-6 \leq n \leq 8$ and T_2 corresponding to a sampling rate of 2 Hz.

(d) Determine $x(t)$ for $t = 1/8$ from (3.9), using $x(nT_2)$ as just determined, with $-6 \leq n \leq 8$, and compare this with the exact value of $\sin(2\pi/8)$. What do you notice?

3.2 With signals of a band-pass nature, i.e. where both $X(\omega) = 0$ for $|\omega| > |\omega_h|$ and $X(\omega) = 0$ for $|\omega| < |\omega_1|$, direct application of the sampling theorem leads to an unnecessarily high sampling frequency $\omega_s = 2\omega_h$. For signals of this type it is possible to work with a lower sampling frequency.

Given a signal $x(t)$ with the spectrum $X(\omega)$ shown in Fig. 3.10:

(a) Draw the spectrum $Y_1(\omega)$ produced if the signal $x(t)$ is sampled at the minimum frequency $\omega_s = \omega_{min}$ given by the sampling theorem.

(b) Draw the spectrum $Y_2(\omega)$ produced if the signal $x(t)$ is sampled at $\omega_s = 2\omega_0$.

(c) How can we recover the original signal $x(t)$ from the spectrum $Y_2(\omega)$?

3.3 For band-limited signals as defined in Exercise 3.2 the minimum sampling frequency is given by:

$$\omega_{min} = \frac{2\omega_h}{\text{largest integer no greater than} \left\{ \dfrac{\omega_h}{\omega_h - \omega_1} \right\}}$$

(a) Determine the minimum sampling frequencies $\omega_{min,1}$ and $\omega_{min,2}$ for the signals $x_1(t)$ and $x_2(t)$ whose spectra $X_1(\omega)$ and $X_2(\omega)$ are given in Fig. 3.11(a) and (b).

(b) Draw the spectra produced after sampling $x_1(t)$ and $x_2(t)$ at these minimum sampling frequencies.

(c) Draw the spectra produced after sampling $x_1(t)$ and $x_2(t)$ at a frequency slightly *lower* than the minimum sampling frequency.

(d) Draw the spectra produced after sampling $x_1(t)$ and $x_2(t)$ at a frequency slightly *higher* than the minimum sampling frequency.

4

Discrete-time signals and systems

4.1 INTRODUCTION

In the previous chapter we saw how we can derive a discrete-time signal from a continuous-time signal. It was also mentioned that it is possible to generate a discrete-time signal without there ever having been an original continuous-time version ('discrete-time signal generator'). The origin of discrete-time signals is of no interest for their processing and we shall not consider it further.

It may perhaps help to list here a number of the conventions that we have mentioned in the earlier chapters:

1. 'Discrete' with no other indication always means 'discrete-time'.
2. The successive values or samples of a discrete signal are represented graphically by 'pins with round heads' (see Figs 3.1 and 3.9).
3. A discrete signal can be expressed both as a function of n and of nT; e.g. as $x[n]$ or $x[nT]$. Both notations are completely equivalent if we put $T = 1$ second, i.e. if we normalize the sampling rate to $1\,\mathrm{Hz}$. The advantage of the first notation is the simplicity of the expressions and we shall therefore make considerable use of it. A disadvantage is that the relationships to the 'true' frequencies (in Hz or rad/s) become less clear. We also get into difficulties if signals with different sampling rates occur in a system. Then there are advantages in changing over to the second notation. We shall encounter examples of this later in this book.
4. If it has not been explicitly stated otherwise, the discrete signals can assume all possible values; i.e. amplitude-quantization effects (such as those that occur in digital systems) will for the present be neglected. This is dealt with later in Chapter 10.

Later on in this book we shall see that there are a number of essential differences between discrete and continuous signals or systems. A number of clear analogies exist, however, and we shall use these wherever possible.

4.2 DISCRETE SIGNALS – DESCRIPTION IN THE TIME DOMAIN

Discrete signals are only defined for discrete values of time. The successive values are separated by a distance of T seconds (the sampling interval). The quantity $1/T$ represents the sampling rate in Hz. We usually work with the quantity $2\pi/T$, i.e. the sampling frequency in rad/s. For the moment we shall put $T = 1$.

4.2.1 Examples

A number of examples of a discrete signal $x[n]$ are given in Fig. 4.1.

- The signal $x_1[n]$ can be formally described by:

$$x_1[n] = \begin{cases} 0 & \text{for } n<0 \text{ and } n>3 \\ n & \text{for } 0\leq n\leq 3 \end{cases} \qquad (4.1)$$

This is a discrete signal of *finite duration*; i.e. below a certain value of n and above a certain value of n, the signal is equal to zero. The signal is defined, however, for all values of n.

- The signal $x_2[n]$ can be expressed formally as:

$$x_2[n] = \begin{cases} 0 & \text{for } n<0 \\ (0.9)^n & \text{for } n\geq 0 \end{cases} \qquad (4.2)$$

This is a discrete signal of *infinite duration*.

- The signal $x_3[n]$ cannot be directly represented by a mathematical expression, but could for example be a discrete representation of a short segment of analog speech.

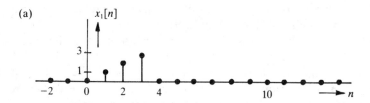

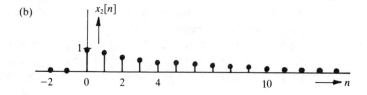

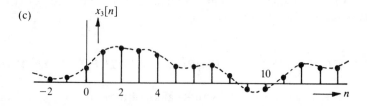

FIGURE 4.1 Examples of discrete signals.

4.2.2 Important discrete signals

There are a number of discrete signals that are important because we shall use them frequently. These are shown in Fig. 4.2. They are, in order:

(a) The discrete unit pulse $\delta[n]$, which is defined as:

$$\delta[n] = \begin{cases} 1 & \text{for } n = 0 \\ 0 & \text{for } n \neq 0 \end{cases} \tag{4.3}$$

This unit pulse fulfils the same role in discrete systems as the Dirac pulse (section 2.4) in continuous systems. In discrete systems, however, we have no difficulties in giving a good definition (we do not have to consider 'generalized functions').

(b) The unit pulse shifted through a time interval i, $\delta[n - i]$, defined as:

$$\delta[n - i] = \begin{cases} 1 & \text{for } n = i \\ 0 & \text{for } n \neq i \end{cases} \tag{4.4}$$

We can use $\delta[n - i]$ to express discrete signals in another way, for example (see Fig. 4.1(c)):

$$x_3[n] = \sum_{i = -\infty}^{\infty} x_3[i]\delta[n - i] \tag{4.5}$$

It will be found that this is a very useful notation, which is generally valid and which we shall encounter frequently. (Equation (4.5) is the exact discrete counterpart of (2.24) for continuous systems.)

(c) The discrete unit-step function $u[n]$:

$$u[n] = \begin{cases} 0 & \text{for } n < 0 \\ 1 & \text{for } n \geq 0 \end{cases} \tag{4.6}$$

We can also express $u[n]$ as:

$$u[n] = \sum_{i = -\infty}^{\infty} u[i]\delta[n - i] = \sum_{i = 0}^{\infty} \delta[n - i] \tag{4.7}$$

(d) and (e) Discrete sine functions $x_4[n]$ and $x_5[n]$. In its most general form the definition of this kind of function is:

$$x[n] = A \sin(n\theta + \varphi) \tag{4.8}$$

with: A = amplitude
θ = relative frequency (the absolute frequency $\omega = \theta/T$)
φ = phase.

Two of these discrete sine functions are drawn here (for $A = 1$, $\varphi = 0$ and with $\theta = \pi/4$ and $\theta = 1$). We can see that there is an interesting difference between these two signals. The function $\sin(n\pi/4)$ is a periodic discrete function of period 8, since:

$$\sin(n\pi/4) = \sin(n\pi/4 + 2\pi) = \sin\{(n + 8)\pi/4\} \tag{4.9}$$

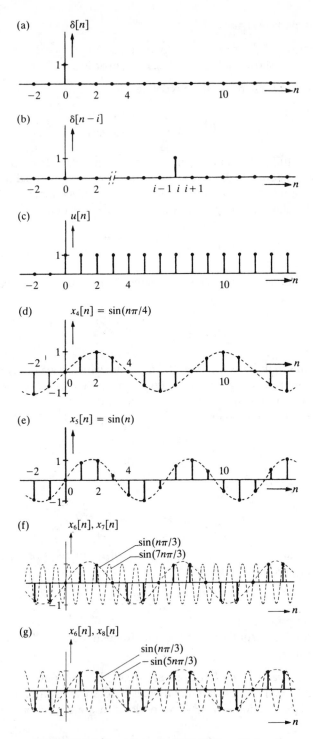

FIGURE 4.2 Some frequently used discrete signals.

The function $\sin(n)$ on the other hand is non-periodic, because no *integer* n_0 can be found for which:

$$\sin(n) = \sin(n + n_0) \tag{4.10}$$

We see here that a discrete sine function does not always represent a periodic signal, whereas a continuous sine function always does.

(f) Another observation that we can make is that two discrete sine functions with relative frequencies θ_1 and θ_2, where $\theta_2 = \theta_1 + i2\pi$ (where i is any integer), are completely identical and therefore indistinguishable, since:

$$\sin(\theta_2 n) = \sin\{(\theta_1 + i2\pi)n\} = \sin(\theta_1 n + ni2\pi) = \sin(\theta_1 n) \tag{4.11}$$

In Fig. 4.2(f) we show this for the signals $x_6[n] = \sin(n\pi/3)$ and $x_7[n] = \sin(n\pi/3 + 2\pi n) = \sin(7n\pi/3)$. These consist of exactly the same sequences of samples. (The dashed lines indicate the curves the functions would follow if n also represented all non-integer numbers.)

(g) There are in fact even more discrete sine functions with other frequencies that give the same sequence of samples as $\sin(n\theta_1)$. These are all the signals given by:

$$x[n] = -\sin(-n\theta_1 + 2\pi i n) \tag{4.12}$$

As an example the signals $x_6[n] = \sin(n\pi/3)$ and $x_8[n] = -\sin(-n\pi/3 + 2\pi n) = -\sin(5n\pi/3)$ are shown in Fig. 4.2(g).

Figure 4.2(f) and (g) can also be considered to be an illustration of the sampling theorem from the previous chapter: for the three hypothetical continuous (= dashed) sine functions that correspond to $x_6[n]$, $x_7[n]$ and $x_8[n]$ it is only for the case of $x_6[n]$ that more than two samples come within one sine period and only this case would satisfy the sampling-theorem condition.

In addition to the real discrete functions that we have encountered so far, we shall also make considerable use of the complex exponential function:

$$e^{jn\theta} = \cos(n\theta) + j\sin(n\theta) \tag{4.13}$$

This function plays a comparable role to the function $e^{j\omega t}$ in continuous-time systems.

4.3 DISCRETE SIGNALS – DESCRIPTION IN THE FREQUENCY DOMAIN

In section 2.2 we saw how we can describe continuous signals in the frequency domain with the aid of the Fourier transform. The same thing can be done for discrete signals, if we take into account the characteristic property that they are only defined at discrete times (and not between them). The Fourier transform $X(e^{j\omega T})$ of an arbitrary discrete signal $x[nT]$, or the FTD of $x[nT]$, is defined as:

$$\boxed{X(e^{j\omega T}) = \sum_{n=-\infty}^{\infty} x[nT]e^{-jn\omega T}} \qquad \text{(FTD)} \tag{4.14}$$

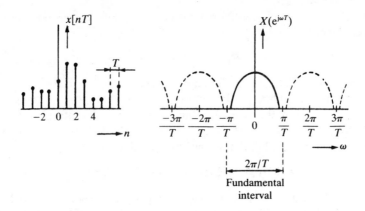

FIGURE 4.3 Schematic representation of an FTD pair as given by equation (4.14); ω is the absolute frequency (in rad/s).

We see two differences from the Fourier transform for continuous signals (the FTC) described by (2.1):

- The integral has been replaced by a summation.
- The frequency variable ω on the left-hand side has been replaced by $e^{j\omega T}$. This has been done to emphasize that the function X is periodic with a period of $2\pi/T$; replacing ω by $\omega + 2k\pi/T$ on the right-hand side gives the same result from the summation. This means that in the graphical representation of X we only have to draw an interval of width $2\pi/T$. The interval is generally taken as $-\pi/T \leq \omega < \pi/T$ (*the fundamental interval*); (see Fig. 4.3).

The function X (also called the spectrum of $x[nT]$) is a complex function, which we can divide as we choose into a real part (R) and an imaginary part (I) or into an amplitude (A) and a phase (φ):

$$X(e^{j\omega T}) = R(e^{j\omega T}) + j I(e^{j\omega T}) = A(e^{j\omega T}) e^{j\varphi(e^{j\omega T})} \tag{4.15a}$$

or more briefly:

$$X(e^{j\omega T}) = R + jI = A e^{j\varphi} \tag{4.15b}$$

The inverse transformation (IFTD) is given by:

$$\boxed{x[nT] = \frac{T}{2\pi} \int_{-\pi/T}^{\pi/T} X(e^{j\omega T}) e^{jn\omega T} d\omega} \quad \text{(IFTD)} \tag{4.16}$$

The functions $X(e^{j\omega T})$ and $x[nT]$ form a Fourier pair, which we represent symbolically by:

$$\boxed{x[nT] \;\circ\!\!-\!\!-\!\!\circ\; X(e^{j\omega T})} \tag{4.17}$$

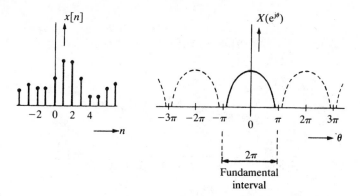

FIGURE 4.4 Schematic representation of an FTD pair as given by equation (4.18); $\theta = \omega T$ is the relative frequency (in radians).

If we make use of the relative frequency $\theta = \omega T$, we can also write (4.14), (4.16) and (4.17) as:

$$X(e^{j\theta}) = \sum_{n=-\infty}^{\infty} x[n]e^{-jn\theta} \qquad \text{(FTD)} \qquad (4.18)$$

$$x[n] = \frac{1}{2\pi} \int_{-\pi}^{\pi} X(e^{j\theta})e^{jn\theta}\,d\theta \qquad \text{(IFTD)} \qquad (4.19)$$

$$x[n] \;\circ\!\!-\!\!-\!\!\circ\; X(e^{j\theta}) \qquad (4.20)$$

For the relative frequency θ the fundamental interval extends from $-\pi \le \theta < \pi$ (see Fig. 4.4). Figures 4.3 and 4.4 are completely equivalent; it is a matter of taste which is preferable in a given situation.

4.3.1 Examples

Example 1
The FTD of the discrete unit pulse $\delta[n]$ gives:

$$\delta[n] \;\circ\!\!-\!\!-\!\!\circ\; X(e^{j\theta}) = \sum_{n=-\infty}^{\infty} \delta[n]e^{-jn\theta} = 1 \qquad (4.21)$$

In Fig. 4.5(a) $\delta[n]$ and the fundamental interval of $X(e^{j\theta}) = 1$ are shown.

Example 2
For the FTD of the signal $x[n]$ that is defined as:

$$x[n] = \begin{cases} 0 & \text{for } n < 0 \\ a^n & \text{for } n \ge 0 \text{ with } |a| < 1 \end{cases} \qquad (4.22a)$$

or:

$$x[n] = a^n u[n] \text{ with } |a| < 1 \qquad (4.22b)$$

we find:[†]

$$X(e^{j\theta}) = \sum_{n=0}^{\infty} a^n e^{-jn\theta} = \sum_{n=0}^{\infty} (ae^{-j\theta})^n$$

$$= \frac{1}{1 - ae^{-j\theta}} = \frac{1}{1 - a\cos(\theta) + ja\sin(\theta)} = A e^{j\varphi} \qquad (4.23a)$$

where:

$$A(e^{j\theta}) = \sqrt{\frac{1}{1 - 2a\cos(\theta) + a^2}} \qquad (4.23b)$$

$$\varphi(e^{j\theta}) = -\arctan\left(\frac{a\sin(\theta)}{1 - a\cos(\theta)}\right) \qquad (4.23c)$$

The modulus $A(e^{j\theta})$ and the phase $\varphi(e^{j\theta})$ are shown in Fig. 4.5(c) (for $a = 0.65$).

4.3.2 Properties of the FTD

Many of the properties (but not all of them) that we give in Appendix I for the FTC have a counterpart in the FTD. A number of these properties will be briefly presented here. We shall make use of the following Fourier pairs, where $x[n]$, $x_1[n]$ and $x_2[n]$ are arbitrary discrete signals:

$$x[n] \quad \circ\!\!-\!\!-\!\!\circ \quad X(e^{j\theta})$$
$$x_1[n] \quad \circ\!\!-\!\!-\!\!\circ \quad X_1(e^{j\theta}) \qquad (4.24)$$
$$x_2[n] \quad \circ\!\!-\!\!-\!\!\circ \quad X_2(e^{j\theta})$$

A. Linearity

$$\boxed{ax_1[n] + bx_2[n] \quad \circ\!\!-\!\!-\!\!\circ \quad aX_1(e^{j\theta}) + bX_2(e^{j\theta})} \qquad (4.25)$$

where a and b are arbitrary constants.

[†]Here we make use of the most elementary property of the geometrical progression:

$$\sum_{n=0}^{\infty} \alpha^n = 1 + \alpha + \alpha^2 + \ldots = \frac{1}{1-\alpha} \text{ for } |\alpha| < 1$$

where α can be a complex number. Later on we shall regularly make use of the property:

$$\sum_{n=0}^{N-1} \alpha^n = 1 + \alpha + \alpha^2 + \ldots + \alpha^{N-1} = \frac{1 - \alpha^N}{1 - \alpha} \text{ for all } \alpha \neq 1$$

B. Time shift

$$\boxed{x[n-i] \ \circ\!\!-\!\!-\!\!\circ \ e^{-ji\theta}X(e^{j\theta})}$$ (4.26)

where i is any integer.

C. Frequency shift

$$\boxed{x[n]e^{jn\theta_0} \ \circ\!\!-\!\!-\!\!\circ \ X(e^{j(\theta-\theta_0)})}$$ (4.27)

A single shift of the spectrum $X(e^{j\theta})$ leads to a complex time function. But if we apply a frequency shift twice, we can again obtain a real function of time (this happens in modulation, for example):

$$2x[n]\cos(n\theta_0) \ \circ\!\!-\!\!-\!\!\circ \ X(e^{j(\theta-\theta_0)}) + X(e^{j(\theta+\theta_0)})$$

$$2x[n]\sin(n\theta_0) \ \circ\!\!-\!\!-\!\!\circ \ \frac{1}{j}\{X(e^{j(\theta-\theta_0)}) - X(e^{j(\theta+\theta_0)})\}$$ (4.28)

D. Convolution in the time domain
The convolution of discrete signals is comparable with the convolution of continuous signals and will be treated at length later in this chapter (section 4.4.4); for the convolution of two discrete signals we have the following property:

$$\boxed{x_1[n] * x_2[n] \ \circ\!\!-\!\!-\!\!\circ \ X_1(e^{j\theta})X_2(e^{j\theta})}$$ (4.29)

Therefore: convolution in the time domain corresponds to multiplication in the frequency domain.

For completeness we shall state here for the moment that the convolution of two discrete signals $x_1[n]$ and $x_2[n]$ is defined as:

$$x_1[n] * x_2[n] = \sum_{i=-\infty}^{\infty} x_1[i]x_2[n-i]$$ (4.30)

E. Convolution in the frequency domain

$$\boxed{x_1[n]x_2[n] \ \circ\!\!-\!\!-\!\!\circ \ \frac{1}{2\pi}X_1(e^{j\theta}) * X_2(e^{j\theta})}$$ (4.31)

Therefore: multiplication in the time domain corresponds to convolution in the frequency domain. The convolution of the two periodic frequency functions is defined here as:

$$X_1(e^{j\theta}) * X_2(e^{j\theta}) = \int_{-\pi}^{\pi} X_1(e^{j\nu})X_2(e^{j(\theta-\nu)})d\nu$$ (4.32)

F. Parseval's theorem

$$\sum_{n=-\infty}^{\infty} |x[n]|^2 = \frac{1}{2\pi} \int_{-\pi}^{\pi} |X(e^{j\theta})|^2 \, d\theta \qquad (4.33)$$

This theorem shows that the 'energy' of a signal can be calculated both in the time domain and in the frequency domain.

G. Arbitrary real time functions

For arbitrary discrete real functions $x[n]$ the real part of the transform $X(e^{j\theta})$ is even and the imaginary part is odd, i.e.:

$$X(e^{j\theta}) = X^*(e^{-j\theta}) \qquad (4.34a)$$

(where $*$ indicates the complex conjugate)
or expressed as real part R and imaginary part I:

$$R(e^{j\theta}) = R(e^{-j\theta}) \quad \text{and} \quad I(e^{j\theta}) = -I(e^{-j\theta}) \qquad (4.34b)$$

or as modulus A and argument φ:

$$A(e^{j\theta}) = A(e^{-j\theta}) \quad \text{and} \quad \varphi(e^{j\theta}) = -\varphi(e^{-j\theta}) \qquad (4.34c)$$

Note: From this property we can draw an interesting conclusion: the function $X(e^{j\theta})$ is completely known if we know its value in the *half* fundamental interval $0 \le \theta < \pi$. We can then calculate the values of $X(e^{j\theta})$ for $-\pi \le \theta < 0$ directly from one of the relations above

TABLE 4.1 Summary of the properties G, H, I and J of the FTD

Property		$x[n]$	$X(e^{j\theta})$
G	1	Real	Real part even Imaginary part odd
H	2	Real and even	Real and even
I	3	Real and odd	Imaginary and odd
J	4	Imaginary	Real part odd Imaginary part even
	5	Imaginary and even	Imaginary and even
	6	Imaginary and odd	Real and odd
	7	Real part even Imaginary part even	Real part even Imaginary part even
	8	Real part odd Imaginary part odd	Real part odd Imaginary part odd
	9	Real part even Imaginary part odd	Real
	10	Real part odd Imaginary part even	Imaginary

(the same thing also applies of course if we know $X(e^{j\theta})$ in the other half fundamental interval $-\pi \leq \theta < 0$); see for example Fig. 4.5.

H. Even real time functions
If $x[n]$ is an *even* and *real* discrete function, i.e. $x[n] = x[-n]$, then $X(e^{j\theta})$ is an *even* and *real* function.

I. Odd real time functions
If $x[n]$ is an *odd* and *real* discrete function, i.e. $x[n] = -x[-n]$, then $X(e^{j\theta})$ is an *odd* and *imaginary* function.

J. Complex time functions
A number of special rules can also be derived for purely imaginary and complex discrete functions. Properties G, H, I and J are summarized in Table 4.1.

K. Waveform decomposition
An arbitrary discrete function $x[n]$ can always be expressed as the sum of an even function $x_e[n]$ and an odd function $x_o[n]$:

$$x[n] = \frac{x[n]+x[-n]}{2} + \frac{x[n]-x[-n]}{2} \tag{4.35a}$$

$$= x_e[n] + x_o[n]$$

where:

$$x_e[n] = \frac{x[n]+x[-n]}{2} \quad \text{and} \quad x_o[n] = \frac{x[n]-x[-n]}{2} \tag{4.35b and c}$$

4.3.3 Frequently used transform pairs

A number of frequently used FTD pairs have been collected in Fig. 4.5. We have included a number of interesting pairs here that we cannot calculate directly from the FTD definition (4.18), because the summation gives a value of infinity (the summation does not converge). These cases of special interest are:

$$x[n] = 1 \quad \circ\!\!-\!\!-\!\!\circ \quad X(e^{j\theta}) = 2\pi\delta(\theta) \tag{4.36a}$$

$$x[n] = \cos(n\theta_o) \quad \circ\!\!-\!\!-\!\!\circ \quad X(e^{j\theta}) = \pi\delta(\theta - \theta_0) + \pi\delta(\theta + \theta_0) \tag{4.36b}$$

$$x[n] = \sin(n\theta_o) \quad \circ\!\!-\!\!-\!\!\circ \quad X(e^{j\theta}) = -\pi j\delta(\theta - \theta_0) + \pi j\delta(\theta + \theta_0) \tag{4.36c}$$

$$x[n] = e^{jn\theta_o} \quad \circ\!\!-\!\!-\!\!\circ \quad X(e^{j\theta}) = 2\pi\delta(\theta - \theta_0) \tag{4.36d}$$

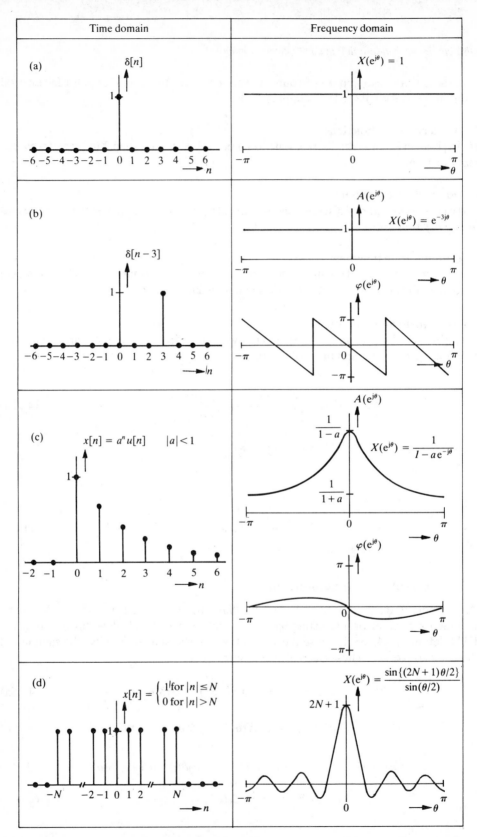

FIGURE 4.5 Examples of FTD pairs.

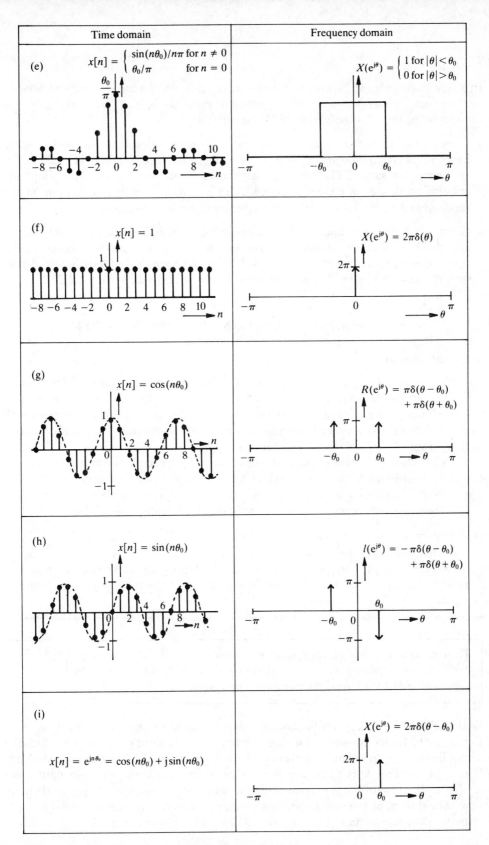

FIGURE 4.5 (*continued*)

In all these we find delta functions in the frequency domain. We can see that we really are dealing with Fourier pairs, because there are no problems in going from the frequency domain to the time domain via the IFTD (verify this).

- *Comment 1.* In section 4.2.2 we saw that two discrete sine functions with relative frequencies θ_1 and θ_2, where $\theta_2 = \theta_1 + i2\pi$ (i an integer), are identical. They therefore also have the same FTD. To avoid ambiguity, from now on we shall adopt the convention that θ_0 in $\cos(n\theta_0)$, $\sin(n\theta_0)$ and $e^{jn\theta_0}$ always refers to the value *in* the fundamental interval (i.e. $-\pi \leq \theta_0 < \pi$), unless expressly stated otherwise.

- *Comment 2.* Clear parallels can be recognized between the FTD pairs (4.36a)–(4.36d) and the FTC pairs of section 2.5. There is however one striking difference: the continuous signals considered in section 2.5 were all periodic, but the discrete signals $\sin(n\theta_0)$, $\cos(n\theta_0)$ and $e^{jn\theta_0}$ do not have to be periodic at all (see section 4.2.2).

4.4 LINEAR TIME-INVARIANT DISCRETE SYSTEMS

4.4.1 Introduction

Now that we have become acquainted with discrete *signals*, we can turn our attention to discrete *systems*. A discrete system is defined as a system that converts one or more discrete input signals $x[n]$ into one or more discrete output signals $y[n]$ in accordance with certain (discrete) rules. In the following we shall mainly be dealing with systems with one real input signal $x[n]$ and one real output signal $y[n]$.

We shall also confine our considerations in the first instance to the important class of linear time-invariant discrete systems (LTD systems). For practical purposes this includes most discrete (and hence digital) filters. A system of this type is shown in Fig. 4.6.

We define the properties *linearity* and *time-invariance* on the basis of $x[n]$ and $y[n]$.

Linearity. A discrete system is linear if the input signal $ax_1[n] + bx_2[n]$ produces an output signal $ay_1[n] + by_2[n]$, where a and b are arbitrary constants. Here $x_1[n]$ and $x_2[n]$ are arbitrary input signals and $y_1[n]$ and $y_2[n]$ the corresponding output signals.

Time-invariance. A discrete system is time-invariant if the input signal $x[n-i]$ produces an output signal $y[n-i]$, where i is an arbitrary integer, $x[n]$ an arbitrary input signal and $y[n]$ the corresponding output signal.

We can clarify the concepts of linearity and time-invariance with the aid of the systems of Fig. 4.7, which each contain just a single multiplier. Depending on the way in which the multiplier is used, completely different properties are obtained. The system H_1, shown in Fig. 4.7(a), is a time-varying system. We can see this in the following way. An input signal $x[n] = \delta[n]$ gives an output signal $y[n] = \cos(n\pi/2)\delta[n] = \cos(0)\,\delta[n] = \delta[n]$. If we now shift this input signal through one sampling interval and thus introduce $\delta[n-1]$ into the system, we find $y[n] = \cos(n\pi/2)\delta[n-1] = \cos(\pi/2)\delta[n-1] = 0$, which is

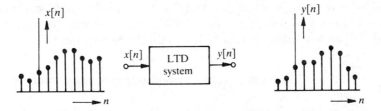

FIGURE 4.6 An LTD system.

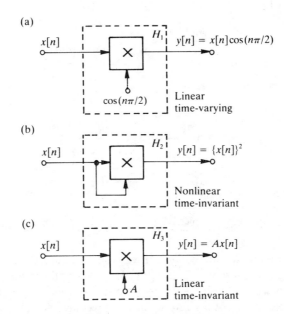

FIGURE 4.7 Linearity and time-invariance in some simple discrete systems.

certainly not a shifted version of the output signal found earlier! The system H_2 of Fig. 4.7(b) is nonlinear because on doubling the input signal the output signal does not double, but quadruples. The other properties stated for the systems of Fig. 4.7 can be verified easily by means of the definitions given earlier (verify them!).

Just as in continuous systems, the concepts of causality and stability are also important in discrete systems in connection with practical realizability. In general we require of feasible systems that they should be *stable* and *causal*. The two formal definitions are:

Stability. A discrete system is stable if any arbitrary input signal $x[n]$ of finite amplitude (i.e. $|x[n]|_{max} \leq A$) produces an output signal $y[n]$ of finite amplitude (i.e. $|y[n]|_{max} \leq B$).

Causality. A discrete system is causal if the output signal at any instant $n = n_0$ is not dependent on the values of the input signal at times *later* than n_0.

(In section 4.4.3 we shall give examples of LTD systems that are stable or non-stable and causal or non-causal.) LTD systems have a number of particularly attractive properties. The first is that all realizable systems of this type can be composed of only three basic elements (Fig. 4.8):

(a) The adder, in which two input signals are added to give one output signal: $y[n] = x_1[n] + x_2[n]$;
(b) The multiplier, in which a signal is multiplied by a constant: $y[n] = Ax[n]$;
(c) The unit-delay element, in which a signal $x[n]$ is delayed by one sampling interval: $y[n] = x[n-1]$.

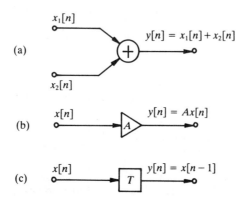

FIGURE 4.8 The three basic elements for LTD systems.

With these elements we can make, for example, a subtracter by combining an adder and a multiplier in which $A = -1$; (see Fig. 4.9(a)). Another simple, but realistic, LTD system in which all the basic elements are present is shown in Fig. 4.9(b). A close relation exists between the theory of discrete signals that we have dealt with earlier in this book and the theory of LTD systems. There is a very good explanation for this, since the most important properties of an LTD system can be derived from only one discrete signal – the output signal that is obtained when the input signal is the discrete unit pulse $\delta[n]$. We call this output signal the *impulse response $h[n]$* of the LTD system.

The impulse response $h[n]$ is a description of the system in the time domain. If we apply the FTD to $h[n]$, we obtain the *frequency response $H(e^{j\theta})$*; this is a description of the system in the frequency domain. It is also possible to describe LTD systems by means of *difference equations*; this is again a description in the time domain.

In the following sections we shall look more closely at these three possibilities, starting with the last. (Later on we shall learn something about yet another important possibility, the *system function $H(z)$*.)

4.4.2 The difference equation as a system description

The discrete system shown in Fig. 4.9(b) is completely described by the two simple equations:

$$v[n] = ax[n] + bx[n-1] + y[n] \tag{4.37a}$$

(a)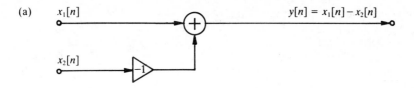

(b)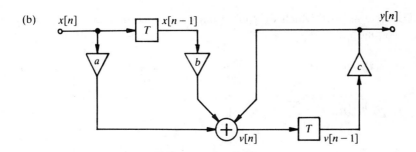

FIGURE 4.9 Two simple LTD systems.

and

$$y[n] = cv[n-1] \tag{4.37b}$$

Equations of this type, in which the value of a discrete signal at a particular moment ($v[n]$ and $y[n]$ here) is given as a function of signal values at earlier instants ($x[n-1]$ and $v[n-1]$ here), are called *difference equations*.

We can combine these two equations to give a single difference equation that contains only delayed versions of the input signal $x[n]$ and delayed versions of the output signal $y[n]$:

$$y[n] = acx[n-1] + bcx[n-2] + cy[n-1] \tag{4.38}$$

With these equations we can in principle calculate the output signal $y[n]$ for any arbitrary input signal $x[n]$. As an example we take $x[n] = \delta[n]$ and we assume that $y[n] = 0$ for $n < 0$; this gives:

$$y[0] = acx[-1] + bcx[-2] + cy[-1] = 0$$

$$y[1] = acx[0] + bcx[-1] + cy[0] = ac$$

$$y[2] = acx[1] + bcx[0] + cy[1] = (b + ac)c$$

$$y[3] = acx[2] + bcx[1] + cy[2] = (b + ac)c^2 \tag{4.39a}$$

$$y[n] = acx[n-1] + bcx[n-2] + cy[n-1] = (b + ac)c^{n-1}$$

Resuming, we find for the output signal:

$$y[n] = \begin{cases} 0 & \text{for } n \leq 0 \\ ac & \text{for } n = 1 \\ (b + ac)c^{n-1} & \text{for } n \geq 2 \end{cases} \qquad (4.39b)$$

The LTD systems with which we shall usually be concerned can be described by the difference equation:

$$y[n] = b_0 x[n] + b_1 x[n-1] + \ldots + b_N x[n-N] + a_1 y[n-1] + \ldots + a_M y[n-M]$$

$$= \sum_{i=0}^{N} b_i x[n-i] + \sum_{i=1}^{M} a_i y[n-i] \qquad (4.40)$$

This is called a linear difference equation with constant coefficients. We can do a number of things with this difference equation:

- For a given $x[n]$ and coefficients a_i and b_i we can calculate the values $y[0]$, $y[1]$, . . . step by step, as we did for the system of Fig. 4.9(b). This is only practical for simple systems and simple signals, however.
- We can use the general theory that exists for the solution of difference equations to find $y[n]$ directly as a function of n. This method is highly analogous to the way in which differential equations are solved. We shall not consider it further here, however, since in general it is a rather roundabout approach.
- The difference equation can be used as the starting point for other system descriptions (in the frequency domain; just like the differential equation for continuous systems). We are then no longer concerned with difference equations but with algebraic equations, which are much easier to work with.
- The difference equation gives a direct indication of a possible structure of a system.

Example
A system that is described by the difference equation:

$$y[n] = \sum_{i=0}^{3} b_i x[n-i] + \sum_{i=1}^{3} a_i y[n-i] \qquad (4.41)$$

can be realized with the circuit of Fig. 4.10. We shall see later, however, that there are many other structures that can be described by the same difference equation.

4.4.3 The impulse response as a system description

As stated, an LTD system can be completely described by its impulse response, i.e. by the output signal $h[n]$ that corresponds to the input signal $x[n] = \delta[n]$; (see Fig. 4.11).
    Sometimes the impulse response can be determined in a very simple way. In other

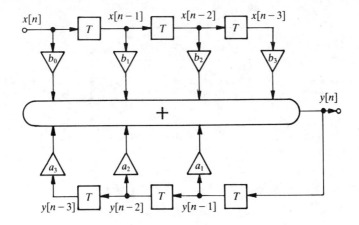

FIGURE 4.10 LTD system corresponding to equation (4.41).

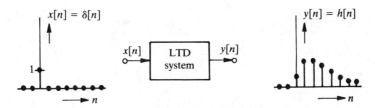

FIGURE 4.11 An LTD system and its impulse response $h[n]$.

cases more complicated methods are necessary, e.g. based on a description in terms of a difference equation.

Example 1
It can be seen directly that the impulse response of the system of Fig. 4.12 is given by:

$$h_A[n] = 2\delta[n] - 0.5\delta[n-1] \tag{4.42}$$

This is a *finite-duration* impulse response and such a system is called a finite impulse response (FIR) filter.

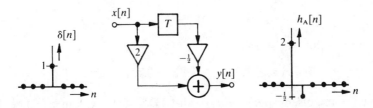

FIGURE 4.12 LTD system A.

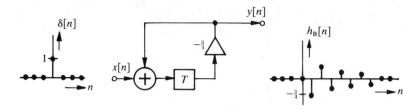

FIGURE 4.13 LTD system B.

Example 2
The impulse response of the system of Fig. 4.13 is given by:

$$h_B[n] = (-3/4)^n u[n-1] \tag{4.43}$$

This is an *infinite-duration* impulse response and such a system is called an infinite impulse response (IIR) filter.

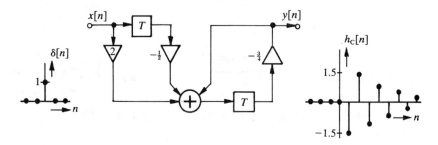

FIGURE 4.14 The cascade connection of the LTD systems A and B.

Example 3
The impulse response of the system of Fig. 4.14 (this is the cascade connection of the systems of Examples 1 and 2) is not quite so easily identified.

However, since we have already encountered the same system in Fig. 4.9(b), with $\delta[n]$ as the input signal, we can use (4.39b). We now substitute $a = 2, b = -\frac{1}{2}$ and $c = -\frac{3}{4}$ and find:

$$h_C[n] = \begin{cases} 0 & \text{for } n \leq 0 \\ -1.5 & \text{for } n = 1 \\ -2(-0.75)^{n-1} & \text{for } n \geq 2 \end{cases} \tag{4.44}$$

We can also write this (verify!) as:

$$h_C[n] = -1.5\delta[n-1] - 2(-0.75)^{n-1} u[n-2] \tag{4.45}$$

If we know the impulse response $h[n]$ of an LTD system, we can calculate the output signal $y[n]$ corresponding to an arbitrary input signal $x[n]$. This can be seen as follows (we use (4.5) in step 4):

1. By definition the input signal $\delta[n]$ produces the output signal $h[n]$.
2. Because of time-invariance the input signal $\delta[n-i]$ then produces the output signal $h[n-i]$.
3. Because of linearity the input signal $x[i]\delta[n-i]$ then produces the output signal $x[i]h[n-i]$
4. Therefore (again because of linearity) the input signal

$$x[n] = \sum_{i=-\infty}^{\infty} x[i]\delta[n-i] \qquad \text{produces the output signal}$$

$$y[n] = \sum_{i=-\infty}^{\infty} x[i]h[n-i]$$

Summarizing, we thus find that an LTD system with impulse response $h[n]$ and input signal $x[n]$ provides an output signal $y[n]$ given by:

$$y[n] = \sum_{i=-\infty}^{\infty} x[i]h[n-i] \tag{4.46}$$

We can see directly from the impulse response $h[n]$ of an LTD system whether the system is stable or causal or both.

For a *stable* LTD system we have:

$$\sum_{i=-\infty}^{\infty} |h[i]| < \infty \tag{4.47}$$

For a *causal* LTD system we have:

$$h[n] = 0 \qquad \text{for } n < 0 \tag{4.48}$$

As an example the impulse responses of four systems are shown in Fig. 4.15, with the conclusions that can be drawn from (4.47) and (4.48).

4.4.4 The discrete convolution

Equation (4.46) gives the 'convolution' of $x[n]$ and $h[n]$ and is expressed in a shortened form as:

$$y[n] = x[n] * h[n] \tag{4.49}$$

Substituting $i' = n - i$ in (4.46) and then replacing i' by i we find:

$$y[n] = \sum_{i=-\infty}^{\infty} x[i]h[n-i] = \sum_{i=-\infty}^{\infty} h[i]x[n-i] \tag{4.50}$$

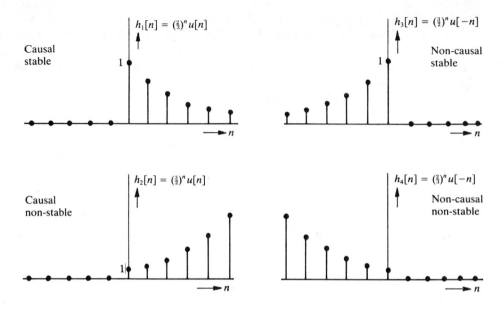

FIGURE 4.15 Stability and causality.

or

$$y[n] = x[n] * h[n] = h[n] * x[n] \tag{4.51}$$

As an illustration the calculation of the convolution of two signals $x[n]$ and $h[n]$ is shown in steps in Fig. 4.16. We start with $h[i]$ and $x[i]$ and perform the following operations:

I.   'Fold $h[i]$ over' in time: this gives $h[-i]$.
II. Shift $h[-i]$ through a distance n: this gives $h[n-i]$ (in Fig. 4.16 we have chosen $n = 2$).
III. Multiply $x[i]$ by $h[n-i]$: this gives $x[i]h[n-i]$.
IV. Sum this product over all i: this gives the signal value $y[2]$ in Fig. 4.16.
V.   Vary n from $-\infty$ to ∞: this gives $y[n]$.

From Fig. 4.16 it can be seen that the signal $y[n]$ resulting after the convolution has a longer duration than either of the original signals $h[n]$ and $x[n]$. In general the convolution of two discrete signals $h[n]$ and $x[n]$ of finite duration (N_1 and N_2) gives a discrete signal $y[n]$ of finite duration of length $N_1 + N_2 - 1$.

If at least one of the signals $h[n]$ and $x[n]$ is of infinite duration, $y[n]$ can also be of infinite duration.

Example

We determine the convolution $y[n] = h[n] * x[n]$ for

$$h[n] = a^n u[n] \quad \text{and} \quad x[n] = b^n u[n] \tag{4.52}$$

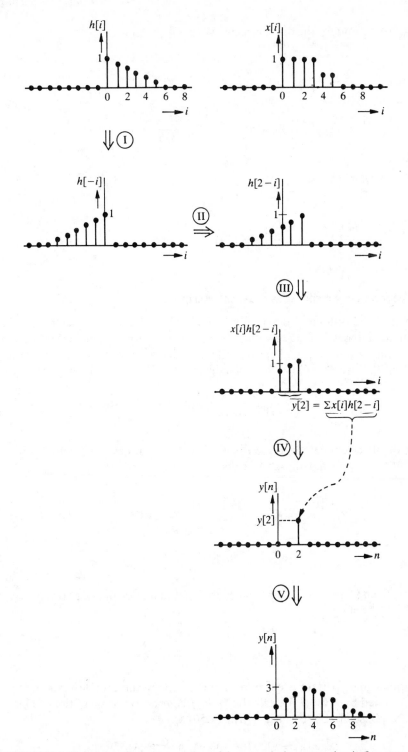

FIGURE 4.16 Schematic representation of the convolution operation in five steps.

Using (4.50) we find for the discrete signal $y[n]$:

$$y[n] = \sum_{i=-\infty}^{\infty} h[i]x[n-i] = \sum_{i=-\infty}^{\infty} a^i u[i] b^{n-i} u[n-i] = \left\{ \sum_{i=0}^{n} a^i b^{n-i} \right\} u[n]$$

$$= \left\{ b^n \sum_{i=0}^{n} \left(\frac{a}{b}\right)^i \right\} u[n] = b^n \frac{1-(a/b)^{n+1}}{1-a/b} u[n]$$

$$= \frac{b^{n+1}\{1-(a/b)^{n+1}\}}{b-a} u[n]$$

$$= \frac{b}{b-a} b^n u[n] - \frac{a}{b-a} a^n u[n]$$

(4.53)

4.4.5 The frequency response as a system description

Just as with LTC systems, a description of LTD systems in the frequency domain often has great advantages compared with a description in the time domain. For a given LTD system of impulse response $h[n]$ we shall now calculate the response $y[n]$ for a cosine input signal $x[n] = \cos(n\theta)$ (so this is *not* the impulse response!). Now since:

$$x[n] = \cos(n\theta) = \frac{e^{j\theta n}}{2} + \frac{e^{-j\theta n}}{2}$$

(4.54)

we first of all calculate the response $y_1[n]$ to the complex exponential input signal $x_1[n] = e^{j\theta n}$. With the aid of (4.50) we find:

$$y_1[n] = \sum_{i=-\infty}^{\infty} h[i]x_1[n-i] = \sum_{i=-\infty}^{\infty} h[i]e^{j\theta(n-i)}$$

$$= e^{j\theta n} \sum_{i=-\infty}^{\infty} h[i]e^{-j\theta i}$$

(4.55)

However, from (4.18), the summation on the right-hand side of (4.55) is exactly equal to the FTD $H(e^{j\theta})$ of $h[n]$.

Therefore:

$$y_1[n] = e^{j\theta n} H(e^{j\theta})$$

(4.56)

The output signal $y_1[n]$ turns out to be none other than the complex exponential signal $x_1[n]$ multiplied by $H(e^{j\theta})$. We call $H(e^{j\theta})$ the *frequency response* of the LTD system of impulse response $h[n]$. We can also write (4.56) as:

$$y_1[n] = e^{j\theta n} H(e^{j\theta}) = e^{j\theta n} A e^{j\varphi} = A e^{j(\theta n+\varphi)}$$

(4.57)

with

$$A = |H(e^{j\theta})| \quad \text{and} \quad \varphi = \arg\{H(e^{j\theta})\}$$

In the same way we find the output signal $y_2[n]$ for the input signal $x_2[n] = e^{-j\theta n}$:

$$y_2[n] = e^{-j\theta n} H(e^{-j\theta})$$

As a result of the linearity property we now also know the response $y[n]$ to the signal $x[n] = \cos(n\theta) = (x_1[n] + x_2[n])/2$:

$$y[n] = \tfrac{1}{2}\{e^{j\theta n} H(e^{j\theta}) + e^{-j\theta n} H(e^{-j\theta})\}$$

The impulse response $h[n]$ is practically always a real function. With the aid of (4.34c) we then find:

$$y[n] = \tfrac{1}{2}A[e^{j(\theta n + \varphi)} + e^{-j(\theta n + \varphi)}] = A\cos(\theta n + \varphi) \tag{4.58}$$

This equation shows that a cosine input signal is converted by an LTD system into a cosine output signal of the same frequency, but with an amplitude and phase that depend on the frequency response $H(e^{j\theta})$ at that particular frequency. This can be considered to be one of the most essential properties of an LTD system.

We have shown above how $H(e^{j\theta})$ can be used if we have an input signal that consists of one or two complex exponentials. But we can do even more with $H(e^{j\theta})$. In (4.19) in section 4.3 about the FTD we found that for an arbitrary signal $x[n]$:

$$x[n] = \frac{1}{2\pi} \int_{-\pi}^{\pi} X(e^{j\theta}) e^{jn\theta} \, d\theta \tag{4.59}$$

This means that $x[n]$ can be taken as an infinite sum of complex exponentials with a 'frequency' θ between $-\pi$ and π, each with an amplitude and phase determined by $X(e^{j\theta})$. On the basis of the linearity property, an LTD system converts this input signal $x[n]$ into an output signal $y[n]$, in which every complex exponential is multiplied by the corresponding value of $H(e^{j\theta})$. Therefore:

$$y[n] = \frac{1}{2\pi} \int_{-\pi}^{\pi} X(e^{j\theta}) H(e^{j\theta}) e^{jn\theta} \, d\theta \tag{4.60}$$

The left-hand side of this equation, by definition, forms a Fourier pair with $Y(e^{j\theta})$ and the right-hand side is the IFTD of the product $X(e^{j\theta}) H(e^{j\theta})$. We therefore also have:

$$Y(e^{j\theta}) = X(e^{j\theta}) H(e^{j\theta}) \tag{4.61}$$

We also know that:

$$y[n] = x[n] * h[n] \tag{4.62}$$

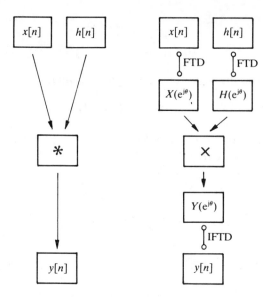

FIGURE 4.17 Two ways of determining the output signal $y[n]$ of an LTD system.

We have thus derived property (4.29), which states that convolution in the time domain corresponds to multiplication in the frequency domain.

This again underlines the significance of the frequency response $H(e^{j\theta})$: we can substitute the fairly easy operation of multiplication in the frequency domain for the relatively difficult operation of convolution in the time domain. We have shown this again schematically in Fig. 4.17.

- *Comments*
  1. $H(e^{j\theta})$ is the FTD of $h[n]$.
  2. Therefore $H(e^{j\theta})$ (like every FTD) is periodic in θ with a period of 2π (see Fig. 4.4).
  3. For real $h[n]$, which is usually the case, the real part of H has even symmetry and the imaginary part of H has odd symmetry; see (4.34b).

The theory discussed so far offers us a choice of three methods of determining the frequency response of a given system.

Method 1: as the FTD of the impulse response $h[n]$.

Method 2: by applying an input $x[n] = e^{j\theta n}$; the output signal is then $y[n] = H(e^{j\theta})e^{j\theta n}$; the frequency response $H(e^{j\theta})$ can be derived from this.

Method 3: from the difference equation of the system with the aid of the time-shift property (4.26); see Example 1 below.

- *Comment.* Method 1 is based on the use of the unit pulse $\delta[n]$ as input signal; in Method 2 we use the complex exponential as input signal and in Method 3 we do not specify the input signal at all. We shall now illustrate these three methods in the following example.

Example 1

We determine the frequency response $H(e^{j\theta})$ of the system given in Fig. 4.18 with $|a| < 1$.[†]

- *Method 1*

The impulse response $h[n]$ is given by:

$$h[n] = a^n u[n]$$

The frequency response is then:

$$H(e^{j\theta}) = \sum_{n=-\infty}^{\infty} h[n]e^{-jn\theta} = \sum_{n=0}^{\infty} a^n e^{-jn\theta} = \sum_{n=0}^{\infty} (ae^{-j\theta})^n$$

$$= \frac{1}{1 - ae^{-j\theta}} \quad (\text{because } |a| < 1) \tag{4.63}$$

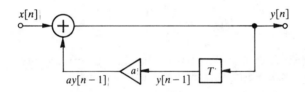

FIGURE 4.18 Simple LTD system.

- *Method 2*

Take $x[n] = e^{j\theta n}$ as input signal. The output signal $y[n]$ is then $H(e^{j\theta})e^{j\theta n}$ and therefore $y[n-1] = H(e^{j\theta})e^{j\theta(n-1)}$. The output signal $y[n]$ can be expressed as:

$$y[n] = x[n] + ay[n-1]$$

From the above it follows that:

$$H(e^{j\theta})e^{j\theta n} = e^{j\theta n} + aH(e^{j\theta})e^{j\theta(n-1)}$$

so that:

$$H(e^{j\theta}) = \frac{e^{j\theta n}}{e^{j\theta n} - ae^{j\theta(n-1)}} = \frac{1}{1 - ae^{-j\theta}} \tag{4.64}$$

- *Method 3*

The difference equation that describes the system has already occurred in Method 2:

$$y[n] = x[n] + ay[n-1] \tag{4.65}$$

---

†Figure 4.18 only represents a *stable* system for $|a| < 1$, since for other values of a the impulse response $h[n]$ increases without limit for increasing n (verify this!). The FTD of $h[n]$ cannot be calculated then either (it does not 'converge'), and the frequency response $H(e^{j\theta})$ therefore does not exist. In Methods 2 and 3 it is tacitly assumed that $H(e^{j\theta})$ does exist. In fact we have to make an explicit check for this. Later we shall see that we can check the stability of a system easily with the aid of 'poles and zeros'.

We use the following Fourier pairs:

$$x[n] \;\circ\!\!\!-\!\!\!-\!\!\!\circ\; X(e^{j\theta})$$
$$y[n] \;\circ\!\!\!-\!\!\!-\!\!\!\circ\; Y(e^{j\theta})$$

From the time-shift rule (4.26) we find:

$$y[n-1] \;\circ\!\!\!-\!\!\!-\!\!\!\circ\; e^{-j\theta}Y(e^{j\theta})$$

Therefore (4.65) is equivalent to:

$$Y(e^{j\theta}) = X(e^{j\theta}) + ae^{-j\theta}Y(e^{j\theta})$$

and since $Y(e^{j\theta}) = H(e^{j\theta})X(e^{j\theta})$, we find:

$$H(e^{j\theta}) = \frac{Y(e^{j\theta})}{X(e^{j\theta})} = \frac{1}{1-ae^{-j\theta}} \tag{4.66a}$$

The modulus $A(e^{j\theta})$ (expressed in dB) and the phase $\varphi(e^{j\theta})$ of H are plotted in Fig. 4.19 for various values of a. Verify that the following relations apply:

$$A(e^{j\theta}) = 1/\sqrt{(1+a^2 - 2a\cos\theta)} \tag{4.66b}$$

$$\varphi(e^{j\theta}) = \arctan\left(\frac{-a\sin\theta}{1-a\cos\theta}\right) \tag{4.66c}$$

Example 2
We want to determine the output signal $y[n]$ of Fig. 4.18 when $x[n] = b^n u[n]$. From Example 1 and Fig. 4.5(c) we find:

$$H(e^{j\theta}) = \frac{1}{1-ae^{-j\theta}} \quad \text{and} \quad X(e^{j\theta}) = \frac{1}{1-be^{-j\theta}}$$

Therefore the FTD of the output signal is:

$$Y(e^{j\theta}) = X(e^{j\theta})H(e^{j\theta}) = \frac{1}{(1-be^{-j\theta})(1-ae^{-j\theta})}$$

$$= \frac{b}{b-a}\frac{1}{1-be^{-j\theta}} - \frac{a}{b-a}\frac{1}{1-ae^{-j\theta}} \tag{4.67}$$

After inverse transformation (in which we use the Fourier pair of Fig. 4.5(c) twice) we find:

$$y[n] = \frac{b}{b-a}b^n u[n] - \frac{a}{b-a}a^n u[n] \tag{4.68}$$

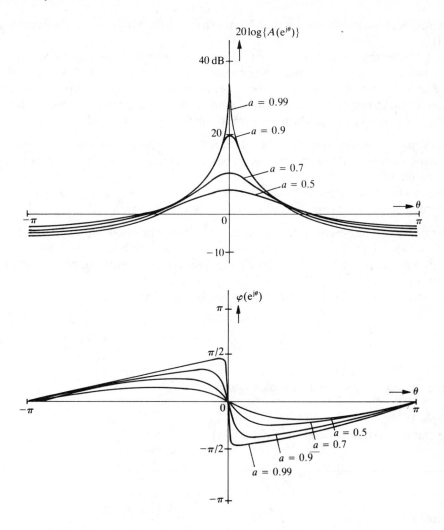

FIGURE 4.19 Modulus $A(e^{j\theta})$ and argument $\varphi(e^{j\theta})$ of $H(e^{j\theta}) = 1/(1 - ae^{-j\theta})$ for various values of a.

Compare 4.68 with the result of (4.53), which was found by convolution in the time domain.

4.5 THE z-TRANSFORM

4.5.1 Definition and examples

From LTC systems (see section 2.10) we know that there are sometimes considerable advantages in turning from the *real* frequency ω to the *complex* frequency p. This can be used to give a unique representation of a system (and of most signals) with the aid of poles and zeros. Many conclusions can then be drawn about the properties of the system (or signal).

For LTD systems we shall now introduce a *complex* frequency variable, usually denoted by the symbol z, in addition to the *real* frequency ω (and the normalized real frequency θ). So that we can go from the discrete-time domain to the complex-frequency domain we make use of the 'z-transform'. After a short formal treatment we shall show that there are many cases in which the z-transform is very easy to handle:

- For example, we can often write the z-transform of the impulse response of a discrete system directly from an inspection of its block diagram. We call this z-transform the *system function*.
- We can also find the FTD discussed in section 4.3 via the z-transform by inserting particular values for z (we shall see later what these values are).

All this makes the z-transform a universal tool for the study of discrete systems. In the theory, we shall confine ourselves to those parts necessary to obtain a sufficient understanding for its application. More detailed theoretical treatments can be found in the existing literature.[8,9,29]

The z-transform (ZT) of a discrete signal $x[n]$ is defined as:

$$X(z) = \sum_{n=-\infty}^{\infty} x[n]z^{-n} \qquad \text{(ZT)} \qquad (4.69)$$

where z can take any arbitrary complex value.

Example 1
The z-transform of the discrete unit pulse $x_1[n] = \delta[n]$ is:

$$X_1(z) = \sum_{n=-\infty}^{\infty} x_1[n]z^{-n} = \sum_{n=-\infty}^{\infty} \delta[n]z^{-n} = z^{-0} = 1 \qquad (4.70)$$

Example 2
The z-transform of the unit pulse shifted through k positions $x_2[n] = \delta[n-k]$ is:

$$X_2(z) = \sum_{n=-\infty}^{\infty} x_2[n]z^{-n} = \sum_{n=-\infty}^{\infty} \delta[n-k]z^{-n} = z^{-k} \qquad (4.71)$$

Example 3
The z-transform of a finite sequence of unit pulses, as shown in Fig. 4.20 and defined as:

$$x_3[n] = \begin{cases} 1 & \text{for } |n| \leq N \\ 0 & \text{for } |n| > N \end{cases} \qquad (4.72)$$

is:

$$X_3(z) = \sum_{n=-N}^{N} z^{-n} = z^{N} + z^{N-1} + \ldots + z^{-N} \qquad (4.73)$$

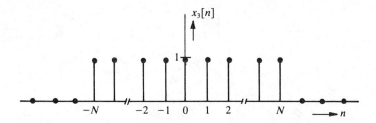

FIGURE 4.20 Finite sequence of unit pulses.

Example 4

The z-transform of $x_4[n] = a^n u[n]$, as shown in Fig. 4.21, with a an arbitrary constant, is:

$$X_4(z) = \sum_{n=-\infty}^{\infty} a^n u[n] z^{-n} = \sum_{n=0}^{\infty} (az^{-1})^n$$

$$= 1 + az^{-1} + a^2 z^{-2} + a^3 z^{-3} + \ldots \tag{4.74}$$

An interpretation of $X(z)$ that is sometimes very useful can be based on the above examples: if $X(z)$ takes the form of a polynomial in z^{-1} then the factor that relates to z^{-i} corresponds exactly to the value of $x[n]$ at the instant $n = i$. Verify this (this interpretation can also be derived directly from the definition of (4.69)).

4.5.2 Convergence properties of the z-transform

Let us consider $X_4(z)$ from the last example a little more closely. For some values of z we find a finite value for $X_4(z)$; we say that $X_4(z)$ then converges. For other values of z, $X_4(z)$ becomes infinite; we say that $X_4(z)$ then does not converge. For $z = 2a$, for example, we find:

$$X_4(2a) = \sum_{n=0}^{\infty} (a/2a)^n = \sum_{n=0}^{\infty} \left(\frac{1}{2}\right)^n = \frac{1}{1 - \frac{1}{2}} = 2 \quad \text{(i.e. } X_4 \text{ converges)}$$

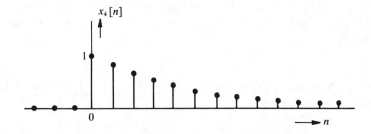

FIGURE 4.21 The discrete function $x_4[n] = a^n u[n]$.

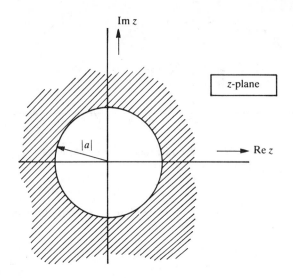

FIGURE 4.22 Region of convergence (hatched) for $X_4(z)$.

and $z = a/2$ gives:

$$X_4(a/2) = \sum_{n=0}^{\infty} (2a/a)^n = \sum_{n=0}^{\infty} 2^n = \infty \quad \text{(i.e. } X_4 \text{ does not converge)}$$

It is easy to show that $X_4(z)$ converges for all values of z for which $|az^{-1}| < 1$, so that $|z| > |a|$, and $X_4(z)$ can then also be expressed, not as an infinite sum as in (4.74), but as:

$$X_4(z) = \frac{1}{1 - az^{-1}} = \frac{z}{z - a} \quad \text{for } |z| > |a| \tag{4.75}$$

The condition $|z| > |a|$ corresponds to the hatched area in the z-plane of Fig. 4.22.

We have to bear in mind that it is only useful to talk about $X(z)$ for the values of z that lie in the region of convergence. Therefore we must formally state the values of z to which the region of convergence corresponds. In practical situations this seldom leads to problems, and indeed we are not always aware of these convergence conditions. However, the convergence conditions are indispensable if we wish to recover the original time function from the z-transform by formal methods. We can show this with the following example.

Example 5
Consider the time functions $x_5[n] = u[n]$ and $y_5[n] = -u[-n-1]$ as shown in Fig. 4.23.
For the z-transforms we find:

$$X_5(z) = \sum_{n=0}^{\infty} z^{-n} = \frac{1}{1 - z^{-1}} = \frac{z}{z - 1} \quad \text{for } |z| > 1 \tag{4.76}$$

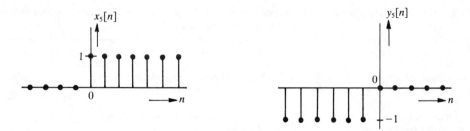

FIGURE 4.23 The discrete functions $x_5[n] = u[n]$ and $y_5[n] = -u[-n-1]$.

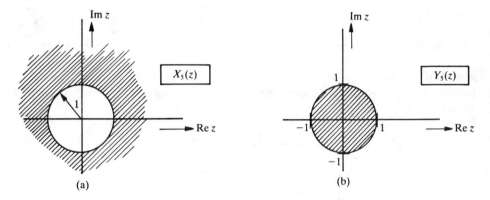

(a)

(b)

FIGURE 4.24 (a) Region of convergence for $X_5(z)$. (b) Region of convergence for $Y_5(z)$.

and

$$Y_5(z) = \sum_{n=-\infty}^{-1} -z^{-n} = 1 - \sum_{n=-\infty}^{0} z^{-n} = 1 - \sum_{n=0}^{\infty} z^n$$

$$= 1 - \frac{1}{1-z} = \frac{z}{z-1} \quad \text{for } |z| < 1 \tag{4.77}$$

The regions of convergence for $X_5(z)$ and $Y_5(z)$ are shown hatched in Fig. 4.24. We see that the same expression applies for the z-transforms of $x_5[n]$ and $y_5[n]$; however, they converge for quite different values of z.

To recover each of the two original sequences from the expression $X(z) = z/(z-1)$, we must know the region of convergence. In practice we shall only rarely encounter such ambiguities, since we almost always work with signals $x[n]$ that have a value of zero below a particular value of n. This determines the region of convergence uniquely (it is the entire z-plane outside a circle of a particular radius with its center at the origin). We shall therefore experience few difficulties in relation to the convergence conditions, and indeed we shall often leave them unstated.

Without stating the mathematical proofs, we have collected a number of general guidelines to convergence properties in Table 4.2.

TABLE 4.2 Some general guidelines to convergence properties (the regions of convergence in the z-plane are shown hatched)

$x[n]$	$X(z)$	z-plane
$x[n]$ is right-sided, i.e. $x[n] = 0$ for all n smaller than a particular n_0	$X(z)$ converges for all z *outside* a circle of a particular radius	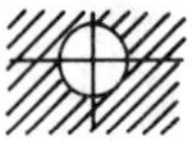
$x[n]$ is left-sided, i.e. $x[n] = 0$ for all n greater than a particular n_0	$X(z)$ converges for all z *inside* a circle of a particular radius	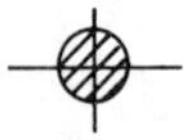
$x[n]$ is of finite length, i.e. $x[n] = 0$ for all $n < n_1$ and all $n > n_2$, where $n_1 < n_2$	$X(z)$ converges for *all values* of z, except possibly for $z = 0$ and $z = \infty$	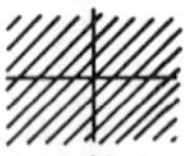
$x[n]$ is two-sided	*If $X(z)$ converges*, then the region of convergence lies *between* two circles	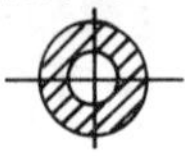

Example 6

The function $x_6[n] = a^n \cos(n\xi)u[n]$ is shown in Fig. 4.25 (for $a = 0.9$ and $\xi = \pi/4$). The z-transform $X_6(z)$ is:

$$X_6(z) = \sum_{n=-\infty}^{\infty} a^n \cos(n\xi)u[n]z^{-n} = \frac{1}{2}\sum_{n=0}^{\infty} a^n(e^{jn\xi} + e^{-jn\xi})z^{-n}$$

$$= \frac{1}{2}\sum_{n=0}^{\infty} (ae^{j\xi}z^{-1})^n + \frac{1}{2}\sum_{n=0}^{\infty} (ae^{-j\xi}z^{-1})^n$$

$$= \frac{1}{2}\left(\frac{1}{1 - ae^{j\xi}z^{-1}} + \frac{1}{1 - ae^{-j\xi}z^{-1}}\right) \quad \text{for } |z| > |a|$$

Further manipulation gives:

$$X_6(z) = \frac{1}{2}\left[\frac{2 - az^{-1}(e^{j\xi} + e^{-j\xi})}{1 - az^{-1}(e^{j\xi} + e^{-j\xi}) + a^2z^{-2}}\right] = \frac{1 - az^{-1}\cos(\xi)}{1 - 2az^{-1}\cos(\xi) + a^2z^{-2}}$$

$$= \frac{z\{z - a\cos(\xi)\}}{z^2 - 2az\cos(\xi) + a^2} \quad \text{for } |z| > |a| \tag{4.78}$$

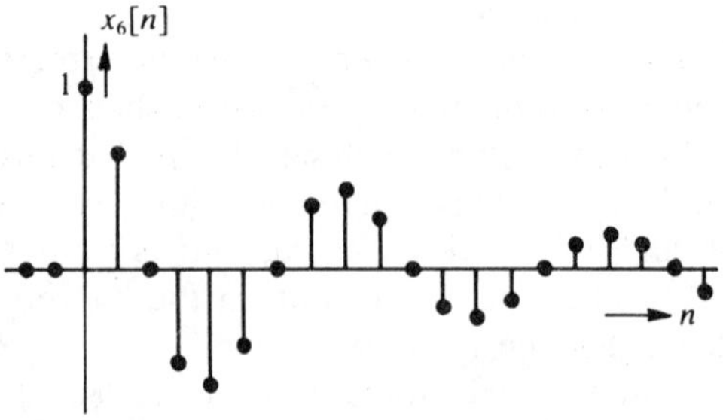

FIGURE 4.25 The discrete function $x_6[n] = a^n \cos(n\xi)u[n]$ for $a = 0.9$ and $\xi = \pi/4$.

Example 7

In the same way as in the previous example we can calculate the z-transform of $x_7[n] = a^n \sin(n\xi)u[n]$; it follows (check it!) from this that:

$$X_7(z) = \frac{az^{-1}\sin(\xi)}{1 - 2az^{-1}\cos(\xi) + a^2 z^{-2}} = \frac{az\sin(\xi)}{z^2 - 2az\cos(\xi) + a^2} \quad \text{for } |z| > |a| \quad (4.79)$$

4.5.3 The inverse z-transform

To recover the original discrete function from a given z-transform we can make use of the inverse z-transform (IZT), which is given formally by:

$$\boxed{x[n] = \frac{1}{2\pi j} \oint_C X(z) z^{n-1} \, dz} \qquad \text{(IZT)} \qquad (4.80)$$

This represents a contour integral in the z-plane, over a counterclockwise arbitrary closed path in the region of convergence and enclosing the origin $z = 0$. That is rather a mouthful, and in practice this formal inverse transform is not often used. An example of the use of (4.80) is given in Appendix III.

TABLE 4.3 Some common z-transform pairs

	$x[n]$	$X(z) = \sum\limits_{n=-\infty}^{\infty} x[n]z^{-n}$
1	$\delta[n]$	1
2	$\delta[n-i]$	z^{-i}
3	$u[n] = \begin{cases} 1 & n \geq 0 \\ 0 & n < 0 \end{cases}$	$z/(z-1)$
4	$a^n u[n]$	$z/(z-a)$
5	$nu[n]$	$z/(z-1)^2$
6	$n^2 u[n]$	$z(z+1)/(z-1)^3$
7	$x[n]$	$X(z)$
7(a)	$x[n-i]$	$z^{-i}X(z)$
7(b)	$a^n x[n]$	$X(z/a)$
7(c)	$nx[n]$	$-z\,d\{X(z)\}/dz$
8(a)	$a^n \cos(n\xi)u[n]$	$\dfrac{z^2 - az\cos(\xi)}{z^2 - 2az\cos(\xi) + a^2}$
8(b)	$a^n \sin(n\xi)u[n]$	$\dfrac{az\sin(\xi)}{z^2 - 2az\cos(\xi) + a^2}$
8(c)	$a^n \sin(n\xi + \psi)u[n]$	$\dfrac{z^2 \sin(\psi) + az\sin(\xi - \psi)}{z^2 - 2az\cos(\xi) + a^2}$
9	$\dfrac{1}{n}u[n-1]$	$\ln\left(\dfrac{z}{z-1}\right)$

Since in our applications we shall practically always be concerned with z-transforms that can be written as:

$$X(z) = \frac{N(z)}{D(z)} = \frac{b_0 + b_1 z^{-1} + b_2 z^{-2} + \ldots + b_N z^{-N}}{1 - a_1 z^{-1} - a_2 z^{-2} - \ldots - a_M z^{-M}} \quad (4.81)$$

we can also find the inverse transform by making use of *expansion in partial fractions* and reduction to known z-transforms.

In certain cases a simple *long division* is sufficient; we can also use this long-division method to find, for a given function in the z-domain, a limited number of values (say the first five) of the corresponding discrete function.

Finally, there is the possibility of making *use of tables* of z-transform pairs, which can be found in the literature.[29] Some frequently occurring z-transform pairs have been collected in Table 4.3.

The three methods mentioned (partial-fraction expansion, long division and using tables) for finding the inverse transform will now be explained with the aid of examples.

Example 8 (*long division*)
Long division of the function $X_8(z) = (1 - z^{-1} + 5z^{-2} - 3z^{-3})/(1 - 3z^{-1})$ gives:

$$
\begin{array}{r}
1 + 2z^{-1} + z^{-2} \\
1 - 3z^{-1} \overline{\smash{\big)}\ 1 - z^{-1} - 5z^{-2} - 3z^{-3}} \\
\underline{1 - 3z^{-1}} \\
2z^{-1} - 5z^{-2} - 3z^{-3} \\
\underline{2z^{-1} - 6z^{-2}} \\
z^{-2} - 3z^{-3} \\
\underline{z^{-2} - 3z^{-3}} \\
0
\end{array}
$$

Therefore $X_8(z) = 1 + 2z^{-1} + z^{-2}$, and using pairs (1) and (2) from Table 4.3 we find (see Fig. 4.26):

$$x_8[n] = \delta[n] + 2\delta[n-1] + \delta[n-2] \quad (4.82)$$

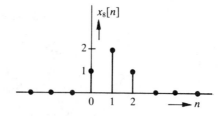

FIGURE 4.26 The discrete function $x_8[n]$.

Example 9 (*long division*)
Long division of the function $X_9(z) = 1/(1 - az^{-1})$ gives:

$$
\begin{array}{r}
1 + az^{-1} + (az^{-1})^2 + \ \cdots \\
1 - az^{-1} \overline{\vert \quad 1 \qquad\qquad\qquad\qquad} \\
-\dfrac{1 - az^{-1}}{az^{-1}} \\
-\dfrac{az^{-1} - (az^{-1})^2}{(az^{-1})^2} \\
-\dfrac{(az^{-1})^2 - (az^{-1})^3}{(az^{-1})^3} \ \text{etc.}
\end{array}
$$

Therefore $X_9(z) = 1 + az^{-1} + a^2 z^{-2} + \ \cdots$ and from Table 4.3 we find:

$$
x_9[n] = \delta[n] + a\delta[n-1] + a^2 \delta[n-2] + \ \cdots \ = a^n u[n] \tag{4.83}
$$

The discrete function $x_9[n]$ corresponds to the function given in Fig. 4.21.

Example 10 (*partial-fraction expansion*)
The z-transform of $x_{10}[n]$ is $X_{10}(z)$, where:

$$
X_{10}(z) = \frac{z}{(z - \alpha)(z - \beta)} \tag{4.84}
$$

By expanding in partial fractions (see Appendix II for details) we can rewrite (4.84) as:

$$
X_{10}(z) = \frac{Az}{z - \alpha} + \frac{Bz}{z - \beta} = \frac{z^2(A + B) - z(A\beta + B\alpha)}{(z - \alpha)(z - \beta)}
$$

therefore: $A + B = 0$ and $A\beta + B\alpha = -1$, or:

$$
A = \frac{1}{\alpha - \beta} \quad \text{and} \quad B = \frac{-1}{\alpha - \beta}
$$

This gives for $X_{10}(z)$:

$$
X_{10}(z) = \frac{1}{\alpha - \beta} \left(\frac{z}{z - \alpha} - \frac{z}{z - \beta} \right)
$$

The discrete function $x_{10}[n]$ is then obtained after inverse transformation of $X_{10}(z)$; see Table 4.3:

$$
x_{10}[n] = \frac{1}{\alpha - \beta}(\alpha^n - \beta^n)u[n] \tag{4.85}
$$

Example 11 (*use of the table*)
We start with the function:

$$X_{11}(z) = \frac{z^2 + 2z/3}{z^2 - 2z/3 + 4/9} \tag{4.86}$$

If we compare (4.86) with pair 8(c) of Table 4.3 we see that $X_{11}(z)$ is a special case of the general form:

$$X_{11}(z) = \frac{Az^2 \sin(\psi) + Aaz \sin(\xi - \psi)}{z^2 - 2az \cos(\xi) + a^2} \tag{4.87a}$$

and therefore:

$$x_{11}[n] = Aa^n \sin(n\xi + \psi)u[n] \tag{4.87b}$$

With a suitable choice of A, a, ξ and ψ we can make the expressions (4.86) and (4.87a) equal to one another; it follows immediately that:

$$\left.\begin{array}{l} A\sin(\psi) = 1 \\ Aa\sin(\xi - \psi) = 2/3 \\ 2a\cos(\xi) = 2/3 \\ a^2 = 4/9 \end{array}\right\} \rightarrow \left\{\begin{array}{l} a = 2/3 \\ \xi = \pi/3 \\ A = 2 \\ \psi = \pi/6 \end{array}\right.$$

Substituting the above values in (4.87b) gives:

$$x_{11}[n] = 2(2/3)^n \sin(n\pi/3 + \pi/6)u[n] \tag{4.88}$$

4.5.4 Properties of the z-transform

Many of the properties of the FTD (see section 4.3.2) have a counterpart in the z-transform. A number of these are given below. We make use of the following z-transform pairs, where $x[n]$, $x_1[n]$ and $x_2[n]$ are arbitrary discrete signals:

$$x[n] \circ\!\!-\!\!-\!\!\circ X(z)$$
$$x_1[n] \circ\!\!-\!\!-\!\!\circ X_1(z) \tag{4.89}$$
$$x_2[n] \circ\!\!-\!\!-\!\!\circ X_2(z)$$

A. Linearity

$$\boxed{ax_1[n] + bx_2[n] \circ\!\!-\!\!-\!\!\circ aX_1(z) + bX_2(z)} \tag{4.90}$$

where a and b are arbitrary constants.

B. Time shift

$$\boxed{x[n-i] \; \circ\!\!-\!\!-\!\!\circ \; z^{-i}X(z)}$$ (4.91)

where i is any integer.

We shall use this property frequently to translate a system description in the form of a difference equation directly into a description in the z-domain.

C. Multiplication by an exponential series

$$\boxed{a^n x[n] \; \circ\!\!-\!\!-\!\!\circ \; X(z/a)}$$ (4.92)

where a is an arbitrary constant.

This property is related to the frequency-shift property of the FTD.

D. Convolution in the time domain

$$\boxed{x_1[n] * x_2[n] \; \circ\!\!-\!\!-\!\!\circ \; X_1(z)X_2(z)}$$ (4.93)

Therefore: convolution in the discrete-time domain corresponds to multiplication in the z-domain.

E. Convolution in the z domain

$$\boxed{x_1[n]x_2[n] \; \circ\!\!-\!\!-\!\!\circ \; \frac{1}{2\pi \mathrm{j}} \oint_C X_1(\nu)X_2(z/\nu)\nu^{-1}\,\mathrm{d}\nu}$$ (4.94)

This relation indicates that convolution in the z-domain corresponds to multiplication in the discrete-time domain. We shall not in fact require to use it in the remainder of this book, but we state it for completeness.

F. Parseval's theorem

$$\boxed{\sum_{n=-\infty}^{\infty} |x[n]|^2 = \frac{1}{2\pi \mathrm{j}} \oint_C X(z)X^*(1/z)z^{-1}\,\mathrm{d}z}$$ (4.95)

where * indicates the complex conjugate.

This theorem, which shows that the 'energy' of a discrete signal can be calculated both in the time domain and in the z-domain, is stated here only for completeness.

4.5.5 The relation between the z-transform and the FTD

We repeat here for an arbitrary discrete signal $x[n]$ the definitions of the z-transform (4.69) and the FTD (4.18):

$$X(z) = \sum_{n=-\infty}^{\infty} x[n]z^{-n} \tag{4.96a}$$

and

$$X(e^{j\theta}) = \sum_{n=-\infty}^{\infty} x[n]e^{-j\theta n} \tag{4.96b}$$

We see at once that we can find $X(e^{j\theta})$ from $X(z)$ by substituting $z = e^{j\theta}$ with $-\pi \le \theta \le \pi$, i.e. by considering all the values of z for which $|z| = 1$ (the unit circle in the z-plane). We see moreover that $\theta = 0$ corresponds to $z = 1$ and $\theta = \pm\pi$ to $z = -1$. This relation is also shown graphically for an arbitrarily chosen function $X(e^{j\theta})$ in Fig. 4.27.

- *Comment.* We now see the broader background to the choice of the rather more complicated notation $X(e^{j\theta})$ for the FTD, instead of the much simpler notation $X(\theta)$. Because of this choice the FTD and the ZT of $x[n]$ give exactly the same mathematical function and we can use the symbol X for both.

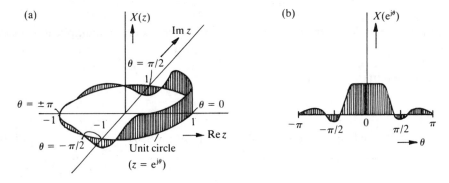

FIGURE 4.27 Relation between the z-transform $X(z)$ and the FTD $X(e^{j\theta})$.

4.5.6 The system function as a system description

If we have an LTD system whose impulse response is $h[n]$, its z-transform $H(z)$ is called the *system function*. We saw earlier that the FTD $H(e^{j\theta})$ of the same impulse response $h[n]$ is called the *frequency response* of that system.

On the basis of the previous section we can formulate a direct relation between $H(z)$ and $H(e^{j\theta})$:

> The frequency response $H(e^{j\theta})$ of an LTD system corresponds to the system function $H(z)$ for $|z| = 1$.

One of the important applications of the system function $H(z)$ is the replacement of convolution in the time domain by multiplication in the z-domain (see property 4.93).

In order words: if we apply an input signal $x[n]$ to a system with impulse response $h[n]$, then the output signal $y[n]$ is:

$$y[n] = \sum_{i=-\infty}^{\infty} x[i]h[n-i] = x[n]*h[n] \tag{4.97a}$$

and in the z-domain we then have:

$$Y(z) = X(z)H(z) \tag{4.97b}$$

This can be understood as follows. From (4.97a) we have:

$$Y(z) = \sum_{n=-\infty}^{\infty} y[n]z^{-n} = \sum_{n=-\infty}^{\infty} \left\{ \sum_{i=-\infty}^{\infty} x[i]h[n-i] \right\} z^{-n}$$

We now change the order of summation and substitute $k = n - i$; this gives successively:

$$Y(z) = \sum_{i=-\infty}^{\infty} x[i] \sum_{n=-\infty}^{\infty} h[n-i]z^{-n} = \sum_{i=-\infty}^{\infty} x[i] \sum_{n=-\infty}^{\infty} h[n-i]z^{-(n-i)} z^{-i}$$

$$= \sum_{i=-\infty}^{\infty} x[i]z^{-i} \sum_{k=-\infty}^{\infty} h[k]z^{-k} = X(z)H(z) \tag{4.98}$$

This is exactly what we wished to prove.

There are three methods available for finding the system function $H(z)$ of a system with impulse response $h[n]$.

Method 1: As the z-transform of the impulse response $h[n]$.

Method 2: By driving the system with the complex input signal $x[n] = z^n$. The output signal $y[n]$ is then given by:

$$y[n] = \sum_{i=-\infty}^{\infty} h[i]x[n-i] = \sum_{i=-\infty}^{\infty} h[i]z^{n-i}$$

$$= z^n \sum_{i=-\infty}^{\infty} h[i]z^{-i} = z^n H(z) \tag{4.99}$$

Therefore a complex input signal $x[n] = z^n$ corresponds to a complex output signal $y[n] = z^n H(z)$. We find $H(z)$ by dividing $y[n]$ by z^n:

Method 3: From the difference equation of the system and using the time-shift property (4.91).

We shall explain these methods with the aid of the two following examples.

Example 1
We determine the system function $H(z)$ of the system shown in Fig. 4.28.

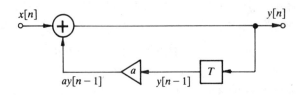

FIGURE 4.28 Simple LTD system.

- *Method 1*

The impulse response $h[n]$ is:

$$h[n] = a^n u[n] \tag{4.100}$$

From Table 4.3 the system function is then:

$$H(z) = \frac{z}{z-a} \quad \text{for } |z| > |a| \tag{4.101}$$

- *Method 2*

Take $x[n] = z^n$ as the input signal. The output signal $y[n]$ is then $H(z)z^n$ and therefore $y[n-1] = H(z)z^{n-1}$. The output signal $y[n]$ of Fig. 4.28 can be written as:

$$y[n] = x[n] + ay[n-1] \tag{4.102}$$

On substituting we find:

$$H(z)z^n = z^n + aH(z)z^{n-1}$$

therefore:

$$H(z) = \frac{z^n}{z^n - az^{n-1}} = \frac{z}{z-a} \tag{4.103}$$

- *Method 3*

Use the difference equation (4.102) and the two z-transform pairs:

$$x[n] \; \circ\!\!-\!\!-\!\!\circ \; X(z)$$

$$y[n] \; \circ\!\!-\!\!-\!\!\circ \; Y(z)$$

The time-shift rule (4.91) says:

$$y[n-1] \; \circ\!\!-\!\!-\!\!\circ \; z^{-1}Y(z)$$

This gives:

$$Y(z) = X(z) + az^{-1}Y(z)$$

therefore:

$$H(z) = \frac{Y(z)}{X(z)} = \frac{1}{1 - az^{-1}} = \frac{z}{z - a} \qquad (4.104)$$

- *Comment.* In this third method, instead of writing out the difference equation in the time domain, we often draw a block diagram in which the signals are given directly in the form of their z-transforms and every delay element is denoted by its system function z^{-1}. For the system of Fig. 4.28 this gives the diagram of Fig. 4.29.

From Fig. 4.29 we find directly:

$$Y(z) = X(z) + az^{-1}Y(z)$$

or:

$$H(z) = \frac{Y(z)}{X(z)} = \frac{z}{z - a} \qquad (4.105)$$

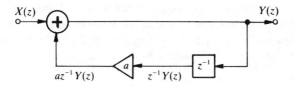

FIGURE 4.29 Block diagram in the z-domain for the circuit of Fig. 4.28.

Example 2
We wish to determine the system function of the system of Fig. 4.30.

Here we only use Method 3 (Method 1 is too cumbersome and for Method 2 we refer the reader to Exercise 4.14 in section 4.6). The circuit diagram is replaced by Fig. 4.31.

We see at once that:

$$Y(z) = b_0 X(z) + b_1 z^{-1} X(z) + b_2 z^{-2} X(z) + a_1 z^{-1} Y(z) + a_2 z^{-2} Y(z)$$

It follows that:

$$H(z) = \frac{Y(z)}{X(z)} = \frac{b_0 + b_1 z^{-1} + b_2 z^{-2}}{1 - a_1 z^{-1} - a_2 z^{-2}} \qquad (4.106)$$

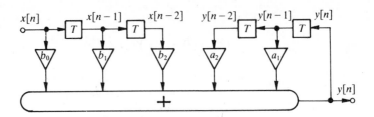

FIGURE 4.30 Second-order LTD system.

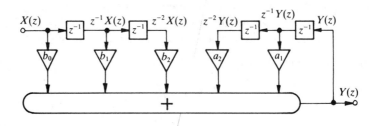

FIGURE 4.31 Alternative circuit diagram for the circuit of Fig. 4.30.

- *Comment.* In this example we have determined the system function of a system that we shall frequently encounter: a 'second-order section'. From the system function, as mentioned earlier, we can derive the frequency response $H(e^{j\theta})$. For every combination of the parameters b_0, b_1, b_2, a_1 and a_2 we then find another function. For simplicity we consider here the special case where $b_0 = 1$ and $b_1 = b_2 = 0$. We then obtain:

$$H(e^{j\theta}) = \frac{1}{1 - a_1 e^{-j\theta} - a_2 e^{-2j\theta}} = A e^{j\varphi} \qquad (4.107)$$

The modulus A (in dB) and the argument φ (in radians) in the fundamental interval $-\pi \leq \theta < \pi$ are plotted in Fig. 4.32 for various combinations of a_1 and a_2.

4.5.7 Poles and zeros as a system description

In the section about the difference equation (section 4.4.2) we have already stated that the causal LTD system, with which we shall mainly be concerned, can be described by the linear difference equation:

$$y[n] = \sum_{i=0}^{N} b_i x[n-i] + \sum_{i=1}^{M} a_i y[n-i] \qquad (4.108)$$

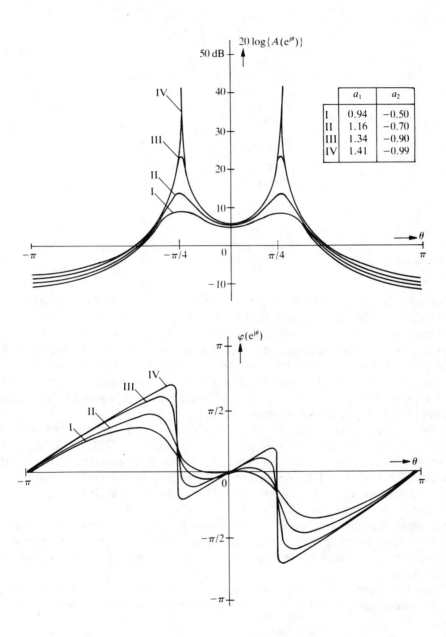

	a_1	a_2
I	0.94	−0.50
II	1.16	−0.70
III	1.34	−0.90
IV	1.41	−0.99

FIGURE 4.32 The modulus and the argument of the frequency response $H(e^{j\theta})$ for a second-order section with $b_0 = 1$ and $b_1 = b_2 = 0$ and a_1 and a_2 as in the table.

The corresponding system function $H(z)$ is:

$$H(z) = \frac{Y(z)}{X(z)} = \frac{\sum\limits_{i=0}^{N} b_i z^{-i}}{1 - \sum\limits_{i=1}^{M} a_i z^{-i}} \tag{4.109}$$

This can also be written as:

$$H(z) = b_0 \frac{(z-z_1)(z-z_2) \cdots (z-z_N)}{(z-p_1)(z-p_2) \cdots (z-p_M)} z^{M-N}$$

$$= b_0 \frac{\prod\limits_{i=1}^{N} (z-z_i)}{\prod\limits_{j=1}^{M} (z-p_j)} z^{M-N} \tag{4.110}$$

The complex quantities z_i are called the *zeros* of $H(z)$ and the complex quantities p_j are called the *poles* of $H(z)$, for if $z = z_i$ then $H(z) = 0$ and if $z = p_j$ then $H(z) = \infty$.

We thus see that $H(z)$ is completely determined, except for the constant b_0, by the values of z_i and p_j. We can represent these poles and zeros graphically in the complex z-plane. For the four second-order sections whose frequency response is shown in Fig. 4.32, the poles-and-zeros plot looks as in Fig. 4.33(a). In these four cases we are dealing with two poles and two zeros (at the origin). The poles-and-zeros plot for a completely arbitrary choice of system is shown in Fig. 4.33(b).

We have also encountered the idea of poles and zeros with LTC systems. We saw there that the vertical axis in the complex p-plane (see for example Fig. 2.11) had an important part to play in formulating a relation between the poles-and-zeros description and the frequency response of the LTC system. For LTD systems the unit circle in the complex z-plane plays a similar part: for the system function $H(z)$ on the unit circle is exactly equal to the frequency response $H(e^{j\theta})$.

Since we shall always be concerned with systems for which the coefficients a and b of $H(z)$ are real, the zeros can only be real or occur as complex conjugate pairs; the same is true for the poles. Figures 4.33(a) and (b) therefore have mirror symmetry about the horizontal axis.

We can draw a number of immediate conclusions from the poles-and-zeros description of all practically realizable LTD systems. We state the most important ones here, without proof:

1. In a stable system all the poles are *inside* the unit circle $|z| = 1$ (the system of Fig. 4.33(b) is thus not stable); zeros may lie inside, on, or outside the unit circle.
2. We can find the relation between the poles-and-zeros description and the frequency response $H(e^{j\theta}) = A e^{j\varphi}$ for the frequency $\theta = \theta_0$ in the following way. The amplitude $A(e^{j\theta_0}) = |H(e^{j\theta_0})|$ is determined by the product of all the distances between the point

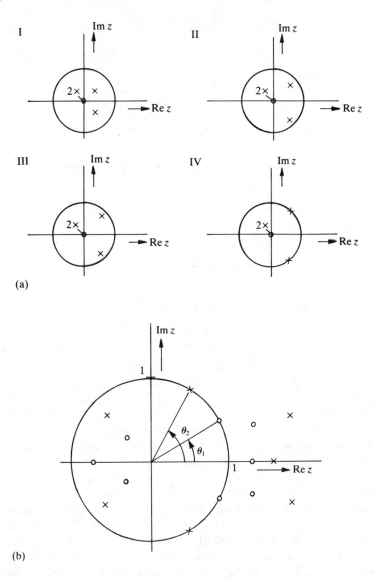

FIGURE 4.33 (a) Poles (x) and zeros (o) of the four second-order sections corresponding to Fig. 4.32. (b) Poles-and-zeros plot for an arbitrary choice of LTD system.

$z = e^{j\theta_0}$ and the zeros, divided by the product of all the distances between $z = e^{j\theta_0}$ and the poles. We find the phase $\varphi(e^{j\theta})$ by adding all the phase contributions from the zeros and subtracting all the phase contributions from the poles from this sum. This is shown graphically for a second-order system (i.e. $M = 2$) with two zeros ($N = 2$) in Fig. 4.34:

$$A(e^{j\theta_0}) = \frac{R_1 R_2}{R_3 R_4} \quad \text{and} \quad \varphi(e^{j\theta_0}) = \beta_1 + \beta_2 - \beta_3 - \beta_4 \qquad (4.111)$$

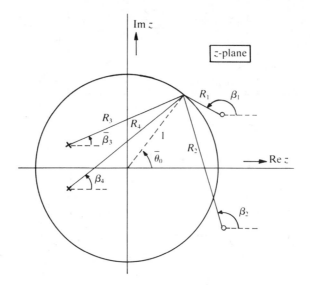

FIGURE 4.34 Graphical determination of $A(e^{j\theta_0})$ and $\varphi(e^{j\theta_0})$.

(For simplicity we have assumed here that $b_0 = 1$.) We see that each pole and each zero has the most effect in the frequency range that corresponds to the nearest part of the unit circle to that pole or zero. Furthermore, the effect of a pole or a zero on the frequency response increases the closer it is to the unit circle. In the extreme case of a zero actually on the unit circle, e.g. at $z = e^{j\theta_1}$ in Fig. 4.33(b), the amplitude of the frequency response is zero at $\theta = \theta_1$ and there is a jump in phase of π radians at that point. On the other hand, a pole on the unit circle, e.g. at $z = e^{j\theta_2}$ in Fig. 4.33(b), gives an infinite amplitude at $\theta = \theta_2$, again with a phase jump of π radians.

3. If all the zeros are inside the unit circle, we have a 'minimum-phase network'.[†] By 'reflecting' (some of) the zeros in the unit circle we change the phase characteristic of the system but not the amplitude characteristic (except perhaps for a constant). 'Reflection' in the unit circle means that a zero $z_i = r_i e^{j\theta_i}$ is replaced by a zero $z_i = (1/r_i)e^{-j\theta_i}$. In Fig. 4.35(a) and (b) an example is given of such a reflection for a minimum-phase system that has three poles at the origin and three zeros. The corresponding amplitude, phase and group-delay characteristics are also shown in Fig. 4.35. We see clearly that the phase characteristic of the minimum-phase system varies far less strongly and the group delay is therefore smaller than in the non-minimum-phase system. This is a general rule.

4. A stable phase-shifter or 'all-pass' network (i.e. $|H(e^{j\theta})| = 1$ for $-\pi \le \theta < \pi$) has only poles inside the unit circle and zeros outside the unit circle. Poles and zeros always occur in pairs that are 'reflected' in the unit circle. An example is given in Fig. 4.36(a) (showing both the poles-and-zeros plot of the system and the amplitude and phase characteristics).

[†]A minimum-phase network is a circuit in which for a given amplitude characteristic the *group delay* $\tau_g = -d\varphi/d\theta$ has the lowest possible value for *all* frequencies.

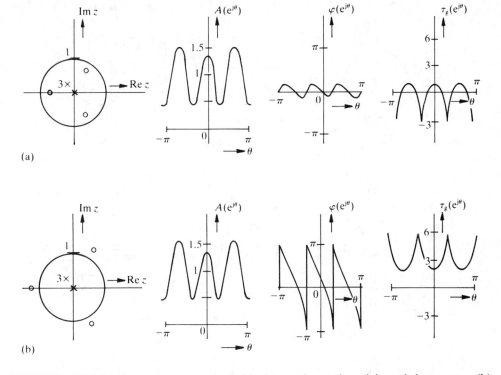

FIGURE 4.35 (a) Minimum-phase system with three poles at the origin and three zeros. (b) Non-minimum-phase system obtained from (a) by reflection of the zeros in the unit circle. This gives a larger group delay τ_g.

5. If a system (except for the origin) has poles only, we call it an 'all-pole' system (Fig. 4.36(b)).

6. If a system (except for the origin) has zeros only, and these are reflected in pairs in the unit circle (possibly also with zeros on the unit circle) then the phase characteristic is linear and we have a linear-phase system; Fig. 4.36(c) (see also Chapter 8 'Design methods for discrete filters').

7. Poles-and-zeros plots of systems connected in cascade can be added together; if a pole and a zero *coincide*, they cancel one another.

8. The number of poles of a causal LTD system is greater than or equal to the number of zeros (poles or zeros at $z = 0$ are included). To show this we write the general expression for the system function $H(z)$ in such a way that only *positive* powers of z occur in both the numerator and the denominator (i.e. including z^3, for example, but not z^{-3}). This can always be done:

$$H(z) = \frac{c_0 z^N + c_1 z^{N-1} + \ldots + c_N}{z^M + d_1 z^{M-1} + \ldots + d_M}.$$

(4.112)

where $c_0 \neq 0$ and c_N and d_M are not both equal to zero. The other c_i's and d_i's can have

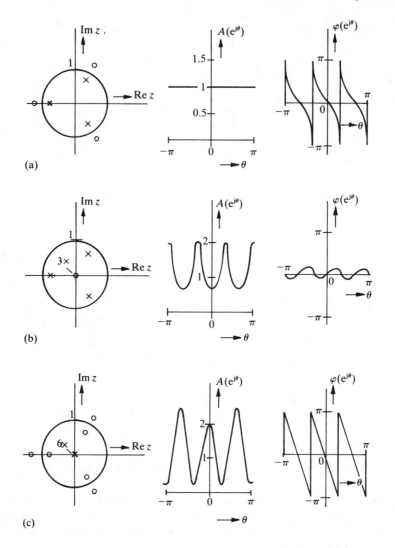

FIGURE 4.36 (a) 'All-pass' network with three poles and zeros. (b) 'All-pole' network. (c) Linear-phase network.

an arbitrary value (thus including 0). $H(z)$ is now said to describe a system of the Mth order; the order is determined by the highest power of z in the denominator. $H(z)$ has M poles and N zeros. Carrying out the long division for $H(z)$ gives a result of the form:

$$H(z) = e_0 z^{N-M} + e_1 z^{N-M-1} + e_2 z^{N-M-2} + \ldots \qquad (4.113)$$

The corresponding impulse response is:

$$h[n] = e_0 \delta[n + N - M] + e_1 \delta[n + N - M - 1] + e_2 \delta[n + N - M - 2] + \ldots \quad (4.114)$$

This signal is only causal if $M \geq N$ (verify this!).

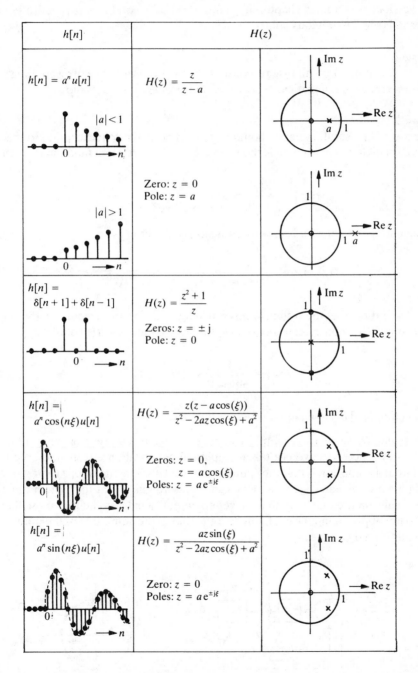

FIGURE 4.37 Examples of impulse responses $h[n]$ and corresponding system functions $H(z)$.

9. If an LTD signal is derived from an existing LTC system (and this is not at all unusual, as we shall see in later chapters), then the causality condition is practically always satisfied (see the previous point).

Example 1
In Fig. 4.37 a number of impulse responses $h[n]$ are given with the corresponding system functions $H(z)$ as mathematical expressions and as poles-and-zeros plots.

Example 2
If they are to be useful, systems should in general be stable. An exception to this is the 'discrete oscillator'; an example is shown in Fig. 4.38. The system function is given by:

$$H(z) = \frac{Y(z)}{X(z)} = \frac{z^{-1}}{1 - 2\cos(\theta_0)z^{-1} + z^{-2}} \tag{4.115}$$

For calculating the poles and zeros this function is rewritten as:

$$H(z) = \frac{z}{z^2 - 2z\cos(\theta_0) + 1} = \frac{z}{(z - e^{-j\theta_0})(z - e^{j\theta_0})} \tag{4.116}$$

The poles and zeros of the system are shown in Fig. 4.38(b). From equation (8b) in Table 4.3 the impulse response corresponding to the system function $H(z)$ of (4.115) is:

$$h[n] = \frac{1}{\sin(\theta_0)} \cdot \sin(n\theta_0)u[n] \tag{4.117}$$

This function $h[n]$ is shown in Fig. 4.39 for $\theta_0 = 0.6$.

- *Comment.* We see in this example a number of clear differences from continuous-time oscillators; since we have not taken amplitude-quantization effects into account and our initial assumption is that the content of the delay elements is zero for $n < 0$, we are dealing with a linear system (and we can therefore talk about the system function!). This means on the one hand that we get no output signal (and hence no oscillation) if we do not apply an input signal. On the other hand, the amplitude of the output signal depends directly on the input signal.

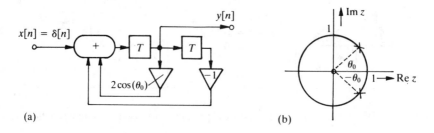

(a)   (b)

FIGURE 4.38 A discrete oscillator. (a) Block diagram. (b) Poles-and-zeros plot.

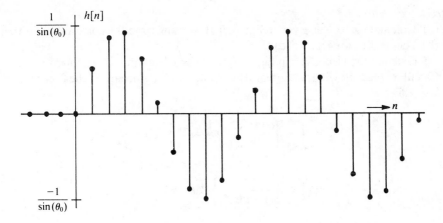

FIGURE 4.39 Impulse response of the discrete oscillator.

4.6 EXERCISES

Exercises to section 4.2

4.1 In (4.9) it was shown that the discrete sine function $\sin(n\pi/4)$ is periodic with a period of 8, while in equation (4.10) it was shown that $\sin(n)$ is not periodic. Determine the value(s) of θ for which the function $A\sin(n\theta + \varphi)$ is periodic and determine the corresponding period(s).

4.2 Draw the following discrete signals, for $-5 \le n \le 5$:

(a) $x[n] = 1$
(b) $x[n] = \cos(2\pi n)$
(c) $x[n] = \cos(\pi n)$
(d) $x[n] = \sin(\pi n/2 + \pi/4)$

(e) $x[n] = \sum_{i=-\infty}^{\infty} \sin(i\pi/2)\delta[n-i]$

(f) $x[n] = \begin{cases} 0 & \text{for } n < 0 \\ 1 & \text{for } n = 0 \\ 0.8 \cdot x[n-1] & \text{for } n > 0 \end{cases}$

4.3 If $x[n] = e^{j\pi n/2}$ we can split $x[n]$ into a real part $\text{Re}\{x[n]\}$ and an imaginary part $\text{Im}\{x[n]\}$ so that $x[n] = \text{Re}\{x[n]\} + j\,\text{Im}\{x[n]\}$. Draw $\text{Re}\{x[n]\}$ and $\text{Im}\{x[n]\}$ for $-5 \le n \le 5$.

4.4 Calculate the values, for $-5 \le n \le 5$, of:

(a) $x[n] = e^{j\pi n/4}$
(b) $x[n] = e^{jn}$

Are these signals periodic and, if so, what is the period?

Exercises to section 4.3

4.5 (a) Calculate $X(e^{j\theta})$ as given in Fig. 4.5(b), starting from the corresponding $x[n]$.
   (b) As (a), but for Fig. 4.5(d).
   (c) Calculate $x[n]$ as given in Fig. 4.5(e), starting from the corresponding $X(e^{j\theta})$.

4.6 Calculate the following two integrals; use Parseval's theorem and the Fourier pairs of Fig. 4.5:

$$I_1 = \int_{-\pi}^{\pi} \left| \frac{\sin\{(2N+1)\theta/2\}}{\sin(\theta/2)} \right|^2 d\theta$$

and

$$I_2 = \int_{-\pi}^{\pi} \left| \frac{1}{1-ae^{-j\theta}} \right|^2 d\theta \quad \text{with} \quad |a| < 1$$

4.7 Calculate the following summation (use Parseval's theorem):

$$S_1 = \sum_{n=-\infty}^{\infty} |x[n]|^2$$

$$\text{with } x[n] = \begin{cases} \sin(n\theta_0)/n\pi & \text{for } n \neq 0 \\ \theta_0/\pi & \text{for } n = 0 \end{cases}$$

Exercises to section 4.4

4.8 Consider an arbitrary linear system with input signal $x[n]$ and output signal $y[n]$. Show that $y[n] = 0$ for all n if $x[n] = 0$ for all n.

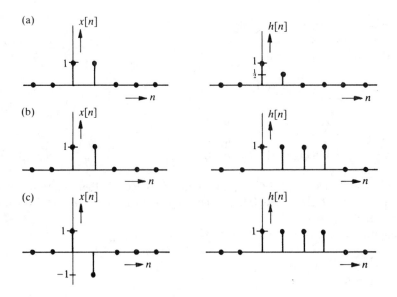

FIGURE 4.40 Exercise 4.10.

4.9 A system has input signal $x[n]$ and output signal $y[n]$. It is described by the difference equation $y[n] = x[n] - ay[n-1]$ with $y[-1] = 1$.
(a) Determine $y[n]$ for the following input signals $x[n]$:

$$x_1[n] = 0$$
$$x_2[n] = \delta[n]$$
$$x_3[n] = 2\delta[n]$$
$$x_4[n] = \delta[n-1]$$

(b) Determine, using the results in (a), whether the system is linear or time-invariant or both.
(c) Can you explain the results in (b)?

4.10 Determine the convolution $y[n] = x[n] * h[n]$ for the signals of the Figs 4.40(a), (b) and (c).

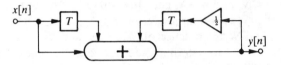

FIGURE 4.41 Exercise 4.11.

4.11 (a) Determine the impulse response $h[n]$ of the system shown in Fig. 4.41.
(b) Show that $h[n]$ can be taken as the convolution of two discrete-time functions $h_1[n]$ and $h_2[n]$: $h[n] = h_1[n] * h_2[n]$

$$\text{with } h_1[n] = \begin{cases} 0 & \text{for } n<0 \text{ and } n>1 \\ 1 & \text{for } 0 \leq n \leq 1 \end{cases}$$

$$\text{and } h_2[n] = \begin{cases} 0 & \text{for } n<0 \\ (\tfrac{1}{2})^n & \text{for } n \geq 0 \end{cases}$$

4.12 A system H, shown in Fig. 4.42, is formed by cascading two LTD systems H_1 and H_2.
(a) Show that if H_1 and H_2 are stable, then H is stable.
(b) Show that if H_1 and H_2 are causal, then H is causal.
(c) The converse of (a) and (b) is *not* generally true! Thus if H is stable, H_1 and H_2 are not necessarily both stable, and if H is causal, H_1 and H_2 are not necessarily both causal. Show this with the aid of the following impulse responses:

$$h_1[n] = \begin{cases} 1 & \text{for } n \geq -1 \\ 0 & \text{for } n < -1 \end{cases}$$

and

$$h_2[n] = \begin{cases} 1 & \text{for } n = 1 \\ -1 & \text{for } n = 2 \\ 0 & \text{for } n \le 0 \quad \text{and } n \ge 3 \end{cases}$$

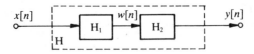

FIGURE 4.42 Exercise 4.12.

4.13 If the impulse response $h[n]$ and the input signal $x[n]$ of an LTD system satisfy $h[n] = 0$ for $n < N_0$ and $n > N_1$ and $x[n] = 0$ for $n < N_2$ and $n > N_3$, where $N_0 \le N_1$ and $N_2 \le N_3$, then $y[n] = 0$ for $n < N_4$ and $n > N_5$.
Determine N_4 and N_5 as a function of N_0, N_1, N_2 and N_3.

4.14 Use methods 2 and 3 to determine the frequency response $H(e^{j\theta})$ of the system shown in Fig. 4.43.

4.15 (For the enthusiast!) Determine whether each of the following systems, with input signal $x[n]$ and output signal $y[n]$, is stable, causal, linear and time-invariant.

(a) $y[n] = \cos(n)x[n]$

(b) $y[n] = \sum_{i=0}^{n} x[i]$

(c) $y[n] = \sum_{i=n-1}^{n+1} x[i]$

(d) $y[n] = x[n-1]$

(e) $y[n] = e^{x[n]}$

(f) $y[n] = ax[n] + b$

Exercises to section 4.5

4.16 Determine the z-transform of the following functions:

(a) $x_a[n] = u[n] - u[n-3]$

(b) $x_b[n] = \alpha^n(u[n] - u[n-3])$

(c) $x_c[n] = \alpha^n u[n] - \alpha^{n-3} u[n-3]$

$$
(d) \ x_d[n] = \begin{cases} n+1 & \text{for } 0 \le n \le 2 \\ 5-n & \text{for } 2 < n \le 4 \\ 0 & \text{for } n < 0 \quad \text{and } n > 4 \end{cases}
$$

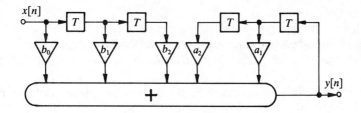

$x[n]$

b_0 b_1 b_2 a_2 a_1

$+$

$y[n]$

FIGURE 4.43 Exercise 4.14.

4.17 Determine the inverse transforms of the following functions and draw them:

(a) $X_e(z) = \dfrac{1 - 0.5z^{-1}}{1 - 0.25z^{-1}}$

(b) $X_f(z) = \dfrac{1 - 0.5z^{-1}}{1 - 0.25z^{-2}}$

(c) $X_g(z) = \dfrac{z^2}{(z - 0.5)(z - 0.75)}$

(d) $X_h(z) = \dfrac{1}{(z - 0.5)^2}$

4.18 Determine the poles and zeros for all the functions given in Exercises 4.16 and 4.17. Sketch for yourself the corresponding poles-and-zeros plots.

4.19 A system has a pole and zeros as given in Fig. 4.44:
  (a) Determine $H(z)$, $h[n]$ and $H(e^{j\theta})$ for this system.
  (b) Draw $h[n]$. Is the system causal?

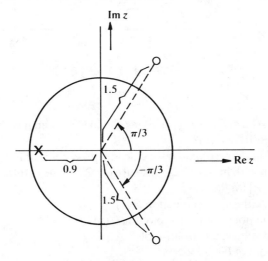

FIGURE 4.44 Exercise 4.19.

(c) Is the system stable?

(d) Is this a minimum-phase system?

If it is not a minimum-phase system, answer the following questions:

(e) Draw the poles-and-zeros plot of a minimum-phase system with the same $|H(e^{j\theta})|$.

(f) Determine the system function, the impulse response and the frequency response of this minimum-phase system.

(g) Draw the impulse response. Is this minimum-phase system causal?

4.20 For the system of Fig. 4.45:

(a) Use Methods 2 and 3 to determine the system function $H(z)$ of this system.

(b) Take $b_0 = b_2 = 1$, $b_1 = 2$, $a_1 = 1.4$ and $a_2 = -0.5$. Determine the poles and zeros of $H(z)$. Is the system stable? Is this a minimum-phase system?

(c) Now take $b_0 = b_2 = 1$, $b_1 = 2.5$, $a_1 = 1$ and $a_2 = -2$. Do the same as in (b).

(d) Now take $b_1 = b_2 = 0$, $b_0 = 1$, $a_1 = 1$ and $a_2 = -0.99$. Determine $H(z)$.

(e) Determine the response of the system defined in (d) to the input signal $x[n] = A\cos(n\pi/3)$.

(f) Take $b_0 = 1$, $b_1 = -0.35$, $b_2 = 0$, $a_1 = 0.7$ and $a_2 = -0.49$. Determine $H(z)$ and the impulse response $h[n]$ again.

(g) Determine the values of the impulse response $h[n]$ for $-5 \le n \le 9$.

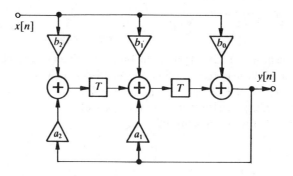

FIGURE 4.45 Exercise 4.20.

4.21 (a) Which system from Fig. 4.35 or 4.36 do we obtain if we cascade the systems of Fig. 4.36(b) and (c)?

(b) And what do we obtain if we cascade the systems of Fig. 4.35(a) and 4.36(a)?

4.22 For a system with impulse response $h[n] = 2^n \cos(n\pi/4)u[n]$:

(a) Draw $h[n]$ for $0 \le n \le 5$.

(b) Calculate the system function $H(z)$.

(c) Draw the poles-and-zeros plot for $H(z)$.

(d) Is this system stable or unstable?

4.23 For the system of Fig. 4.46:

(a) Express $y_1[n]$ in terms of $y_1[n-1]$, $y_2[n-1]$ and $x[n]$; do the same for $y_2[n]$.

(b) Take $A = \cos(\theta_0)$
      $B = \sin(\theta_0)$
      $y_1[-1] = \cos(-\theta_0)$
      $y_2[-1] = \sin(-\theta_0)$
Show that (if $x[n] = 0$), then:
$y_1[n] = \cos(n\theta_0)$ and $y_2[n] = \sin(n\theta_0)$.
Note: For these values of A, B, $y_1[n]$ and $y_2[n]$ this circuit thus represents a discrete oscillator that supplies a sine signal and a cosine signal at the same frequency (verify this!).

(c) Calculate (for arbitrary A and B) the system functions $H_1(z) = Y_1(z)/X(z)$ and $H_2(z) = Y_2(z)/X(z)$.

(d) Take $A = B = \frac{1}{2}\sqrt{2}$. Draw the poles-and-zeros plot for $H_1(z)$ and $H_2(z)$.

(e) Take $A = B = \frac{1}{2}\sqrt{2}$ and $x[n] = \delta[n]$. Calculate and draw the impulse responses $h_1[n]$ and $h_2[n]$ for $-2 \le n \le 10$.

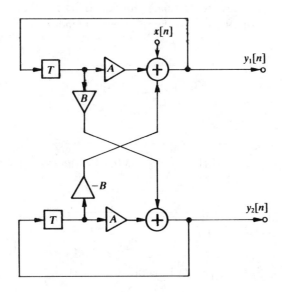

FIGURE 4.46 Exercise 4.23.

5

The DFT and the FFT

5.1 THE DFT FOR PERIODIC DISCRETE SIGNALS

In section 4.3 we saw how the Fourier transform (the FTD) of an arbitrary discrete signal $x[n]$ is defined, and we now recapitulate the equations for the forward and inverse transforms, (4.18) and (4.19). We use the relative frequency $\theta = \omega T$:

$$X(e^{j\theta}) = \sum_{n=-\infty}^{\infty} x[n]e^{-jn\theta} \qquad \text{(FTD)} \qquad (5.1)$$

$$x[n] = \frac{1}{2\pi} \int_{-\pi}^{\pi} X(e^{j\theta})e^{jn\theta}\,d\theta \qquad \text{(IFTD)} \qquad (5.2)$$

Now we should like to look at the case where we are dealing with a periodic function $x_p[n]$ of period N, so that for all n:

$$x_p[n] = x_p[n + lN] \quad \text{with} \quad l = 0, \pm 1, \pm 2 \ldots \qquad (5.3)$$

We cannot simply apply (5.1) and (5.2) to this signal directly, since we run into problems with convergence in evaluating the sum and the integral. So with periodic discrete signals we set about finding a frequency representation in another way. We make use of what we already know about $x_p[n]$. Now, because of the periodic nature of $x_p[n]$, we know:

1. That in a description in the frequency domain contributions will only occur at frequencies that are multiples of the fundamental frequency θ_0, and possibly at $\theta = 0$. The value of the fundamental frequency is given by $\theta_0 = 2\pi/N$. We can see this as follows. A cosine signal at this frequency can be expressed formally as $\cos(2\pi n/N + \varphi)$ where φ is an arbitrary phase. This signal has a period of length exactly equal to N and therefore represents the fundamental of $x_p[n]$.
2. That we can get all the frequency information from a single period of the signal $x_p[n]$.

These two points bring us to the definition of the (N-point) discrete Fourier transform (DFT):

$$X_p[k] = \sum_{n=0}^{N-1} x_p[n]e^{-j(2\pi/N)kn} \qquad \text{(DFT)} \qquad (5.4)$$

and the (N-point) inverse discrete Fourier transform (IDFT):

$$x_p[n] = \frac{1}{N}\sum_{k=0}^{N-1} X_p[k]e^{j(2\pi/N)kn} \qquad \text{(IDFT)} \qquad (5.5)$$

If we are given the N values $x_p[0]$, $x_p[1]$, ..., $x_p[N-1]$, or the *fundamental interval* of the sequence $x_p[n]$ in the time domain, we can use the DFT to find the frequency function $X_p[k]$ for *every* integer value of k. The function $X_p[k]$ is also periodic with a period N, since:

$$X_p[k+lN] = \sum_{n=0}^{N-1} x_p[n]e^{-j(2\pi/N)(k+lN)n} = \sum_{n=0}^{N-1} x_p[n]e^{-j(2\pi/N)kn}\,e^{-j2\pi ln}$$

$$= \sum_{n=0}^{N-1} x_p[n]e^{-j(2\pi/N)kn} = X_p[k]$$

Therefore with the N values $X_p[0]$, $X_p[1]$, ..., $X_p[N-1]$, i.e. with the *fundamental interval* of $X_p[k]$, we have sufficient informaton to describe $X_p[k]$ completely (see Fig. 5.1).

From the values of $X_p[k]$ in the fundamental interval we find $x_p[n]$ again with the aid of (5.5). In a graphical representation of the functions $x_p[n]$ and $X_p[k]$ it is sufficient to draw the fundamental interval. In Fig. 5.1 the values are indicated by continuous lines inside this interval and by dashed lines outside it.

The factor $1/N$ that precedes the summation sign of the IDFT is chosen such that after applying the forward and inverse transforms we recover the original function $x_p[n]$ exactly. The functions $x_p[n]$ and $X_p[k]$ form a Fourier pair, which we represent symbolically as:

$$x_p[n] \; \circ\!\!-\!\!-\!\!\circ \; X_p[k] \qquad (5.6)$$

Example 1

The 4-point DFT of the periodic function $x_1[n]$ defined as:

$$x_1[n] = \begin{cases} 1 & \text{for } n = 4l \text{ with } l = 0, \pm 1, \pm 2 \ldots \\ 0 & \text{elsewhere} \end{cases} \qquad (5.7)$$

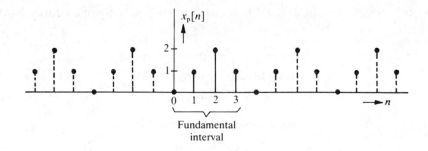

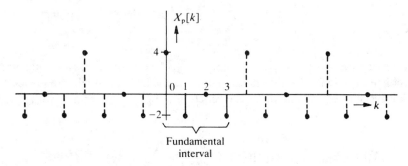

FIGURE 5.1 Example of a DFT pair $x_p[n]$ and $X_p[k]$ with $N = 4$.

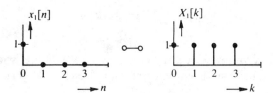

FIGURE 5.2 The 4-point DFT pair $x_1[n]$ and $X_1[k]$.

follows directly from (5.4):

$$X_1[k] = \sum_{n=0}^{3} x_1[n] e^{-j(2\pi/4)kn} = e^{-j(2\pi/4)k0} = e^{-j0} = 1 \qquad (5.8)$$

The fundamental intervals of $x_1[n]$ and $X_1[k]$ are shown in Fig. 5.2. We find the inverse transform by substituting (5.8) in (5.5):

$$x_1[n] = \frac{1}{4} \sum_{k=0}^{3} X_1[k] e^{j(2\pi/4)kn} = \frac{1}{4}(1 + e^{j\pi n/2} + e^{j2\pi n/2} + e^{j3\pi n/2})$$

Thus:

$$x_1[0] = \frac{1}{4}(1 + 1 + 1 + 1) = 1; \quad x_1[1] = \frac{1}{4}(1 + j - 1 - j) = 0$$

$$x_1[2] = \frac{1}{4}(1 - 1 + 1 - 1) = 0; \quad x_1[3] = \frac{1}{4}(1 - j - 1 + j) = 0$$

Example 2

The 4-point DFT of the periodic function $x_2[n]$ defined as:

$$x_2[n] = \begin{cases} 1 & \text{for } n = 4l+1 \text{ with } l = 0, \pm 1, \pm 2, \ldots \\ 0 & \text{elsewhere} \end{cases} \tag{5.9}$$

is found from (5.4):

$$X_2[k] = \sum_{n=0}^{3} x_2[n] e^{-j(2\pi/4)kn} = e^{-j2\pi k/4} = e^{-j\pi k/2} \tag{5.10}$$

The fundamental interval for both $x_2[n]$ and $X_2[k]$ is given in Fig. 5.3. The complex function $X_2[k]$ is shown here not only as modulus $A_2[k]$ and phase $\varphi_2[k]$, but also as real part $R_2[k]$ and imaginary part $I_2[k]$.

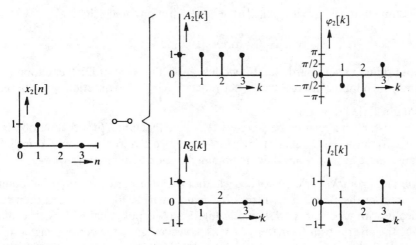

FIGURE 5.3 The 4-point DFT pair $x_2[n]$ and $X_2[k] = A_2[k]e^{j\varphi_2[k]} = R_2[k] + jI_2[k]$.

Example 3

The 16-point IDFT of $X_3[k]$ defined as:

$$X_3[k] = \begin{cases} 1 & \text{for } k = 2+16l \text{ and } k = 14+16l \\ 0 & \text{elsewhere} \end{cases} \tag{5.11}$$

is found by substituting $X_3[k]$ in (5.5):

$$x_3[n] = \frac{1}{16} \sum_{k=0}^{15} X_3[k] e^{j2\pi kn/16} = \frac{1}{16} (e^{j4\pi n/16} + e^{j28\pi n/16})$$

$$= \frac{1}{16} (e^{j\pi n/4} + e^{j7\pi n/4}) = \frac{1}{16} (e^{j\pi n/4} + e^{-j\pi n/4} e^{j2\pi n})$$

$$= \frac{1}{16} (e^{j\pi n/4} + e^{-j\pi n/4}) = \frac{\cos(\pi n/4)}{8} \tag{5.12}$$

The fundamental intervals of $x_3[n]$ and of $X_3[k]$ are shown in Fig. 5.4.

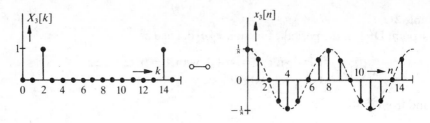

FIGURE 5.4 The 16-point DFT pair $X_3[k]$ and $x_3[n] = 0.125\cos(\pi n/4)$.

5.2 THE DFT FOR FINITE-DURATION DISCRETE SIGNALS

Let us assume that we have a signal $x[n]$ of finite duration for which:

$$x[n] = 0 \quad \text{for } n<0 \quad \text{and } n\geq N \tag{5.13}$$

If without thinking very hard about it we just apply the N-point DFT equation (5.4) to this signal, we obtain a discrete frequency function $X[k]$. Two questions arise here:

1. What does $X[k]$ represent?
2. Can we establish a relation between $X[k]$ and the function $\tilde{X}(e^{j\theta})$ that we obtain if we fill in $x[n]$ in the FTD relation (5.1)? (We use two different symbols X and $\tilde{X}$ here because these are functions that are obtained from $x[n]$ in different ways.)

Question (1): From $X[k]$ we do not see whether we have started from $x[n]$ as defined in (5.13), or from the function $x_p[n]$ that is equal to $x[n]$ for $0\leq n<N$ and then repeats periodically with a period N as shown in Fig. 5.5 (N is put equal to 4 as an example). Applying the inverse transform (5.5) to $X[k]$, however, gives the periodic signal $x_p[n]$ by definition. We could say that in applying the DFT to a signal of finite duration we are thinking of it as being repeated periodically. After forward and inverse transformation we find the original signal $x[n]$ again by taking the samples of $x_p[n]$ in the fundamental interval $0\leq n<N$ and putting $x[n] = 0$ outside it.

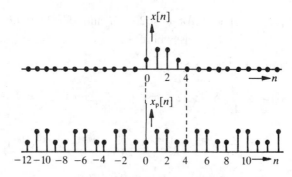

FIGURE 5.5 The discrete finite-duration function $x[n]$ and the function $x_p[n]$ that is a periodic continuation of $x[n]$.

Question (2): Applying the FTD equation (5.1) to the signal $x[n]$ gives:

$$\tilde{X}(e^{j\theta}) = \sum_{n=-\infty}^{\infty} x[n]e^{-j\theta n} = \sum_{n=0}^{N-1} x[n]e^{-j\theta n} \tag{5.14}$$

Applying the DFT equation (5.4) gives:

$$X[k] = \sum_{n=0}^{N-1} x[n]e^{-j(2\pi k/N)n} \tag{5.15}$$

Comparison of (5.14) and (5.15) shows that:

$$X[k] = \tilde{X}(e^{j2\pi k/N}) \tag{5.16}$$

i.e. that for the frequencies $\theta_k = 2\pi k/N$ the function $\tilde{X}$ has exactly the same (complex) value as $X[k]$. This is explained graphically in Fig. 5.6.

We could also say that X is a sampled version (sampled in the frequency domain!) of $\tilde{X}$. The values $X[k]$ for $0 \le k < N$ completely describe the signal $x_p[n]$ and therefore $x[n]$ and $\tilde{X}(e^{j\theta})$ as well. This reminds us of the sampling theorem for continuous-time signals that we encountered earlier. We found there that a continuous-time signal can be completely described by samples, if the corresponding frequency spectrum is strictly band limited. Here we see an example of the fact that a continuous frequency spectrum can be completely described by samples, because the corresponding continuous-time signal is strictly duration-limited.

In the foregoing we have seen that the DFT is a useful aid for the description in the frequency domain of periodic discrete signals and discrete signals of finite duration. The great advantage of the DFT is that both functions of the Fourier pair are discrete – in the time domain or the frequency domain, as appropriate – and are completely described by N (complex) numbers $x_p[n]$ or $X_p[k]$. The DFT is therefore very suitable for calculation on a digital computer. It is in fact possible to use the DFT in even more cases than we have mentioned so far. We shall come back to this later in this chapter.

We should also mention here that there is another method of calculation that can be used to obtain the DFT very efficiently (the fast Fourier transform or FFT). This will also be discussed later.

5.3 PROPERTIES OF THE DFT

Many of the properties that we have encountered in the signal transforms treated earlier have a counterpart in the DFT. As we have seen, the DFT was originally defined for periodic signals. But now we have also learned how we can use the DFT for non-periodic signals of finite duration, if we think of them as being continued periodically (this is why we use the subscript 'p' where possible). In the application of some of the properties of the DFT it is extremely important that we remember which of the two kinds of signal we started with. The different consequences and the underlying reasons for them will be discussed with the properties where this effect comes into play.

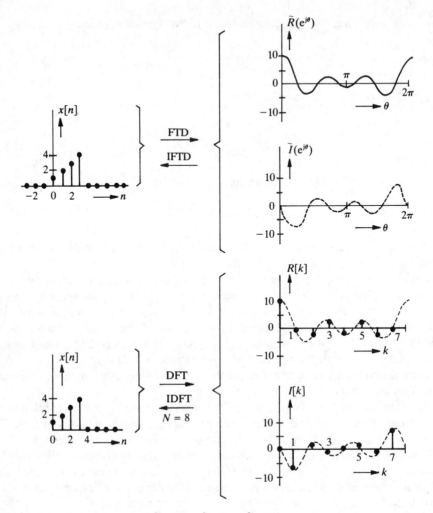

FIGURE 5.6 Relation between $\tilde{X}(e^{j\theta}) = \tilde{R}(e^{j\theta}) + j\tilde{I}(e^{j\theta})$ and $X[k] = R[k] + jI[k]$ if $x[n]$ is a signal of finite duration.

We shall now briefly summarize the most important DFT properties, making use of the following Fourier pairs:

$$f_p[n] \circ\!\!-\!\!-\!\!\circ F_p[k]$$

$$g_p[n] \circ\!\!-\!\!-\!\!\circ G_p[k]$$

(5.17)

where $f_p[n]$ and $g_p[n]$ are arbitrary periodic discrete functions of period N.

A. Linearity

$$\boxed{af_p[n] + bg_p[n] \circ\!\!-\!\!-\!\!\circ aF_p[k] + bG_p[k]}$$

(5.18)

where a and b are arbitrary constants.

B. Time/frequency symmetry

$$\boxed{\frac{1}{N} F_p[n] \; \circ\!\!-\!\!-\!\!-\!\!\circ \; f_p[-k]} \qquad (5.19)$$

Caution: Here F_p represents the time function and f_p represents the frequency function.

C. Circular time shift

$$\boxed{f_p[n-i] \; \circ\!\!-\!\!-\!\!-\!\!\circ \; e^{-j(2\pi/N)ki} F_p[k]} \qquad (5.20)$$

where i is an arbitrary integer.

This property should be interpreted with care if $f_p[n-i]$ has been obtained from the periodic continuation of a signal $f[n]$ of finite duration, i.e. $f[n] = 0$ for $n < 0$ and $n \geq N$. We show this in Fig. 5.7. We see that the normal interpretation applies for a time shift of the function $f_p[n]$ through one sampling interval. But if we consider the fundamental interval of $f_p[n-1]$ in order to come back again to a signal of finite duration we do not find the usual shifted version $f[n-1]$, but a *circularly* shifted version $f_c[n-1]$ (see Fig. 5.7(d)). In circular shift the signal always remains equal to zero outside the fundamental interval. The circular shift has the effect that the samples that disappear from the fundamental interval in the normal shift appear again on the other side. Circular shift is not limited to shifts where $i \leq N$. It can easily be seen that for the signal $f[n]$ from Fig. 5.7:

$$f_c[n-5] = f_c[n-1] \quad \text{and} \quad f_c[n-8] = f_c[n-4] = f_c[n] = f[n]$$

In general for circular shift (with period N):

$$f_c[n-i] = f_c[n-i+lN] \quad \text{with} \quad l = 0, \; \pm 1, \; \pm 2, \; \ldots \qquad (5.21)$$

D. Frequency shift

$$\boxed{e^{j(2\pi/N)ni} f_p[n] \; \circ\!\!-\!\!-\!\!-\!\!\circ \; F_p[k-i]} \qquad (5.22)$$

where i is an arbitrary integer.

We thus see that shifting the spectrum $F_p[k]$ leads to a *complex* time function. We do obtain a *real* time function if we apply a frequency shift to $F_p[k]$ twice. This happens in modulation, for example:

$$2f_p[n]\cos(2\pi ni/N) \; \circ\!\!-\!\!-\!\!-\!\!\circ \; F_p[k-i] + F_p[k+i] \qquad (5.23)$$

- *Comment*: In considering the fundamental interval of $F_p[k]$, $F_p[k-1]$, $\ldots$ we see the same effect as described before for time shift: in the fundamental interval we are always concerned with circular shift. This is illustrated in Fig. 5.8, where the funda-

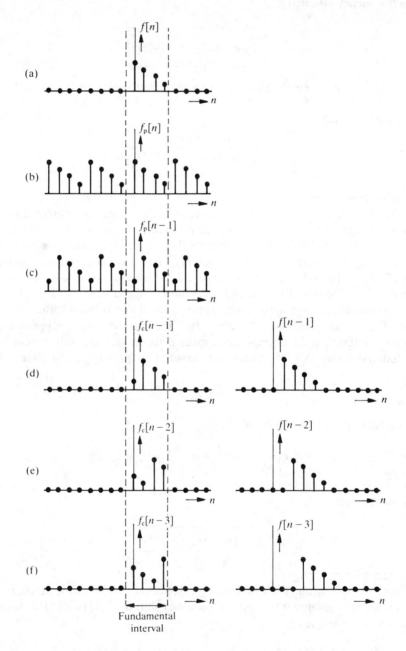

FIGURE 5.7 Circular shift (with $N = 4$) of a signal $f[n]$ of finite duration through one, two or three sampling intervals gives us $f_c[n-1]$, $f_c[n-2]$ and $f_c[n-3]$. The 'linearly' shifted signals $f[n-1]$, $f[n-2]$ and $f[n-3]$ are also shown for comparison.

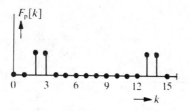

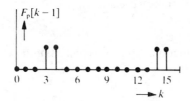

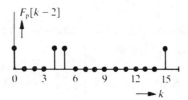

FIGURE 5.8 Circular frequency shift.

mental interval of an arbitrary periodic function $F_p[k]$ of period 16 and its shifted versions $F_p[k-1]$ and $F_p[k-2]$ are shown.

E. Circular convolution[†] in the time domain

$$\boxed{f_p[n] \circledast g_p[n] \;\; \circ\!\!-\!\!-\!\!\circ \;\; F_p[k]G_p[k]} \tag{5.24}$$

Circular convolution in the time domain thus corresponds to multiplication in the frequency domain.

F. Circular convolution[†] in the frequency domain

$$\boxed{f_p[n]g_p[n] \;\; \circ\!\!-\!\!-\!\!\circ \;\; \frac{1}{N}\, F_p[k] \circledast G_p[k]} \tag{5.25}$$

Circular convolution in the frequency domain thus corresponds to multiplication in the time domain.

[†]We shall return to the definition of circular convolution (sometimes called cyclic convolution), designated by ⊛ , in section 5.5.

G. Parseval's Theorem

$$\sum_{n=0}^{N-1} |f_p[n]|^2 = \frac{1}{N} \sum_{k=0}^{N-1} |F_p[k]|^2 \qquad (5.26)$$

This theorem states that the 'energy' of a periodic discrete signal can be calculated both in the time domain and in the frequency domain.

H. Arbitrary real time functions

For arbitrary periodic real functions $f_p[n]$ the real part of the transform $F_p[k]$ is even and the imaginary part is odd:

$$F_p[k] = F_p^*[N-k] \qquad (5.27a)$$

(where * indicates the complex conjugate) or expressed as real part R_p and imaginary part I_p:

$$R_p[k] = R_p[N-k] \quad \text{and} \quad I_p[k] = -I_p[N-k] \qquad (5.27b)$$

or as modulus A_p and argument φ_p:

$$A_p[k] = A_p[N-k] \quad \text{and} \quad \varphi_p[k] = -\varphi_p[N-k] \qquad (5.27c)$$

Caution: From this property, which applies for all real periodic time functions $f_p[n]$, we can draw an interesting conclusion. The function $F_p[k]$ is completely determined if we know its value for $0 \le k \le N/2$ if N is even or for $0 \le k \le (N-1)/2$ if N is odd. The values of $F_p[k]$ in the rest of the fundamental interval can be derived directly from the above equations. (For $N = 4$ we therefore only have to know the first three values of $F_p[k]$; for $N = 5$ also the first three. For $N = 100$ or $N = 101$ we only have to know the first 51 values.) What general conclusion can you draw from (5.27a) in relation to $F_p[N/2]$ for N even? Verify this in Examples 1 to 3 in section 5.1.

I. Even real time functions

If $f_p[n]$ is an *even* and *real* function, i.e. $f_p[n] = f_p[N-n]$, then the frequency function $F_p[k]$ is also *even* and *real*.

J. Odd real time functions

If $f_p[n]$ is an *odd* and *real* function, i.e. $f_p[n] = -f_p[N-n]$, then the frequency function $F_p[k]$ is *odd* and *imaginary*.

K. Complex time functions

The DFT, see (5.4) and (5.5), is also defined for complex signals in time $f_p[n]$. For special complex time functions $f_p[n]$ (e.g. even or odd) $F_p[k]$ has the same properties as given earlier in Table 4.1 for the functions $x[n]$ and $X(e^{j\theta})$.

L. Waveform decomposition

An arbitrary periodic discrete function $f_p[n]$ can always be expressed as the sum of an even function $f_{pe}[n]$ and an odd function $f_{po}[n]$:

$$f_p[n] = \frac{f_p[n] + f_p[N-n]}{2} + \frac{f_p[n] - f_p[N-n]}{2}$$

$$= f_{pe}[n] + f_{po}[n]$$

(5.28)

where

$$f_{pe}[n] = \frac{f_p[n] + f_p[N-n]}{2} \quad \text{and} \quad f_{po}[n] = \frac{f_p[n] - f_p[N-n]}{2}$$

M. Alternative inversion equation

The IDFT (5.5) can also be expressed in another way:

$$f_p[n] = \frac{1}{N} \left[\sum_{k=0}^{N-1} F_p^*[k] e^{-j(2\pi/N)kn} \right]^*$$

(5.29)

(where * denotes the complex conjugate).

In this inversion equation the power of 'e' has the same sign as in the DFT equation (5.4); this means that virtually the same computer program can be used for executing the forward and inverse transformations.

N. Orthogonality property

A property that strictly speaking has nothing to do with the definitions given for the DFT or the IDFT, but is sometimes very useful in calculating them, is the 'orthogonality property':

$$\sum_{n=0}^{N-1} e^{j(2\pi/N)nk} = \begin{cases} N & \text{for } k = lN \quad \text{with integer } l \\ 0 & \text{elsewhere} \end{cases}$$

(5.30a)

The orthogonality property is sometimes formulated:

$$\sum_{n=0}^{N-1} e^{j(2\pi/N)n(k-m)} = \begin{cases} N & \text{for } k = lN + m \quad \text{with integer } l \\ 0 & \text{elsewhere} \end{cases}$$

(5.30b)

The proof of this property forms one of the exercises at the end of this chapter.

5.4 THE CHOICE OF N

When we use the DFT or the IDFT, we have to decide on the number of samples N [see (5.4) and (5.5)] that will be taken into account in the calculation. It might seem obvious that for periodic discrete signals N should always be taken equal to the number of samples occurring in one period, and that for finite-duration discrete signals N should be taken equal to the number of samples that differ from zero. But in fact this is not necessary at all. Nor is it yet clear whether we can use the DFT to calculate anything for signals that are not periodic *and* are not of finite duration. Let us now look at these aspects more closely.

5.4.1 The choice of N for periodic signals

We consider the signal $x[n] = \cos(2\pi n/6)$. This signal is periodic with a period of 6 and is shown in Fig. 5.9.

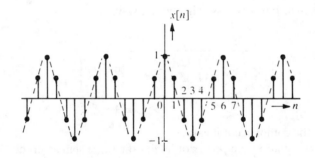

FIGURE 5.9 The signal $x[n] = \cos(2\pi n/6)$.

Since $x[n]$ is a cosine signal (although a discrete one), we should expect a description in the frequency domain to contain a contribution at one particular frequency only. Let us see if the DFT yields a result that fulfils this expectation. We do this by applying the DFT for various choices of N. As an example we take $N = 6$, $N = 12$ and $N = 16$. This gives the following results:

- *For $N = 6$:*

$$X_1[k] = \sum_{n=0}^{5} x[n] e^{-j(2\pi/6)kn} = \sum_{n=0}^{5} \cos(2\pi n/6) e^{-j(2\pi/6)kn}$$

$$= \frac{1}{2} \sum_{n=0}^{5} (e^{j2\pi n/6} + e^{-j2\pi n/6}) e^{-j(2\pi/6)kn}$$

$$= \frac{1}{2} \sum_{n=0}^{5} (e^{j(2\pi/6)n(1-k)} + e^{j(2\pi/6)n(-1-k)})$$

Using property (5.30b) we find for $X_1[k]$:

$$X_1[k] = \begin{cases} 3 & \text{for } k = 6l+1 \text{ with integer } l \\ 3 & \text{for } k = 6l-1 \text{ with integer } l \\ 0 & \text{elsewhere} \end{cases} \quad (5.31)$$

- *For N = 12:*

$$X_2[k] = \sum_{n=0}^{11} x[n] e^{-j(2\pi/12)kn}$$

In the same way as before we find:

$$X_2[k] = \begin{cases} 6 & \text{for } k = 12l+2 \text{ with integer } l \\ 6 & \text{for } k = 12l-2 \text{ with integer } l \\ 0 & \text{elsewhere} \end{cases} \quad (5.32)$$

- *For N = 16:*

$$X_3[k] = \sum_{n=0}^{15} x[n] e^{-j(2\pi/16)kn}$$

$$= \frac{1}{2} \sum_{n=0}^{15} (e^{j2\pi n/6} + e^{-j2\pi n/6}) e^{-j(2\pi/16)kn}$$

$$\quad (5.33)$$

$$= \frac{1}{2} \sum_{n=0}^{15} \left[e^{j(2\pi/48)n(8-3k)} + e^{-j(2\pi/48)n(8+3k)} \right]$$

Caution: This equation cannot be simplified further by using the orthogonality property (5.30), because the summation goes to 15 and the denominator of the exponent is 48 (and *not* 15 + 1).

The fundamental interval of each of the functions X_1, X_2 and X_3 is represented graphically in Fig. 5.10; for X_3 we have only shown the modulus.

In Fig. 5.10(a) we see a contribution for $k = 1$; this is the lowest frequency (except for the frequency zero) that the DFT can reproduce and that corresponds exactly to a single period of $x[n]$. The contribution from $X_1[k]$ at $k = 5$ is a consequence of the fact that we started out from a real time function; see property (5.27).

In Fig. 5.10(b) we see a contribution for $k = 2$; because the 12-point DFT that we use takes exactly two periods of $x[n]$ into account. This means that $x[n] = \cos(2\pi n/6)$ does not represent the lowest frequency for which exactly one period fits into the fundamental interval $0 \le n < 12$. We can say that this 12-point DFT permits a higher frequency resolution than the 6-point DFT.

In Fig. 5.10(c) we see a very interesting effect. Because the number of periods of $x[n]$ that fit into the fundamental interval of the DFT is not an integer number, but 8/3, none of the possible discrete frequencies $\theta_k = 2\pi k/N = 2\pi k/16$ corresponds exactly to the frequency of the cosine signal that we started with. We therefore find a frequency picture that looks very different from Fig. 5.10(a) and (b). We now find contributions at all the possible frequencies of $X_3[k]$; but the contributions from $k \approx 8/3$ and $k \approx 16 - 8/3$ are the largest. This effect is called 'leakage', and we can explain it by a closer look at the time domain.

If the period of $x[n]$ exactly matches the number of samples in the fundamental interval of the DFT, then joining the fundamental intervals together gives the original signal $x[n]$, i.e. $x_p[n] = x[n]$; see Fig. 5.11. If this condition is not satisfied, joining the fundamental intervals together does not produce the original signal, so that $x_p[n] \neq x[n]$. We see irregularities at spacings of N samples (see Fig. 5.12). We should therefore not be surprised if the DFT gives completely different results in these two situations. To avoid the leakage effect in applying the DFT to periodic signals we should preferably make the factor N equal to an integer number of times the number of samples that fit into one period of the signal. There are however good reasons (we shall return to this later in the section on the FFT) for choosing N equal to an integer power of 2 (so that for example $N = 8, 16, 32, \ldots$); we then have to limit the effect of leakage as much as possible in

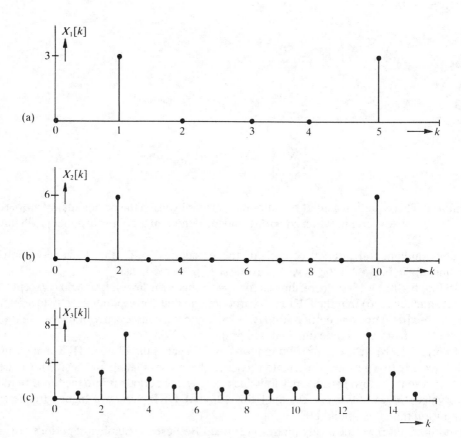

FIGURE 5.10 The fundamental interval of the frequency functions X_1, X_2 and $|X_3|$.

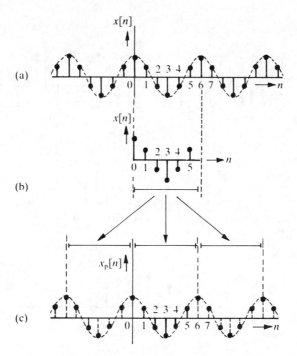

FIGURE 5.11 (a) The signal $x[n] = \cos(2n\pi/6)$. (b) Fundamental interval with $N = 6$. (c) Periodic continuation of the fundamental interval.

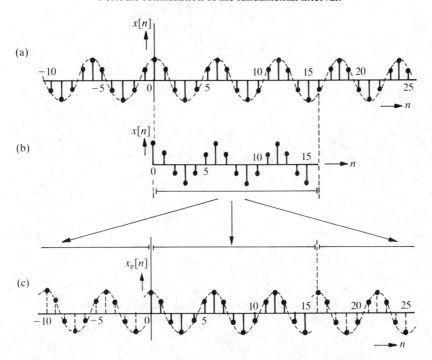

FIGURE 5.12 As Fig. 5.11, but now with $N = 16$; leakage now occurs in the frequency spectrum $X_p[k]$.

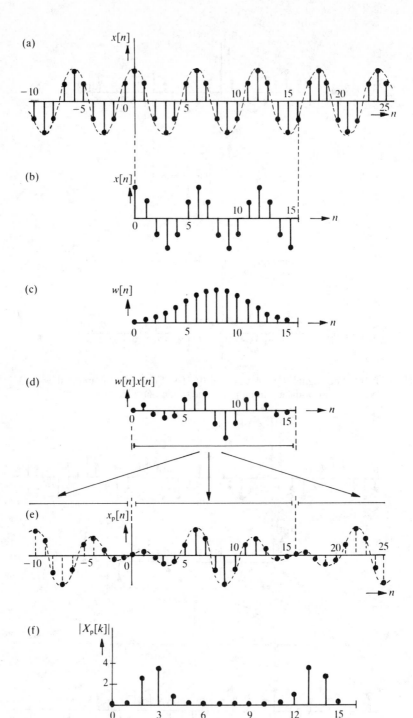

FIGURE 5.13 The use of a window function to reduce leakage.

some other way. One way of doing this is to use a 'window'. This means that we modify the signal values of $x[n]$ in the fundamental interval of the DFT in such a way that irregularities in $x_p[n]$ no longer occur when these intervals are joined together. This is illustrated in Fig. 5.13, where we make use of the Hanning window[†] (with $N = 16$):

$$w[n] = \frac{1 - \cos(2\pi n/N)}{2} \quad \text{with } 0 \le n \le N - 1 \tag{5.34}$$

Multiplying the original values of the samples $x[n]$ in the fundamental interval by the values given by the window function $w[n]$ yields a new function $x[n]w[n]$ that on periodic repetition gives a function $x_p[n]$ with no large irregularities (see Fig. 5.13(e)). Although the function $x_p[n]$ clearly differs from the original function $x[n]$, application of the DFT (see Fig. 5.13(f)) gives a result in which the leakage has been strongly reduced. The DFT is still unable to reproduce the exact frequency of the original cosine function, but all the frequency contributions are more strongly concentrated around $k \approx 8/3$ and $k \approx 16 - 18/3$ (compare the Figs 5.10(c) and 5.13(f)). Later on in this book we shall encounter more situations in which we use window functions and we shall also discuss other kinds of windows there. We shall also consider the background to the use of window functions in more detail, e.g. through an interpretation in the frequency domain.

We should like to close this section with one more comment. In section 4.2.2 we saw that a discrete cosine function is not always periodic; e.g. $x[n] = \cos(n)$. If we want to apply a DFT to such a signal, there is no N we can choose that corresponds to an integer number of periods of $x[n]$. In this case we can never completely avoid leakage, but we can try to minimize it by choosing N such that on periodic repetition of the fundamental interval the irregularities are as small as possible. It is still possible to apply a window function as well, of course.

5.4.2 The choice of N for finite-duration signals

We take a discrete signal $x[n]$ of finite duration for which:

$$x[n] = 0 \quad \text{for} \quad n < 0 \quad \text{and} \quad n \ge 4 \tag{5.35}$$

This signal has only four samples $x[0]$, $x[1]$, $x[2]$ and $x[3]$ that are not zero. How should we choose N if we want to obtain a frequency representation of this signal by means of the DFT? It will at once be clear that N must in any case be equal to at least four, since otherwise we just cannot make use of all the information we have about $x[n]$ in calculating the DFT. But how large exactly should we make N?

Figure 5.14 gives a graphical representation of what we find if we apply an N-point DFT with $N = 4$, $N = 6$ and $N = 16$ to $x[n]$. For every choice of $N \ge 4$ we find an unambiguous frequency representation; only the number of frequency samples is different. All of the frequency information about $x[n]$ can be found just by applying the DFT of the minimum length $N = 4$. For a larger N we obtain a greater frequency

---

[†]Sometimes a different definition of a Hanning window is used. We shall return to this in section 8.2.1.

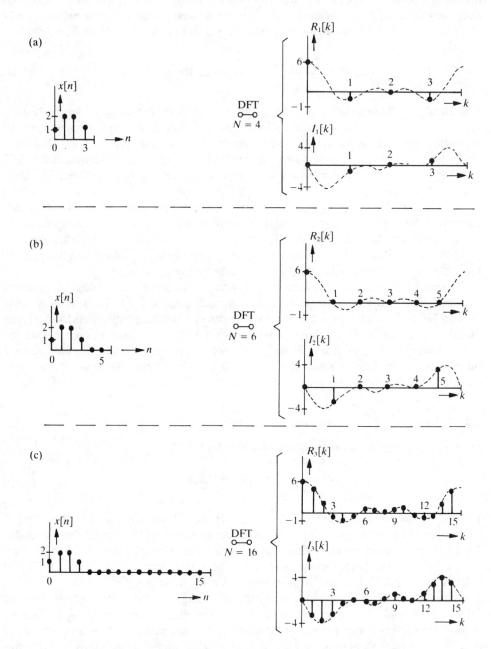

FIGURE 5.14 The effect of the choice of N for finite-duration signals.

resolution, but this alone does not provide any more information about $x[n]$. We can see this a little more clearly if we call to mind the relation that exists between the FTD and the DFT of a discrete signal of finite duration (see (5.16) and Fig. 5.6). The DFT can be considered as a sampled version of the FTD. For the minimum value of N we obtain the minimum number of samples necessary for an unambiguous discrete representation of the FTD. For higher values of N we find more samples of the FTD, but these do not really represent any extra information. (The dashed envelopes in Fig. 5.14 represent the FTD of the signal $x[n]$.) We shall see in section 5.5.1 that for the execution of certain operations (in particular the calculation of convolutions) we are sometimes obliged to apply a DFT for which N is larger than the minimum value, which corresponds to the length of the signal with which we are working.

5.4.3 The choice of N for non-periodic infinite-duration signals

For non-periodic discrete signals of infinite duration we can only obtain an exact description in the frequency domain by applying the FTD, since with the DFT we are forced to take only a finite number of samples of the signal into account. By choosing N large enough, however, we can obtain an approximate description in the frequency domain that is often good enough for practical application.

We should like to illustrate this with an example. Let us consider the signal $x[n] = (\frac{1}{2})^n u[n]$. Figure 5.15(a) shows this signal and the corresponding FTD (just the amplitude). Figure 5.15(b) shows the frequency function $|\tilde{X}[k]|$ that we obtain by applying the DFT

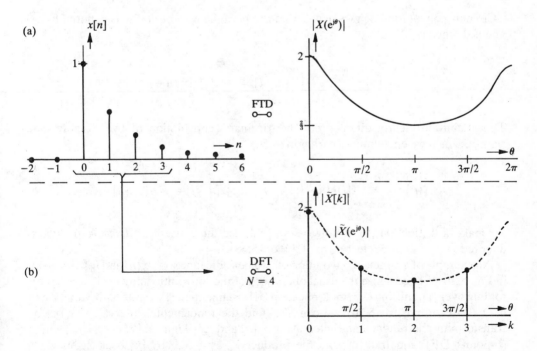

FIGURE 5.15 (a) The FTD of $x[n] = (\frac{1}{2})^n u[n]$. (b) The DFT of $x[n] = (\frac{1}{2})^n u[n]$ with $N = 4$.

to the first four samples of $x[n]$ that are not zero. The values of $|\tilde{X}[k]|$ do not now represent a sampled version of $|X(e^{j\theta})|$ but of $|\tilde{X}(e^{j\theta})|$; i.e. of the FTD of $\tilde{x}[n]$; where:

$$\tilde{x}[n] = \begin{cases} x[n] & \text{for } 0 \le n < 4 \\ 0 & \text{elsewhere} \end{cases}$$

As $\tilde{x}[n]$ becomes more like $x[n]$ (i.e. as we make N larger and thus consider more samples of $x[n]$), $|\tilde{X}(e^{j\theta})|$ becomes more like $X(e^{j\theta})$ and $\tilde{X}[k]$ also represents a better approximation to $X(e^{j\theta})$. In the example of Fig. 5.15(b) the difference between $|X(e^{j\theta})|$ and the values of $|\tilde{X}[k]|$ obtained with the DFT with $N = 4$ is smaller than 7%.

5.5 CIRCULAR CONVOLUTION

One of the important applications of a Fourier transform is the replacement of convolution in the time domain by multiplication in the frequency domain. We have already encountered this with the FTC and the FTD. It would also be desirable to use the DFT for this purpose. In section 5.3 we have already stated the circular (sometimes called cyclic) convolution property of the DFT without further explanation; see (5.24):

$$y_p[n] = x_p[n] \circledast h_p[n] \quad \circ\!\!-\!\!-\!\!\circ \quad Y_p[k] = X_p[k]H_p[k] \tag{5.36}$$

Circular convolution of two signals that are periodic with period N is denoted by $\circledast$ and is defined as:

$$y_p[n] = \sum_{i=0}^{N-1} x_p[i]h_p[n-i] = \sum_{i=0}^{N-1} h_p[i]x_p[n-i] \tag{5.37}$$

This definition clearly differs from the 'ordinary' convolution of two discrete-time signals, which we encountered earlier in (4.50):

$$y[n] = \sum_{i=-\infty}^{\infty} x[i]h[n-i] = \sum_{i=-\infty}^{\infty} h[i]x[n-i] = x[n] * h[n] \tag{5.38}$$

To make a distinction, from now on we shall use the concept of *linear convolution*, denoted by $*$, if we are referring to (4.50) or (5.38).

An example of a circular convolution of two periodic functions $x_p[n]$ and $h_p[n]$ is given in Fig. 5.16. We recognize the usual operations of the convolution process in this figure: folding over (I), shifting (II), multiplication (III), summation (IV) and finally variation of n (V). According to (5.36) we can also find the fundamental interval of $y_p[n]$ by transforming the fundamental intervals of $x_p[n]$ and $h_p[n]$ into $X_p[k]$ and $H_p[k]$ with a (6-point) DFT and transforming the product $Y_p[k] = X_p[k]H_p[k]$ back again via a (6-point) IDFT. This is shown schematically in Fig. 5.17. Since the apparent detour via

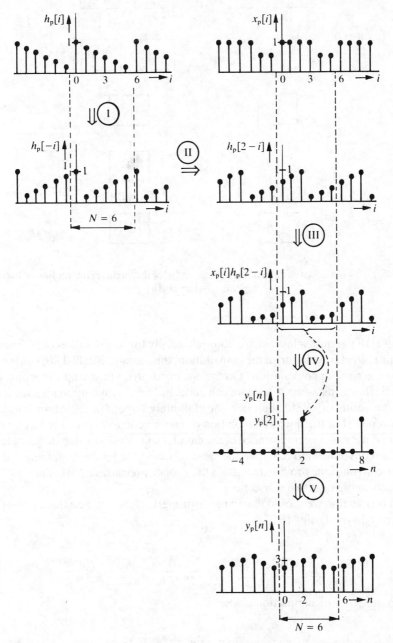

FIGURE 5.16 Graphical representation of the circular convolution $y_p[n] = x_p[n] \circledast h_p[n]$ (the calculation for $n = 2$ is shown).

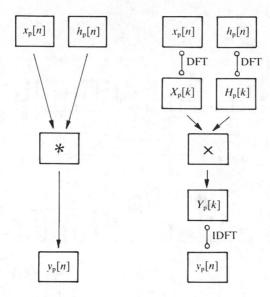

FIGURE 5.17 Two ways of determining the result $y_p[n]$ of the circular convolution of two periodic signals $x_p[n]$ and $h_p[n]$.

DFTs and IDFT requires less calculation – especially for larger values of N (see section 5.7) – than a direct calculation of the convolution, this 'detour' is called *fast convolution*.

What use is circular convolution? On the one hand circular convolution is only defined for periodic functions, while we cannot calculate the linear convolution for such functions because the result of (5.38) then becomes infinitely large. On the other hand we are usually interested in the linear convolution of two functions where at least one of them has a finite duration. The importance of the circular convolution is that in this case we can use circular convolution to calculate the same results in the fundamental interval as with the linear convolution, provided we take the proper precautions. We shall look at this more closely in the next two subsections.

To end this section we should like to prove property (5.36). We do this by determining the DFT of $y_p[n]$ and using (5.37):

$$Y_p[k] = \sum_{n=0}^{N-1} y_p[n]e^{-j(2\pi/N)kn} = \sum_{n=0}^{N-1}\sum_{i=0}^{N-1} x_p[i]h_p[n-i]e^{-j(2\pi/N)kn} \qquad (5.39a)$$

Changing the order of summation gives:

$$Y_p[k] = \sum_{i=0}^{N-1} x_p[i] \sum_{n=0}^{N-1} h_p[n-i]e^{-j(2\pi/N)kn}$$

$$= \sum_{i=0}^{n-1} x_p[i] \sum_{n=0}^{N-1} h_p[n-i]e^{-j(2\pi/N)(n-i)k}e^{-j(2\pi/N)ik} \qquad (5.39b)$$

Substituting $n - i = l$ in the above equation gives:

$$Y_p[k] = \sum_{i=0}^{N-1} x_p[i] e^{-j(2\pi/N)ik} \sum_{l=-i}^{N-1-i} h_p[l] e^{-j(2\pi/N)lk}$$

$$= \sum_{i=0}^{N-1} x_p[i] e^{-j(2\pi/N)ik} \sum_{l=0}^{N-1} h_p[l] e^{-j(2\pi/N)lk}$$

$$= X_p[k] H_p[k] \tag{5.39c}$$

5.5.1 The convolution of two finite-duration signals

Let us assume that we want to determine the linear convolution of two signals $x[n]$ and $h[n]$ of finite duration, which only differ from zero for $0 \le n < 6$ (see Fig. 5.18). These signals exactly represent the fundamental interval of the signals $x_p[i]$ and $h_p[i]$ of Fig. 5.16. *Circular* convolution of $x[n]$ and $h[n]$ by making use of a (6-point) DFT and IDFT therefore gives the same result $y_p[n]$ as in Fig. 5.16. If we now want to form a signal of finite duration from $y_p[n]$ in the usual way by taking the fundamental interval and putting zeros outside it, we obtain the signal $y[n]$, also shown in Fig. 5.18. But this signal is not in the least like the *linear* convolution of $x[n]$ and $h[n]$ that we derived in Fig. 4.16! This is not really surprising, however, because a linear convolution of two signals of finite duration (with lengths N_1 and N_2) gives a signal of length $N_1 + N_2 - 1$. In our example we know that the *linear* convolution of $x[n]$ and $h[n]$ therefore gives a signal of length $6 + 6 - 1 = 11$. Because of the method of calculation, however, we find a signal with the length of the fundamental interval; i.e. of length 6.

The situation thus changes completely if we treat $x[n]$ and $h[n]$ as if they have a fundamental interval with a length of at least 11; then $y[n]$ also obtains a length of at least 11. This is shown in Fig. 5.19, where we have chosen a fundamental interval with a length of $N = 12$. We see that we do now obtain a result that corresponds exactly to the desired linear convolution.

We can now draw the following important conclusion: the *linear* convolution of two signals of finite duration N_1 and N_2 can be obtained from the fundamental interval of a *circular* convolution calculated with a length $N \ge N_1 + N_2 - 1$.

So we now know how we can calculate the convolution of two signals of finite duration quickly by using two DFTs and one IDFT; we only have to make sure we choose N large enough.

5.5.2 The convolution of a finite-duration signal with an infinite-duration signal

In practice it is not unusual to want to calculate the linear convolution of two signals where one has a finite duration N_1 and the other a very long ('infinite') duration. This situation occurs if for example we want to filter a discrete version $x[n]$ of a continuous speech signal with a filter that has an impulse response $h[n]$ of finite duration N_1. We then find the output signal $y[n]$ from (see Fig. 5.20):

$$y[n] = x[n] * h[n] = \sum_{i=-\infty}^{\infty} x[i] h[n-i] \tag{5.40}$$

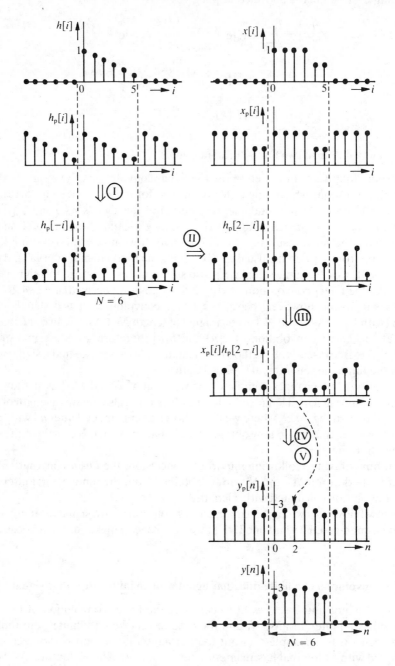

FIGURE 5.18 Circular convolution of two signals with a finite duration of $N = 6$ (the length of the fundamental interval is also equal to 6).

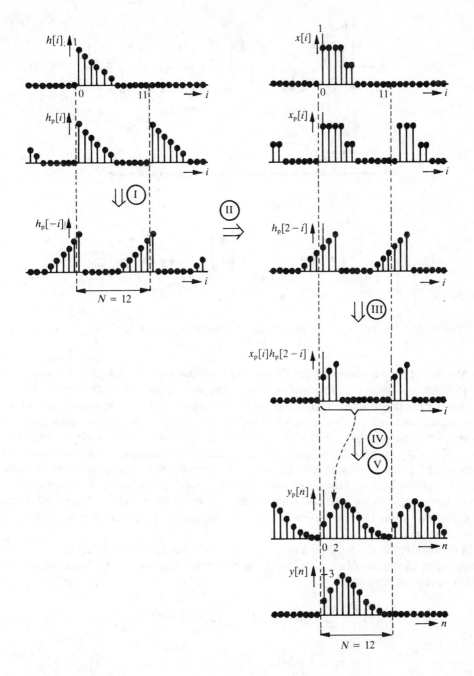

FIGURE 5.19 Circular convolution of two signals with a finite duration of $N = 6$ (the length of the fundamental interval is now made equal to 12).

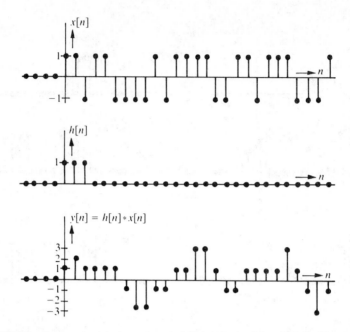

FIGURE 5.20 The linear convolution of an infinite-duration signal $x[n]$ with a finite-duration signal $h[n]$.

We can of course obtain $y[n]$ by calculating the summation of (5.40) for every value of n ('direct calculation'). However, we have already mentioned several times that an indirect calculation of a convolution via two Fourier transforms, multiplication in the frequency domain and inverse transformation often requires much less computation. Is something similar possible in this case too? At first sight it does not seem so. Determining the DFT of $h[n]$ is no real problem, because of the finite duration, but the determination of the DFT of the very long ('infinite') signal $x[n]$ does present difficulties. This is because we should first wait until the entire signal $x[n]$ is at our disposal, which would introduce a very large delay before we had even calculated the first few values of $y[n]$. Also, the calculation of this DFT would require much computation (and a correspondingly large memory capacity).

Fortunately there is a solution to the problem. We find it by considering the signal as a succession of signals $x_1[n]$, $x_2[n]$, . . ., each of finite duration N_2. Equation (5.40) can then be rewritten as:

$$y[n] = \sum_{i=-\infty}^{\infty} x_1[i]h[n-i] + \sum_{i=-\infty}^{\infty} x_2[i]h[n-i] + \ \ldots$$

$$= y_1[n] + y_2[n] + \ \ldots \tag{5.41}$$

where $y_1[n]$, $y_2[n]$, . . . are all convolutions of two signals of finite duration N_1 and N_2. These can be calculated one at a time and then added to give $y[n]$. This is illustrated in Fig. 5.21, where we have divided the signal $x[n]$ into segments of $N_2 = 5$ samples and

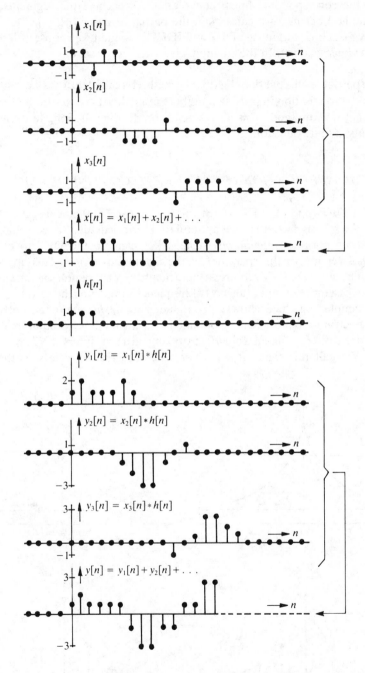

FIGURE 5.21 Calculation of a linear convolution by dividing an infinite-duration signal into a number of signals of finite duration.

performed the corresponding linear convolution for each of the segments. This is a powerful method because we can obtain the partial results $y_1[n]$, $y_2[n]$, . . . without difficulty by fast convolution (via DFT and IDFT) provided we choose N large enough (see the previous section). In the example given here we must have $N \geq N_1 + N_2 - 1 = 3 + 5 - 1 = 7$.

Since the partial results overlap slightly, the method is called the overlap-add method.

We should also mention in passing that there is an alternative to this method, known as the overlap-save method. This also depends on dividing the infinite-duration signal into segments of finite duration.

5.6 THE APPLICATION OF THE DFT TO CONTINUOUS SIGNALS

Although the DFT (and the FFT, which will be discussed in the next section) has been defined for *discrete* signals, it is such a useful tool for calculating spectra that it is often applied to (non-periodic) *continuous* signals. We can use the DFT to calculate an *approximated* version of the spectrum of the continuous signal. This approximation comes about because the DFT requires a finite number (N) of samples in time as the input signal and gives a finite number (also N) of frequency samples as the result.

As an example Fig. 5.22 shows a continuous signal $h(t)$ and the corresponding frequency spectrum $H(\omega)$ that can be obtained by using the FTC. So that we can use the DFT we derive a discrete signal from $h(t)$ by taking, in a time interval NT, N samples at a spacing T. We call this signal $h_p[n]$, where the subscript 'p' refers to the periodic

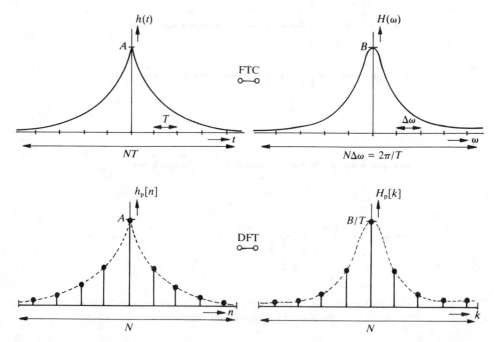

FIGURE 5.22 Approximation of the spectrum $H(\omega)$ of a continuous signal $h(t)$ with the DFT.

continuation that we are used to with the DFT. Applying the DFT to $h_p[n]$ then gives the frequency function $H_p[k]$. We shall show that the N samples of $H_p[k]$ are an *approximation* to N values of $H(\omega)$.

First, there are several important comments we should make. In going from $h(t)$ to $h_p[n]$, we have to choose the sampling interval T and the number of samples N. On the one hand the product NT determines the error that we introduce by considering only a part of $h(t)$. On the other hand, we recognize the two parameters N and T directly in the results of the DFT: the total frequency interval of $H(\omega)$ that we approximate has a width of $2\pi/T$ and the spacing $\Delta\omega$ between successive frequency samples is $2\pi/NT$, so that:

$$\Delta\omega = 2\pi/NT \quad \text{or} \quad T \cdot \Delta\omega = 2\pi/N \tag{5.42}$$

In addition we find a factor $1/T$ by which the amplitudes of $H(\omega)$ and $H_p[k]$ differ. These relationships are illustrated in Fig. 5.22. By making T small enough and N large enough we can approximate the function $H(\omega)$ arbitrarily well with a DFT (although at the expense of an ever-increasing number of calculations).

We now want to look more closely at the sources of the errors that we find in the samples of $H_p[k]$ as compared with the true values of $H(\omega)$ that we want to calculate. Here we use Fig. 5.23. The transition from the function $h(t)$ to the function $h_p[n]$ can be split into a number of successive steps. First of all we derive a finite-duration signal $\hat{h}(t)$ by multiplying $h(t)$ by the window function $w(t)$ whose spectrum is $W(\omega)$ (Fig. 5.23(b)). The spectrum of $\hat{h}(t) = h(t)w(t)$ is given by $\hat{H}(\omega) = H(\omega) * W(\omega)$ (Fig. 5.23(c)). Next we convert $\hat{h}(t)$ into a discrete signal $\hat{h}[n] = \hat{h}(nT)$. This has a periodic spectrum $\tilde{H}(e^{j\omega T})$, which we can find from $\hat{h}[n]$ with the FTD (Fig. 5.23(d)). From Chapter 3 on sampling (3.5c) we know that we can also express the resulting spectrum $\tilde{H}$ in terms of $\hat{H}$ through the following equation:

$$\tilde{H}(e^{j\omega T}) = \frac{1}{T} \sum_{i=-\infty}^{\infty} \hat{H}\left(\omega - \frac{2\pi i}{T}\right) \tag{5.43}$$

Finally, we can consider the N samples of $\tilde{h}[n]$ that differ from zero as the fundamental interval of the periodic discrete signal $h_p[n]$, and apply an N-point DFT to it (Fig. 5.23(e)). The resulting frequency function $H_p[k]$ gives N samples of the function $\tilde{H}(e^{j\omega T})$ in its fundamental interval:

$$H_p[k] = \tilde{H}(e^{j2\pi k/N}) \tag{5.44}$$

There are two reasons why the samples of $H_p[k]$ do not correspond exactly to samples of the function $H(\omega)$ at corresponding frequencies (leaving aside the constant amplitude factor $1/T$ which we could easily correct). The first source of error is the multiplication of $h(t)$ by the window function $w(t)$, which means that we are no longer dealing with $H(\omega)$ but with $\hat{H}(\omega)$, which becomes a better approximation to $H(\omega)$ as the window becomes wider. The second source of error arises through the sampling of $\hat{h}(t)$, which can cause aliasing to occur in the frequency domain (as we saw in section 3.2 and as shown in Fig. 5.23(d)). Since it is physically impossible for both $h(t)$ and $H(\omega)$ to have a finite width, it is by definition impossible to avoid *both* sources of error in this approach. At least one of the two will always occur and in practice both sources of error are of some significance.

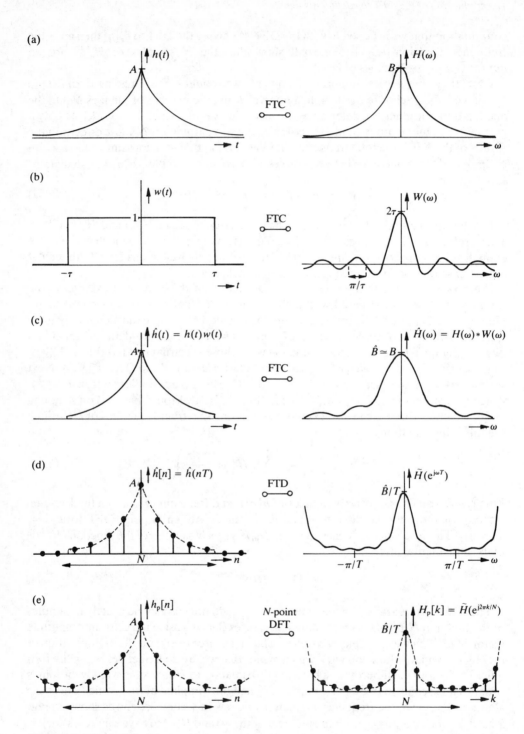

FIGURE 5.23 Sources of error in the approximation of the spectrum $H(\omega)$ of a continuous signal $h(t)$ with the DFT.

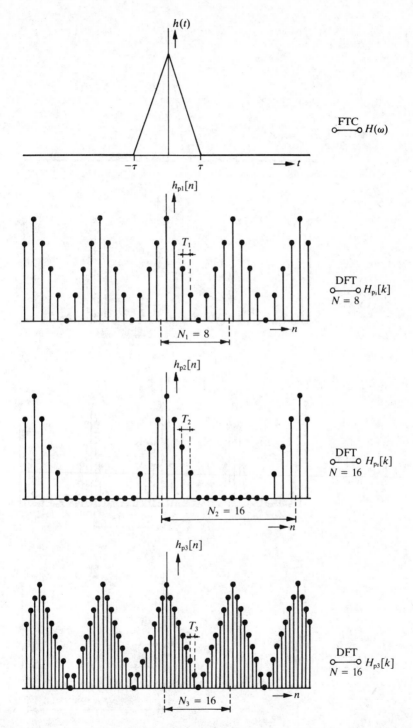

FIGURE 5.24 The effect of the choice of T and N on the DFT approximation of the spectrum $H(\omega)$ of a continuous signal $h(t)$.

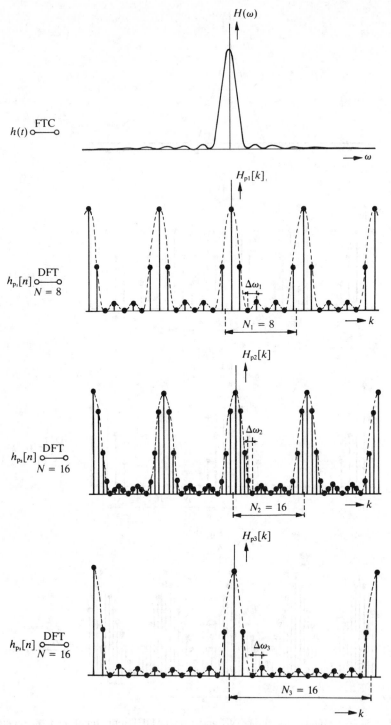

FIGURE 5.24 *(continued)*.

We should now like to consider in more detail with the aid of Fig. 5.24 the consequences of a particular choice of N and T for the approximation of $H(\omega)$ by $H_p[k]$. Here we have a continuous function $h(t)$ of finite duration with a corresponding FTC $H(\omega)$ that extends over the interval $-\infty \leq \omega \leq \infty$. We shall now calculate an approximation to $H(\omega)$ with the aid of the DFT in three different ways. To do this we choose:

Case 1: $N_1 = 8$ and $T_1 = \tau/4$, so that $\Delta\omega_1 = 2\pi/N_1 T_1 = \pi/\tau$;
Case 2: $N_2 = 16$ and $T_2 = \tau/4$, so that $\Delta\omega_2 = \pi/2\tau$;
Case 3: $N_3 = 16$ and $T_3 = \tau/8$, so that $\Delta\omega_3 = \pi/\tau$.

Since $h(t)$ is of finite duration, in this example we can completely eliminate the error resulting from the multiplication by a window function by ensuring that in all three cases $N \cdot T \geq 2\tau$.

It is useful to look at the second source of error, the aliasing in the frequency domain. Both in case 1 and in case 2 the samples of the discrete time function are at a spacing of $\tau/4$, which implies that we obtain the same aliasing. In case 2 we find twice as many frequency samples as in case 1, but these frequency samples do *not* give a better approximation to $H(\omega)$, in spite of the application of a DFT twice as large. In case 3 the samples of the discrete time function are twice as close together, so that the fundamental interval of $H_{p3}[k]$ represents a frequency range twice as large as that of $H_{p1}[k]$. Errors in $H_{p3}[k]$ due to aliasing are greatly reduced with respect to $H_{p1}[k]$ and $H_{p2}[k]$.

- *Comment.* In this section we have always centered the fundamental interval around $n = 0$ and $k = 0$ to facilitate comparison with the continuous functions. In practical calculations we shall usually work with the 'non-centered' values $h_p[0], \ldots, h_p[N-1]$ and $H_p[0], \ldots, H_p[N-1]$. Because of the periodic nature of h_p and H_p this does not introduce any fundamental problems.

5.7 THE FFT

In the preceding sections we have paid considerable attention to the DFT, because of the important part this transform plays in discrete signal processing. In the practical application of the DFT the number of operations required (multiplications and additions) is very important, since it determines the amount of time required or the kind of equipment (e.g. the type of computer) we require to calculate a particular DFT. Let us therefore look more closely at the equation (5.4) that we use to calculate an N-point DFT directly (for simplicity we shall now omit the subscript 'p'):

$$X[k] = \sum_{n=0}^{N-1} x[n]e^{-j(2\pi/N)kn} \tag{5.45}$$

For the calculation of each of the N values $X[k]$ we must perform N multiplications of the type $x[n] \times e^{-j(2\pi/N)kn}$ and $N-1$ additions. In total we therefore have N^2 multiplications and $N(N-1)$ additions. In the most general case, where $x[n]$ is complex, these are complex multiplications and complex additions.

We therefore see that the direct calculation of an N-point DFT requires a number of complex operations of the order of magnitude of N^2. For $N = 4$ this is 16, but for $N = 2048$ we are already at $N^2 = 4\,194\,304$. In recent years especially there has been much research on methods that make it possible to calculate an N-point DFT with considerably fewer operations. This has led to various mathematical tricks, which do indeed produce the desired result. The procedure always followed is first to calculate a number of DFTs of shorter length and then to combine the results appropriately. This can be done in very many different ways. These are denoted by the collective name 'fast Fourier transform' (FFT), because their common characteristic is the reduction in the number of operations and hence in the computer time required.

In the literature on the FFT it is usual to substitute W_N^{kn} for the expression $e^{-j(2\pi/N)kn}$ from the DFT, where W_N is called the *twiddle factor*:

$$W_N = e^{-j(2\pi/N)} \tag{5.46}$$

In the various versions of the FFT that exist, use is always made of a number of characteristic properties of the twiddle factor, such as:

$$W_N^{kn} = W_N^{k(n+N)} = W_N^{(k+N)n} \tag{5.47a}$$

$$W_N^{2kn} = W_{N/2}^{kn} \tag{5.47b}$$

$$W_N^{k(N-n)} = (W_N^{kn})^* \tag{5.47c}$$

where * denotes the complex conjugate.

The most widely known and used are the FFT algorithms, where N is an integer power of two, so that $N = 2^M$.

With these algorithms it is possible to reduce the number of operations required to the order of magnitude of $N \cdot \log_2(N) = N \cdot M$. Table 5.1 gives a summary of N^2, $N \cdot \log_2(N)$ and $N/\log_2(N)$ for various values of N. The last column therefore shows how many times the speed of calculation can be increased by using an FFT instead of a direct calculation of the DFT.

The gain in speed that can be attained by using the FFT instead of a direct calculation of the DFT is shown even more clearly when we plot the results of Table 5.1 as a graph. This is done in Fig. 5.25.

TABLE 5.1 Comparison of the number of operations in a direct calculation of a DFT and in an FFT

N	N^2	$N \cdot \log_2(N)$	$N/\log_2(N)$
2	4	2	2.00
4	16	8	2.00
8	64	24	2.67
16	256	64	4.00
32	1 024	160	6.40
64	4 096	384	10.67
128	16 384	896	18.29
256	65 536	2 048	32.00
512	262 144	4 608	56.89
1024	1 048 576	10 240	102.40
2048	4 194 304	22 528	186.18

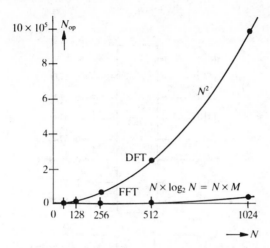

FIGURE 5.25 Comparison of the number of operations required for a direct calculation of a DFT and for the calculation of an FFT.

To give some idea of the kind of techniques that can be used in the FFT, we should like to show how an N-point DFT (with N even) can be executed by calculating two $N/2$-point DFTs and combining the results. We take a discrete signal $x[n]$ of length N (N is even). We then define two new signals $x_1[n]$ and $x_2[n]$ that each have a length of $N/2$ and which consist of the even and odd samples of $x[n]$ respectively:

$$\left.\begin{array}{l} x_1[n] = x[2n] \\ x_2[n] = x[2n+1] \end{array}\right\} \quad \text{with } n = 0, 1, \ldots, (N/2) - 1 \qquad (5.48)$$

Making use of the twiddle factor W_N, we find for the DFT of $x[n]$:

$$X[k] = \sum_{n=0}^{N-1} x[n] W_N^{nk} = \underset{(n \text{ even})}{\sum_{n=0}^{N-1}} x[n] W_N^{nk} + \underset{(n \text{ odd})}{\sum_{n=0}^{N-1}} x[n] W_N^{nk}$$

$$= \sum_{n=0}^{(N/2)-1} x[2n] W_N^{2nk} + \sum_{n=0}^{(N/2)-1} x[2n+1] W_N^{(2n+1)k} \qquad (5.49a)$$

and using (5.47b) and (5.48) we find:

$$X[k] = \sum_{n=0}^{(N/2)-1} x_1[n] W_{N/2}^{nk} + W_N^k \sum_{n=0}^{(N/2)-1} x_2[n] W_{N/2}^{nk}$$

$$= X_1[k] + W_N^k X_2[k] \qquad (5.49b)$$

Here $X_1[k]$ and $X_2[k]$ represent the $N/2$-point DFTs of $x_1[n]$ and $x_2[n]$ exactly!

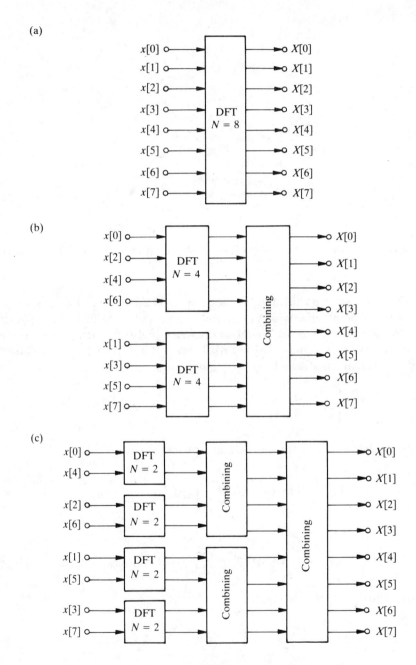

FIGURE 5.26 The principle of the FFT illustrated for $N = 8$. (a) 8-point DFT; (b) 8-point DFT based on two 4-point DFTs; (c) 8-point DFT based on four 2-point DFTs.

- *Comment.* While $X_1[k]$ and $X_2[k]$ are periodic with period $N/2$, W_N is periodic with period N; consequently $X[k]$ is also periodic with period N.

For the direct calculation of each of these smaller DFTs the number of operations required is of the order of magnitude of $(N/2)^2$, and hence in total $2(N/2)^2 = N^2/2$. For the calculation of all the values of $X[k]$ there is also the multiplication of W_N^k by $X_2[k]$ and the addition of $X_1[k]$ and $W_N^k X_2[k]$. For large values of N, however, the total number of operations nevertheless remains of the order of magnitude of $N^2/2$. So we have gained 50% in comparison with a direct calculation of $X[k]$! If now $N/2$ is also an even number, we can again base the calculation of $X_1[k]$ and $X_2[k]$ on smaller DFTs (this time of length $N/4$) and make yet another gain. If in turn $N/4$ is also even, we can repeat the trick again and again, until we finally arrive at 2-point DFTs.

This principle is illustrated in Fig. 5.26 for an 8-point DFT. (It is interesting to note that the repeated splitting of the input samples into even and odd samples has a shuffling effect, and indeed the process is often called 'shuffling'.)

It is not immediately clear from Fig. 5.26 exactly what operations are executed in the blocks designated by 'combining'. We can indicate this most easily with the aid of the *FFT butterfly*. This represents an operation in which, starting with two complex numbers A and B, two new complex numbers Y and Z are calculated such that:

$$Y = A + W_N^p B \quad \text{and} \quad Z = A - W_N^p B \tag{5.50}$$

W_N in (5.50) is the well-known twiddle factor and p a certain integer between 0 and $N-1$. We can represent equation (5.50) schematically as in Fig. 5.27.

The FFT butterfly with $p = 0$ exactly represents the relation between the input samples $x[0] = A$ and $x[1] = B$ and the output samples $X[0] = Y$ and $X[1] = Z$ of a 2-point DFT (verify this by using (5.45) and (5.50) with $N = 2$ and $p = 0$). With the aid of the butterfly we can now draw Fig. (5.26(c)) again in a more detailed form, as in Fig. 5.28. Representations of this kind are often encountered in the extensive existing literature on the FFT.

- *Comment 1.* The version of the FFT shown in Fig. 5.28 is known in the literature as the 'Radix-two decimation-in-time FFT'.
- *Comment 2.* In the 8-point FFT of Fig. 5.28 we can see that there are three successive steps, each containing four butterflies which each have one complex multiplication factor. For an arbitrary N-point FFT, where N is an integer power of 2, we find in the

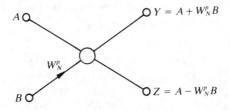

FIGURE 5.27 An FFT butterfly.

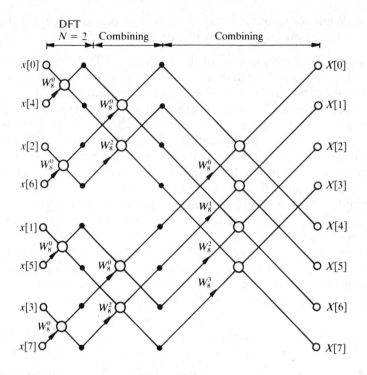

FIGURE 5.28 The 8-point FFT shown in more detail with the FFT butterfly.

same way $\log_2(N)$ steps, each containing $N/2$ butterflies. In total an FFT calculation following this pattern requires $(N/2) \cdot \log_2(N)$ complex multiplications. This corresponds to our earlier assertion that an N-point FFT requires a number of operations of the order of magnitude of $N \cdot \log_2(N)$.

5.8 EXERCISES

Exercises to sections 5.1 and 5.2

5.1 The periodic functions $x_1[n]$ and $x_2[n]$ have a period $N = 4$ and are defined as:

$$x_1[n] = \begin{cases} 0 & \text{for } n = 4l \\ 1 & \text{for } n = 4l+1 \quad \text{and} \quad n = 4l+3 \\ 2 & \text{for } n = 4l+2 \end{cases}$$

$$x_2[n] = \begin{cases} 1 & \text{for } n = 4l \quad \text{and} \quad n = 4l+3 \\ 2 & \text{for } n = 4l+1 \quad \text{and} \quad n = 4l+2 \end{cases}$$

with $l = 0, \pm1, \pm2, \ldots$.
(a) Calculate the DFT of $x_1[n]$ and $x_2[n]$ for $N = 4$.

(b) Calculate the FTD of the signals $\tilde{x}_1[n]$ and $\tilde{x}_2[n]$ defined as:

$$\tilde{x}_1[n] = \begin{cases} x_1[n] & \text{for } 0 \le n < 4 \\ 0 & \text{for } n < 0 \quad \text{and} \quad n \ge 4 \end{cases}$$

$$\tilde{x}_2[n] = \begin{cases} x_2[n] & \text{for } 0 \le n < 4 \\ 0 & \text{for } n < 0 \quad \text{and} \quad n \ge 4 \end{cases}$$

(c) Show that the DFT is a sampled version of the FTD.

5.2 Calculate the IDFT $x_3[n]$ of the periodic function $X_3[k]$ with period $N = 16$:

$$X_3[k] = \begin{cases} 2 & \text{for } k = 16l + 1 \quad \text{and} \quad k = 16l + 15 \\ 1 & \text{for } k = 16l + 3 \quad \text{and} \quad k = 16l + 13 \\ 0 & \text{elsewhere} \end{cases}$$

with $l = 0, \pm 1, \pm 2, \ldots$.

Exercises to section 5.3

5.3 (a) Calculate the DFT $X[k]$ of the function (take $N = 4$):

$$x[n] = \begin{cases} 1 & \text{for } n = 0 \\ 2 & \text{for } n = 1 \\ 0 & \text{elsewhere} \end{cases}$$

(b) Show that $X[k]$ can also be found from (5.8) and (5.10), using the linearity property (5.18).

5.4 (a) Calculate the DFT $X_p[k]$ of the following periodic function with period $N = 4$:

$$x_p[n] = 1 \quad \text{for} \quad 0 \le n < 4$$

(b) Show that $X_p[k]$ can also be found from (5.7) and (5.8), using the time/frequency symmetry property (5.19).

5.5 Show that (5.10) follows immediately from (5.8), using the time-shift property (5.20).

5.6 Show that Parseval's theorem applies for the DFT pairs from Exercises 5.1 and 5.2.

5.7 Prove the orthogonality property (5.30a):

$$\frac{1}{N} \sum_{n=0}^{N-1} e^{j(2\pi/N)nk} = \begin{cases} 1 & \text{for } k = lN \\ 0 & \text{elsewhere} \end{cases}$$

Hint: Prove this property first for $k = lN$ and for $k \ne lN$ use the relation:

$$\sum_{n=0}^{N-1} a^n = \frac{1 - a^N}{1 - a} \quad \text{for } a \ne 1$$

Exercise to section 5.4

5.8 For a finite-duration signal $x[n]$:

$$x[n] = \begin{cases} 1 & \text{for } n = 0 \quad \text{and} \quad n = 3 \\ 2 & \text{for } n = 1 \quad \text{and} \quad n = 2 \\ 0 & \text{elsewhere} \end{cases}$$

Determine the DFT $X[k]$ of this signal $x[n]$ with $N = 4$, $N = 6$ and $N = 16$. Check whether the results correspond to the results shown in Fig. 5.14.

Exercises to section 5.5

5.9 The functions $x_p[n]$ and $h_p[n]$ are periodic with period N and are defined in the fundamental interval as:

$$x_p[n] = \begin{cases} (0.5)^n & \text{for } n = 0, 1, 2, 3 \\ 0 & \text{for } n = 4, 5, \ldots, N-1 \end{cases}$$

and

$$h_p[n] = \begin{cases} 1 & \text{for } n = 0, 1, 2 \\ 0 & \text{for } n = 3, 4, \ldots, N-1 \end{cases}$$

(a) Draw $x_p[n]$ and $h_p[n]$ for $N = 4$, $N = 6$ and $N = 8$.
(b) Determine the circular convolution $y_p[n] = x_p[n] \circledast h_p[n]$ graphically for $N = 4$, $N = 6$ and $N = 8$.

5.10 (a) Calculate the linear convolution $y[n] = x[n] * h[n]$ of the signals $x[n]$ and $h[n]$ as shown in Fig. 5.20.
(b) Calculate the linear convolutions $y_1[n] = x_1[n] * h[n]$, $y_2[n] = x_2[n] * h[n]$ and $y_3[n] = x_3[n] * h[n]$ of the signals $x_1[n]$, $x_2[n]$, $x_3[n]$ and $h[n]$ as shown in Fig. 5.21.
(c) Check to see if $y[n] = y_1[n] + y_2[n] + y_3[n] + \ldots$.

Exercises to section 5.7

5.11 Prove the relations (5.47a), (5.47b) and (5.47c).

5.12 In section 5.7 it was stated that the FFT butterfly (with $p = 0$) of Fig. 5.27 represents a 2-point DFT. This is shown in Fig. 5.29. Show that the butterfly of Fig. 5.29 does in fact represent a 2-point DFT.

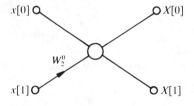

FIGURE 5.29 Exercise 5.12.

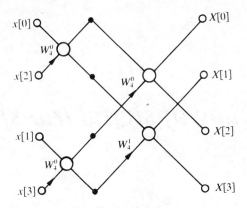

FIGURE 5.30 Exercises 5.13 and 5.14.

5.13 Show that the FFT shown schematically in Fig. 5.30 corresponds to a 4-point DFT.

5.14 A discrete signal $x[n]$ is defined as:

$$x[n] = \begin{cases} n+1 & \text{for } 0 \le n < 4 \\ 0 & \text{elsewhere} \end{cases}$$

(a) Calculate $X[k]$ by making use of (5.4); take $N = 4$.

(b) Calculate $X[k]$ by making use of the diagram in Fig. 5.30.

6

Overview of signal transforms

6.1 INTRODUCTION

In the preceding chapters we have looked at a large number of signal transforms. We can use these to go from a time description of a signal to a frequency description or vice versa. The choice of a particular transform is connected on the one hand with the properties of the signal (periodic or non-periodic, continuous-time or discrete-time), and on the other hand with the objective of the transformation (description with the aid of a real frequency variable or a complex frequency variable).

We have seen how – under certain conditions – we can convert a continuous signal $x(t)$ into a discrete signal $x[n]$ and how – again under certain conditions – we can derive a periodic discrete signal $x_p[n]$ from this signal. For each of these signals there is a specific signal transform. However, there are very clear relationships between these transforms. We often have a preference for going from one signal representation to another (and hence from one signal transform to the other), for example because the computation may be easier. The other side of the bargain is that the result we obtain is only an approximated version or a sampled version – or both of these.

In this chapter we should like to look more closely at the relations that exist between the various kinds of descriptions in terms of time and frequency. To start with we must first decide on a rather clearer notation than the one we have used so far. This is necessary to avoid confusion between the many time and frequency functions that we want to distinguish from one another. We shall therefore adopt the following conventions for the rest of this chapter (see Fig. 6.1):

- $x_c(t)$ is a *non-periodic, continuous-time* signal that gives the spectrum $X_c(\omega)$ via the Fourier transform for continuous signals (FTC).
- $x_d[n]$ is a *non-periodic, discrete-time* signal that gives the spectrum $X_d(e^{j\theta})$ via the Fourier transform for discrete signals (FTD).
- $x_p[n]$ is a *periodic, discrete-time* signal that gives the spectrum $X_p[k]$ via the N-point discrete Fourier transform (DFT).
- $x_h(t)$ is a *periodic, continuous-time* signal that gives the coefficients α_k via the Fourier series (FS); we can consider all of the coefficients α_k taken together as a discrete function of k.

Stylized examples of all these time and frequency functions can be seen with their most characteristic properties in Fig. 6.1. The figure illustrates a useful rule of thumb: if a time

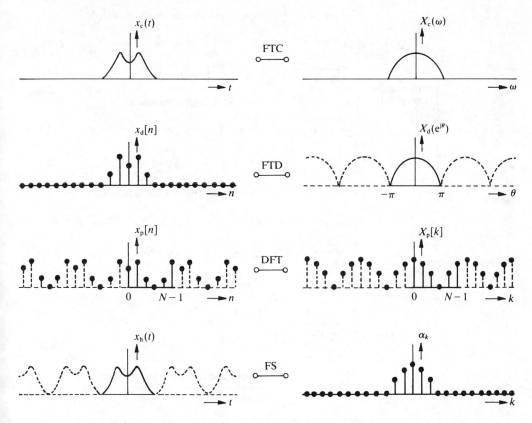

FIGURE 6.1 Four different kinds of signals $x_c(t)$, $x_d[n]$, $x_p[n]$, $x_h(t)$ and the corresponding frequency spectra $X_c(\omega)$, $X_d(e^{j\theta})$, $X_p[k]$ and α_k obtained via various signal transforms.

function and a frequency function form a transform pair, and one of them is discrete, then the other is periodic (and vice versa!).

For completeness Table 6.1 lists all the signal transforms we have dealt with so far. The table shows the most suitable transforms for every type of signal.

6.2 A QUALITATIVE COMPARISON

The eight functions of Fig. 6.1 can be shown in a symbolic representation, as in Fig. 6.2. The four time functions are represented by the four circles at the center, the four corresponding frequency functions by the other circles. The chain-dotted line represents the boundary between the time domain and the frequency domain. We can cross this boundary by using the transforms shown.

The same symbolic representation is again shown in Fig. 6.3(a), where the boundary line now shown separates the functions that are continuous (in time or in frequency) from the functions that are discrete (in time or in frequency). In Chapter 3 'Conversion from

TABLE 6.1 Overview of signal transforms

Fourier transform for continuous signals

Non-periodic signals

Continuous-time signal $x_c(t)$

Real frequency variable ω

$$X_c(\omega) = \int_{-\infty}^{\infty} x_c(t)\,e^{-j\omega t}\,dt \qquad \text{FTC}$$

$$x_c(t) = \frac{1}{2\pi} \int_{-\infty}^{\infty} X_c(\omega)\,e^{j\omega t}\,d\omega \qquad \text{IFTC}$$

(Two-sided) Laplace transform[†]

Complex frequency variable p

$$X_c(p) = \int_{-\infty}^{\infty} x_c(t)\,e^{-pt}\,dt \qquad \text{LT}$$

$$x_c(t) = \frac{1}{2\pi j} \int_{\sigma-j\infty}^{\sigma+j\infty} X_c(p)\,e^{pt}\,dp \qquad \text{ILT}$$

Discrete-time signal $x_d[n]$

Fourier transform for discrete signals

Real frequency variable θ

$$X_d(e^{j\theta}) = \sum_{n=-\infty}^{\infty} x_d[n]\,e^{-jn\theta} \qquad \text{FTD}$$

$$x_d[n] = \frac{1}{2\pi} \int_{-\pi}^{\pi} X_d(e^{j\theta})\,e^{jn\theta}\,d\theta \qquad \text{IFTD}$$

(Two-sided) z-transform

Complex frequency variable z

$$X_d(z) = \sum_{n=-\infty}^{\infty} x_d[n]\,z^{-n} \qquad \text{ZT}$$

$$x_d[n] = \frac{1}{2\pi j} \oint_C X_d(z)\,z^{n-1}\,dz \qquad \text{IZT}$$

Periodic signals

Discrete-time signal $x_p[n]$ — Real frequency variable k

N-point discrete Fourier transform (possibly calculated with the 'fast Fourier transform' or FFT)

$$X_p[k] = \sum_{n=0}^{N-1} x_p[n]\,e^{-j(2\pi/N)kn} \qquad \text{DFT}$$

$$x_p[n] = \frac{1}{N} \sum_{k=0}^{N-1} X_p[k]\,e^{j(2\pi/N)kn} \qquad \text{IDFT}$$

Continuous-time signal $x_h(t)$ — Real frequency variable k

Fourier series

$$\alpha_k = \frac{1}{T} \int_{-T/2}^{T/2} x_h(t)\,e^{-jk\omega_0 t}\,dt$$

$$x_h(t) = \sum_{k=-\infty}^{\infty} \alpha_k\,e^{jk\omega_0 t}$$

$\text{with } \omega_0 = \dfrac{2\pi}{T} \qquad \text{FS}$

[†]The equations for the transforms LT and ILT do not appear again in this book.

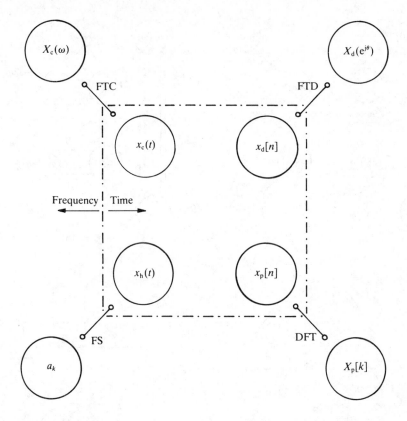

FIGURE 6.2 Symbolic representation of the transform pairs from Fig. 6.1. The chain-dotted line indicates the boundary between the time domain and the frequency domain.

continuous-time signals to discrete-time signals and vice versa' we have already looked in some detail at the transition from a continuous-time function to a discrete-time function ('sampling') and the transition in the reverse direction ('interpolation'). Similarly, we can go from a continuous-frequency function to a discrete-frequency function and vice versa. Here also we speak of sampling (S) and interpolation (I), but now in the frequency domain. In Fig. 6.3(b) the transitions that correspond to interpolation are indicated by a continuous arrow and the transitions that correspond to sampling are indicated by a dashed arrow. Figure 6.3(c) once more gives a schematic representation of what exactly we mean by sampling and interpolation; either time (t or n) or frequency (ω, θ or k) is shown as the variable along the horizontal axis.

Yet another kind of boundary line is shown in Fig. 6.4(a): here the boundary separates periodic from non-periodic functions. We can go from a periodic function to a non-periodic function by choosing a single period (C; continuous arrows in Fig. 6.4(b)) and we can make the transition in the inverse direction by periodically repeating a non-periodic function (R; dashed arrows in Fig. 6.4(b)). The two operations that correspond to the transitions are shown again schematically in Fig. 6.4(c).

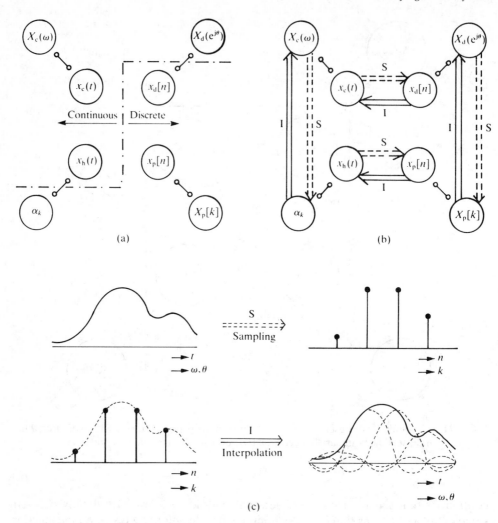

FIGURE 6.3 (a) Symbolic representation of the various transform pairs. The chain-dotted line indicates the boundary between continuous and discrete functions. (b) Functions of one kind can be converted into functions of the other kind by sampling (S) or interpolation (I) with $(\sin x)/x$ functions. (c) Schematic representation of the operations S and I.

   The transitions indicated by continuous arrows in Figs 6.3(b) and 6.4(b) (interpolation and choosing a single period) can always be executed without approximation, i.e. we can always 'get back again' to the situation before the transition. This is *not* automatically so for the transitions indicated by a dashed arrow. We have already seen this in Chapter 3 'The conversion of continuous-time signals into discrete-time signals and vice versa' for the transition from $x_c(t)$ to $x_d[n]$: we have to satisfy the conditions of the sampling theorem, which at the very least means that $X_c(\omega)$ is zero above certain frequencies or (in other words) is band-limited.

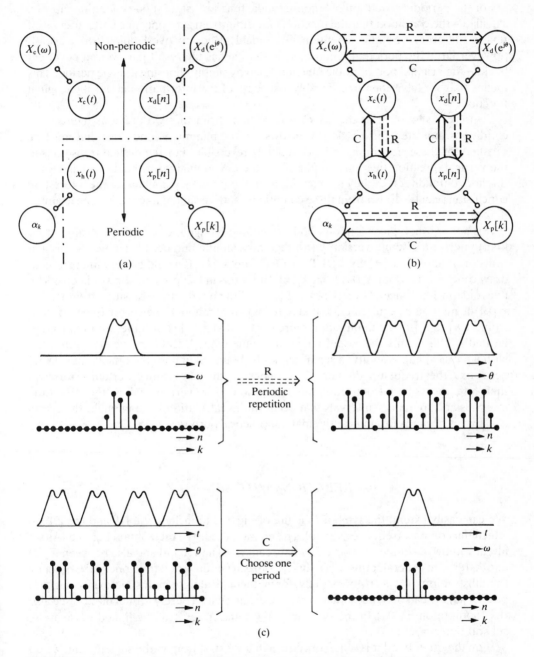

FIGURE 6.4 (a) Symbolic representation of the various transform pairs. The chain-dotted line indicates the boundary between periodic and non-periodic functions. (b) By choosing a single period (C) or periodic repetition (R) functions of one kind can be converted into functions of the other kind. (c) Schematic representation of the operations C and R.

For the periodic repetition of a non-periodic function, at least one condition must be satisfied – the condition that the function may differ from zero only in a finite interval. If this is not true, periodic repetition unavoidably leads to overlapping (or aliasing), however large the period we choose. This aliasing has the result that we can never get back to the original non-periodic function again by simply choosing a single period. This means that we have then irretrievably lost some of the information about the original function.

Figure 6.5 shows all of the possible transitions from the two previous figures; the conditions indicated for the dashed arrows are the minimum conditions that must be satisfied for these transitions to be completely reversible. For the dashed transitions in the vertical direction these are conditions in the time domain; for the dashed arrows in the horizontal direction they are conditions in the frequency domain. Some interesting conclusions can be drawn from these conditions, and we should like to give one example here.

It is not unusual, given a non-periodic (continuous) signal $x_c(t)$ with spectrum $X_c(\omega)$, first to derive a discrete version $x_d[n]$, then take a finite number of samples of this and transform these, as $x_p[n]$, by a DFT (or FFT) into $X_p[k]$. It would be very interesting to draw conclusions about $X_c(\omega)$ from $X_p[k]$. But is this in fact possible, or would errors be unavoidably introduced? Let us look at Fig. 6.5. For an error-free transition from $x_c(t)$ to $x_d[n]$ the minimum requirement is that $X_c(\omega)$ is band-limited. For an error-free transition from $x_d[n]$ to $x_p[n]$ the minimum requirement is that $x_d[n]$ is of finite duration ('time-limited'). This is only the case if $x_c(t)$ is also time-limited. But here we are confronted with a physical impossibility: a signal cannot be bounded in both the time domain $[x_c(t)$ here] *and* the frequency domain $[X_c(\omega)$ here] at the same time. Certain errors will therefore *always* be introduced in this calculation. However, by making the finite time intervals and frequency intervals that occur in the calculation large enough, the errors can always be made insignificantly small for practical purposes.

6.3 RELATIONS WITH LT AND ZT

We can easily extend the symbolic picture of Fig. 6.2 to include the LT and the ZT, in which the complex frequency variables p and z are used. We then obtain Fig. 6.6. As we already know, a simple relation exists between $X_c(p)$ and $X_c(\omega)$ and also between $X_d(z)$ and $X_d(e^{j\theta})$. In general terms, $X_c(p)$ and $X_c(\omega)$ can be converted from one to the other by the substitution $p = j\omega$. (More strictly, for mathematical rigor in this context we ought to put $X_c(j\omega)$ instead of $X_c(\omega)$, but here we are dealing with a purely formal matter, see also the footnote on p. 31.) In the same way $X_d(z)$ and $X_d(e^{j\theta})$ can be linked through the substitution $z = e^{j\theta}$.

If on the other hand it is $X_c(\omega)$ or $X_d(e^{j\theta})$ that we start from in the substitution, we do not normally know the values of p or z for which the new function $X_c(p)$ or $X_d(z)$ converges. We should then really look at this separately. In practical situations there are usually no unexpected problems here, so that we generally do not have to bother with the question of convergence.

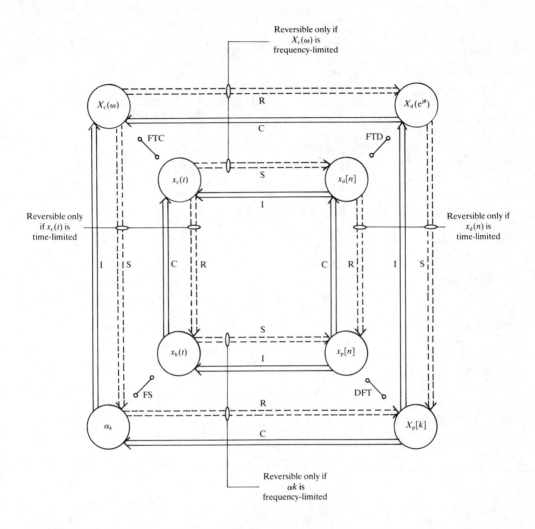

FIGURE 6.5 All the possible transitions between the various functions; the condition indicated for each dashed arrow is the minimum condition that must be satisfied for that transition to be error-free (and therefore reversible).

6.4 THE USE OF THE DELTA FUNCTION

In the description of the transition from continuous time to discrete time we have used the Dirac pulse or delta function $\delta(t)$ as an intermediate step. The delta function itself is a *continuous* function, because the independent variable t can take all possible (real) values. Although the delta function is a function that from its nature can never arise in practice (it is called a 'generalized function'), it is a very convenient theoretical

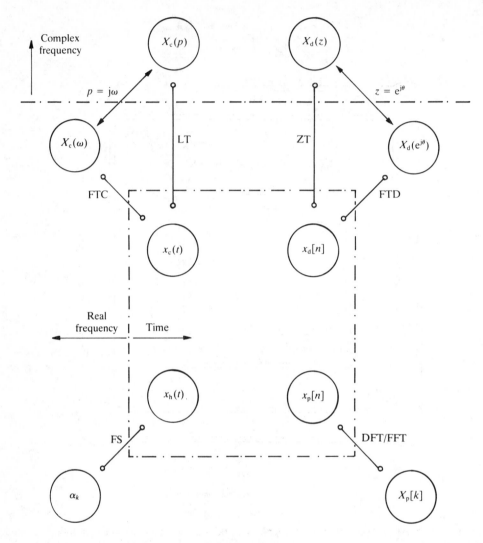

FIGURE 6.6 Extension of Fig. 6.2 to include the complex frequency functions $X_c(p)$ and $X_d(z)$.

mathematical aid to use at the boundary between continuous time and discrete time. Situations in which it has an important part to play are often found in control theory, for example, where we may encounter many kinds of complicated combinations of continuous-time and discrete-time signal-processing operations. Extensive use of the delta function here will often provide a convincing mathematical description entirely in terms of the FTC and the IFTC.

We should like to illustrate this with the aid of Fig. 6.7. Only *continuous* functions occur here, in both the time domain and the frequency domain. The similarity to Fig. 6.1 is very noticeable, however.

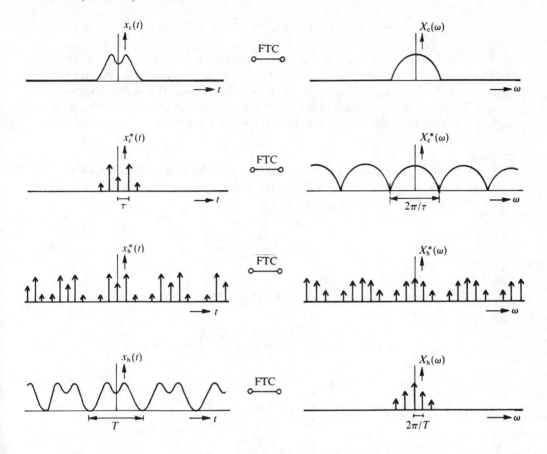

FIGURE 6.7 By making extensive use of delta functions the most divergent time and frequency functions can be linked through the FTC alone. In principle we are then always dealing with *continuous* functions. (Caution: this figure is the only place in this book where the symbol * is used to distinguish sampled time functions from non-sampled time functions.)

In certain situations the use of delta functions leads to problems. A good example of this is that the product of two delta functions is not defined, and cannot therefore be used. The same thing also applies, of course, for the product of two signals consisting of sequences of delta functions. This means that an operation such as modulation, which is essentially the multiplicaton of two signals, cannot be described properly in such cases. With discrete-time signals we do not have these problems; the multiplication of two samples just gives a new sample whose value is the product of the two old values.

In cases where an entirely (or almost entirely) discrete-time description is adequate, it is preferable to use the FTD and DFT, because these are more directly associated with practical circuits (digital circuits and computers) than are the rather more abstract calculations with delta functions. We shall therefore avoid such calculations where we can.

6.5 A QUANTITATIVE COMPARISON

We should now like to illustrate what we have said above with the aid of two examples. In the first example (Fig. 6.8) we choose for $x_c(t)$ a function strictly bounded in time, which takes the form of a triangle of width T. In the second example (Fig. 6.9) we choose for $X_c(\omega)$ a function strictly bounded in frequency with the same shape and width V. We also stipulate that the time functions always have the value A at time 0.

Example 1

The transforms from Table 6.1 can be used to derive the following equations for the various frequency functions of Fig. 6.8 directly, assuming a sampling interval of $\tau = T/8$ for the discrete-time functions $x_d[n]$ and $x_p[n]$:

$$X_c(\omega) = \frac{AT}{2}\left[\frac{\sin(\omega T/4)}{\omega T/4}\right]^2 \tag{6.1}$$

$$X_d(e^{j\theta}) = A + \frac{3A}{2}\cos(\theta) + A\cos(2\theta) + \frac{A}{2}\cos(3\theta) \tag{6.2}$$

$$X_p[k] = A + \frac{3A}{2}\cos\left(\frac{2\pi k}{8}\right) + A\cos\left(\frac{4\pi k}{8}\right) + \frac{A}{2}\cos\left(\frac{6\pi k}{8}\right) \tag{6.3}$$

$$\alpha_k = \begin{cases} A/2 & k = 0 \\ 2A/k^2\pi^2 & k = \pm1,\ \pm3,\ \pm5,\ \ldots \\ 0 & k = \pm2,\ \pm4,\ \pm6,\ \ldots \end{cases} \tag{6.4}$$

These frequency functions are shown in Fig. 6.8 and eight corresponding values for each are also given in Table 6.2. A clear relationship between the various functions can be seen from this. It is noteworthy, however, that there are differences in value that originate directly from the different definitions of the transforms (look at frequency 0, for example). There are also differences due to aliasing, because the conditions of the sampling theorem are not satisfied.

TABLE 6.2 Eight corresponding values for the frequency functions of Example 1

k	ω	θ	$X_c(\omega)$	$X_d(e^{j\theta})$	$X_p[k]$	α_k
0	0	0	$0.5 \times AT$	$0.5 \times 8A$	$0.5 \times 8A$	$0.5 \times A$
1	$2\pi/T$	$\pi/4$	$0.2026 \times AT$	$0.2134 \times 8A$	$0.2134 \times 8A$	$0.2026 \times A$
2	$4\pi/T$	$\pi/2$	0	0	0	0
3	$6\pi/T$	$3\pi/4$	$0.0225 \times AT$	$0.0366 \times 8A$	$0.0366 \times 8A$	$0.0225 \times A$
4	$8\pi/T$	π	0	0	0	0
5	$10\pi/T$	$5\pi/4$	$0.0081 \times AT$	$0.0366 \times 8A$	$0.0366 \times 8A$	$0.0081 \times A$
6	$12\pi/T$	$3\pi/2$	0	0	0	0
7	$14\pi/T$	$7\pi/4$	$0.0041 \times AT$	$0.2134 \times 8A$	$0.2134 \times 8A$	$0.0041 \times A$
			↓	↓	↓	↓
			Continues non-periodically	Repeats periodically	Repeats periodically	Continues non-periodically

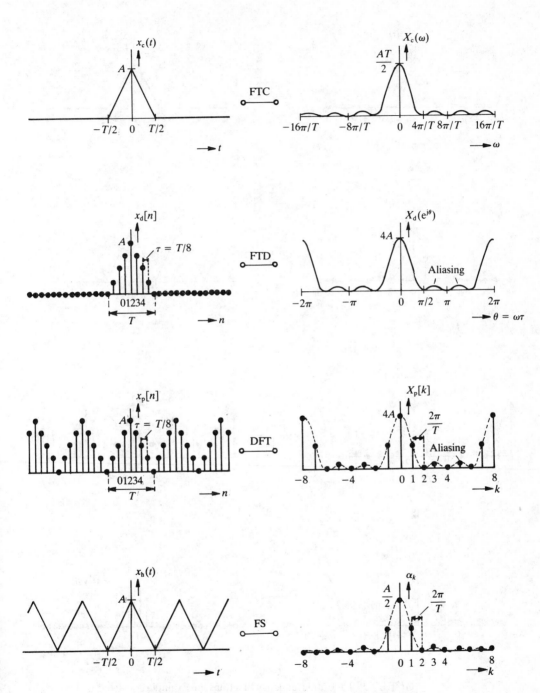

FIGURE 6.8 Time and frequency functions for Example 1.

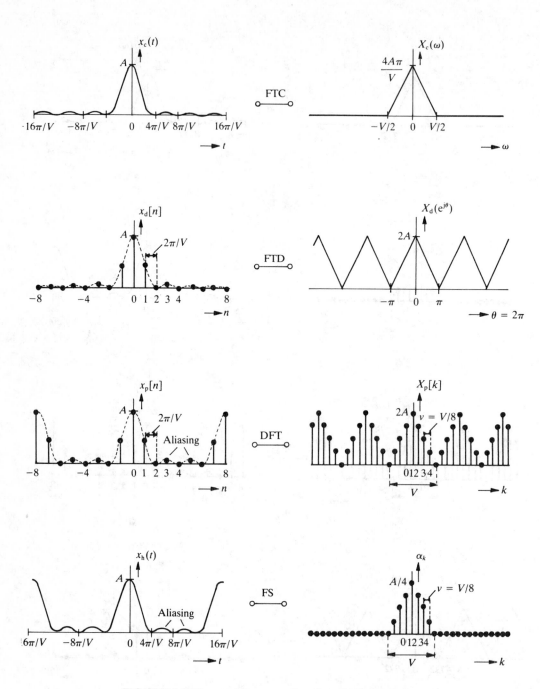

FIGURE 6.9 Time and frequency functions for Example 2.

Example 2

The inverse transforms from Table 6.1 can be used to derive the equations for the various time functions of Fig. 6.9, starting from the fairly simply described frequency functions. We assume here that the sampling interval for the discrete-frequency functions $X_p[k]$ and α_k is $\nu = V/8$. We then find:

$$x_c(t) = A \left[\frac{\sin(Vt/4)}{Vt/4} \right]^2 \qquad (6.5)$$

$$x_d[n] = \begin{cases} A & n = 0 \\ 4A/n^2\pi^2 & n = \pm 1, \pm 3, \pm 5, \ldots \\ 0 & n = \pm 2, \pm 4, \pm 6, \ldots \end{cases} \qquad (6.6)$$

$$x_p[n] = \frac{A}{4} + \frac{3A}{8} \cos\left(\frac{2\pi n}{8}\right) + \frac{A}{4} \cos\left(\frac{4\pi n}{8}\right) + \frac{A}{8} \cos\left(\frac{6\pi n}{8}\right) \qquad (6.7)$$

$$x_h(t) = \frac{A}{4} + \frac{3A}{8} \cos\left(\frac{Vt}{8}\right) + \frac{A}{4} \cos\left(\frac{2Vt}{8}\right) + \frac{A}{8} \cos\left(\frac{3Vt}{8}\right) \qquad (6.8)$$

Eight corresponding values from each of these four time functions are shown in Table 6.3. A comparison with Example 1 reveals the high degree of symmetry that exists between the time and frequency domains. We can even speak of 'aliasing in the time domain' for the functions $x_p[n]$ and $x_h(t)$ in Fig. 6.9.

TABLE 6.3 Eight corresponding values for the time functions of Example 2

n	t	$x_c(t)$	$x_d[n]$	$x_p[n]$	$x_h(t)$
0	0	A	A	A	A
1	$2\pi/V$	$0.4052 \times A$	$0.4052 \times A$	$0.4238 \times A$	$0.4238 \times A$
2	$4\pi/V$	0	0	0	0
3	$6\pi/V$	$0.0450 \times A$	$0.0450 \times A$	$0.0732 \times A$	$0.0732 \times A$
4	$8\pi/V$	0	0	0	0
5	$10\pi/V$	$0.0162 \times A$	$0.0162 \times A$	$0.0732 \times A$	$0.0732 \times A$
6	$12\pi/V$	0	0	0	0
7	$14\pi/V$	$0.0082 \times A$	$0.0082 \times A$	$0.4238 \times A$	$0.4238 \times A$
		↓	↓	↓	↓
		Continues non-periodically	Continues non-periodically	Repeats periodically	Repeats periodically

6.6 EXERCISES

6.1 Derive equations (6.1)–(6.4) for the frequency functions $X_c(\omega)$, $X_d(e^{j\theta})$, $X_p[k]$ and α_k from Fig. 6.8.

6.2 (a) Use the definitions of the Laplace transform and the z-transform to determine the functions $X_c(p)$ and $X_d(z)$ that correspond to the time functions $x_c(t)$ and $x_d[n]$ from Fig. 6.8.

(b) Show that the substitution $j\omega = p$ and $e^{j\theta} = z$ transforms the functions $X_c(\omega)$ and $X_d(e^{j\theta})$ found in Exercise 6.1 into the functions $X_c(p)$ and $X_d(z)$ found in Exercise 6.2(a).

6.3 Derive equations (6.5)–(6.8) for the time functions $x_c(t)$, $x_d[n]$, $x_p[n]$ and $x_h(t)$ from Fig. 6.9.

7

Filter structures

7.1 INTRODUCTION

In the preceding chapters we have regularly talked about linear time-invariant discrete systems. The great majority of these LTD systems are formed by the LTD filters. We shall define the concept of 'filter' here very generally as a circuit (or an algorithm) that converts an input signal into an output signal whose spectrum is related in a certain prescribed way to the spectrum of the input signal (certain frequencies are suppressed or attenuated, for example), without the occurrence of frequency shifts (as is the case in modulation).

In this chapter we should like to deal with the various widely encountered LTD filter structures. We can classify these filters in a number of ways. We have already seen that we can make a distinction between:

(a) finite impulse response filters, or FIR filters, and
(b) infinite impulse response filters, or IIR filters.

This classification tells us little about the structure of the filters, but the next one is more informative in this respect:

(a) non-recursive discrete filters (NRDFs), i.e. filters in which there is no feedback path, and
(b) recursive discrete filters (RDFs), i.e. filters in which there is at least one feedback path.

We should say here that it is often wrongly assumed that FIR filters always have an NRDF structure and that an RDF structure always gives an IIR filter. Although this often is the case, it is not necessarily so (Fig. 7.1). We shall encounter examples of this later.

A concept encountered in the literature on filter structures is that of the *canonic* network. A discrete filter is said to be canonic if it contains the minimum number of delay elements necessary to realize the associated frequency response.

7.2 NON-RECURSIVE DISCRETE FILTERS

Non-recursive discrete filters (NRDFs) are characterized by the fact that they have no feedback paths (and hence no closed loops). Two examples are given in Fig. 7.2.

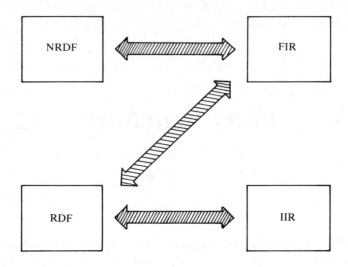

FIGURE 7.1 Usually an FIR filter has an NRDF structure and an RDF structure gives an IIR filter. In special cases the combination of an RDF structure and an FIR filter is possible (if all the poles coincide with zeros).

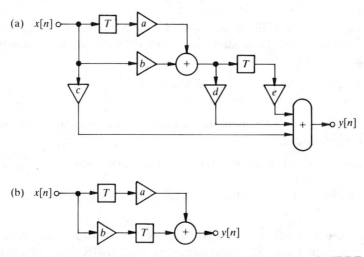

FIGURE 7.2 (a) Arbitrary example of an NRDF. (b) Non-canonic NRDF.

It is easy to see that the impulse response of such a filter can never contain more non-zero samples than the number of delay elements in the longest path that exists between input and output, plus one. The consequence of this is that an NRDF is always of the FIR type. The example of Fig. 7.2(a) contains two delay elements and the length of the impulse response is three. In Fig. 7.2(b) an example is given of a non-recursive filter with two delay elements and an impulse response of length one (this filter is therefore non-canonic). A widely occurring form of NRDF is shown in Fig. 7.3. This type of filter is

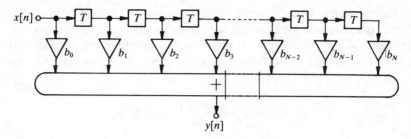

FIGURE 7.3 Transversal filter.

called a *transversal filter*. It is characterized by the fact that the only signal stored in the delay elements is the original input signal $x[n]$.

The system function $H(z)$ of a (causal) NRDF can always be expressed in the form:

$$H(z) = \frac{Y(z)}{X(z)} = b_0 + b_1 z^{-1} + \ldots + b_N z^{-N} = \sum_{i=0}^{N} b_i z^{-i}$$

$$= \frac{\sum_{i=0}^{N} b_i z^{N-i}}{z^N} \tag{7.1}$$

The corresponding impulse response $h[n]$ is:

$$h[n] = b_0 \delta[n] + b_1 \delta[n-1] + \ldots + b_N \delta[n-N] = \sum_{i=0}^{N} b_i \delta[n-i] \tag{7.2}$$

We see from (7.1) that $H(z)$ has only zeros, plus N poles at the origin ($z = 0$). We also see from (7.2) that the corresponding impulse response $h[n]$ has a maximum duration of $(N+1)$ and therefore always has a finite duration. In general, for the creation of an impulse response of infinite duration at least one pole must be outside the origin. (Note that in the transversal filter the impulse response can be identified immediately from the filter coefficients!)

However, a good approximation to almost any system function can be obtained with a filter that has only zeros, merely by making the length of the impulse response of the filter large enough. In practice we can have NRDFs with an impulse response comprising hundreds or thousands of samples (and hence hundreds or thousands of zeros, too!).

Another property of an NRDF is that it will always be stable. We can see this in two ways:

1. Stability requires that there should be no poles outside the unit circle (see Chapter 4 'Discrete-time signals and systems', section 4.5.7, point 1); this condition is automatically satisfied, since there are no poles at all outside the origin.
2. Stability requires that the sum of the absolute values of the samples of the impulse response is finite; see (4.47). Since we have a finite number of samples, each with a finite value, this condition is also automatically satisfied.

Yet another property of NRDFs (which is also generally applicable for FIR filters, even if they are of the recursive type) is that we can make filters with an exactly linear phase characteristic. We say that a discrete filter has a linear phase characteristic $\varphi(e^{j\theta})$ if the derivative of the phase characteristic multiplied by -1 (the *group delay* τ_g) has the following property:

$$\tau_g(e^{j\theta}) = -\frac{d\varphi(e^{j\theta})}{d\theta} = \text{constant} \tag{7.3}$$

A constant group delay means that signal components at different frequencies receive the same delay in the filter, with the result that certain properties of the shape of the input signal as a function of time (such as symmetric transitions from one signal level to another) are preserved. This is mainly of importance in fields where there are special requirements in this respect (as in television and data transmission).

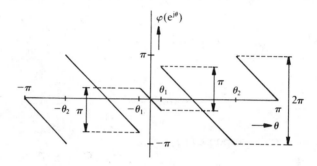

FIGURE 7.4 Example of a linear phase characteristic.

A constant group delay $\tau_g(e^{j\theta})$ means that the phase characteristic $\varphi(e^{j\theta})$ has a constant slope, but can have jumps of π radians. Such a jump occurs only if the system function of the filter has a zero on the unit circle in the z-plane (see section 4.5.7, point 2). At frequencies where the phase jump occurs the frequency response has an amplitude of zero and the fact that $\tau_g(e^{j\theta})$ is not defined at that frequency is of no significance. An example of a linear phase characteristic is given in Fig. 7.4. We see here a jump of π radians at $\theta = \pm\theta_1$ due to the presence of zeros at $z = e^{\pm j\theta_1}$ in the system function. The apparent jumps of 2π at $\theta = \pm\theta_2$ are not true discontinuities but are caused by the fact that we always limit the range of the phase φ (just as in continuous systems!) to the interval $-\pi \leq \varphi \leq \pi$.

TABLE 7.1 The four different types of FIR filters with linear phase

Phase jumps of π radians at		Finite-duration impulse response $h[n]$ (of length L)
$\theta = 0$	$\theta = \pm\pi$	
No	No	$h[n]$ is symmetric; L is odd
No	Yes	$h[n]$ is symmetric; L is even
Yes	Yes	$h[n]$ is antisymmetric; L is odd
Yes	No	$h[n]$ is antisymmetric; L is even

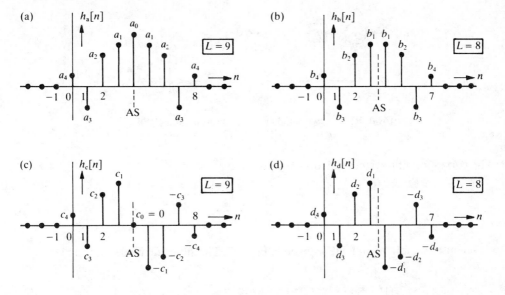

FIGURE 7.5 Characteristic impulse responses of the four possible types of FIR filter with a linear phase and impulse response of length L. (AS = axis of symmetry.) (a) Symmetric impulse response, L is odd; (b) Symmetric impulse response, L is even; (c) Antisymmetric impulse response, L is odd; (d) Antisymmetric impulse response, L is even.

Discrete filters with a linear phase characteristic are characterized by impulse responses with very specific properties. These properties are determined by the presence or absence of jumps in phase at $\theta = 0$ and $\theta = \pm\pi$. Four cases can be distinguished; they are shown in Table 7.1.

Examples of the impulse responses corresponding to these four classes of linear-phase filters are given in Fig. 7.5. We shall return to this again in section 8.2.4.

7.3 RECURSIVE DISCRETE FILTERS

Recursive discrete filters (RDFs) are characterized by the fact that there is at least one feedback path in the filter. An example is given in Fig. 7.6.

This filter can be described by the difference equations:

$$v[n] = x[n] + ay[n]$$
$$y[n] = v[n-1] = x[n-1] + ay[n-1] \tag{7.4}$$

and has the impulse response:

$$h[n] = a^{n-1}u[n-1] \tag{7.5}$$

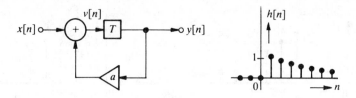

FIGURE 7.6 A simple recursive discrete filter (RDF).

The corresponding system function $H(z)$ is:

$$H(z) = \frac{z^{-1}}{1 - az^{-1}} \tag{7.6}$$

A second example of an RDF is given in Fig. 7.7. This filter can be described by:

$$v[n] = x[n] - a^4 x[n-4] + ay[n] \tag{7.7}$$

$$y[n] = v[n-1] = x[n-1] - a^4 x[n-5] + ay[n-1]$$

and the corresponding impulse response $h[n]$ is (verify this!):

$$h[n] = \begin{cases} 0 & \text{for } n \le 0 \\ a^{n-1} & \text{for } 1 \le n \le 4 \\ 0 & \text{for } n \le 5 \end{cases} \tag{7.8a}$$

or

$$h[n] = \delta[n-1] + a\delta[n-2] + a^2 \delta[n-3] + a^3 \delta[n-4] \tag{7.8b}$$

So what we have here is the rather unusual case of a filter with a *recursive* structure and a *finite-duration* impulse response!

We can use these two examples to make a number of interesting comments about RDFs.

1. Because of the presence of feedback paths closed loops are formed. Each closed loop must contain at least one delay element. Otherwise counters or multipliers would have to perform operations on signals not yet available, since they still have to calculate them.
2. Instabilities can occur for certain values of the coefficients in the closed loops; e.g. for $|a| \ge 1$ in Fig. 7.6.
3. The impulse response of Fig. 7.6 is of infinite duration, while that of Fig. 7.7 is of finite duration. This last result comes about because the contributions from the closed loop and from the non-recursive part of the circuit cancel one another out exactly for $n = 4$ (this requires, incidentally, that the multiplications by a and by $-a^4$ must be per-

formed with infinite accuracy; in a real digital system, where we are only concerned with discrete values, this would require particular attention!).

4. So far we have always tacitly assumed that the delay elements (registers) that occur in LTD systems are 'empty' at the moment the system is switched on (i.e. they represent contents 0), so that the output signal is only determined by the input signal presented to the system. If this condition is not satisfied, the output signal $y[n]$ is partly determined by the initial condition of the registers. This is extremely important, especially in a system that contains feedback, because in principle the initial state can

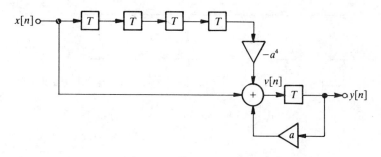

FIGURE 7.7 RDF with a finite-duration impulse response.

continue to affect the output signal for an infinitely long time. (In the system of Fig. 7.7 this means that for $|a| > 1$ even a system that is originally stable will change into an unstable system.) From now on we shall always assume that the initial state of the registers is zero when the system is switched on.

7.3.1 The direct form 1

In general we can describe the relation between the input signal and the output signal of an RDF by the difference equation (4.40):

$$y[n] = \sum_{i=0}^{N} b_i x[n-i] + \sum_{i=1}^{M} a_i y[n-i] \tag{7.9}$$

This difference equation can be directly translated into a structure that is called the *direct form 1* in the literature (Fig. 7.8). We see that this structure contains $N + M$ delay elements and $N + M + 1$ multipliers. In addition, for every output sample $N + M + 1$ signals must be added together in the adders. The system function $H(z)$ of this filter is:

$$H(z) = \frac{Y(z)}{X(z)} = \frac{\displaystyle\sum_{i=0}^{N} b_i z^{-i}}{1 - \displaystyle\sum_{i=1}^{M} a_i z^{-i}} \tag{7.10}$$

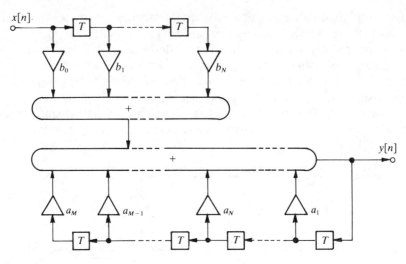

FIGURE 7.8 The direct form 1 structure.

In general $H(z)$ has zeros and poles (see section 4.5.7). The zeros are realized in the non-recursive part of the filter structure with the coefficients b_0 to b_N and the poles in the purely recursive part with the coefficients a_1 to a_M.

The direct form 1, as shown in Fig. 7.8, is non-canonic, because we can realize the same filter with fewer delay elements. In the next section we shall encounter a canonic structure.

7.3.2 The direct form 2

We can think of the structure of Fig. 7.8 as the cascade connection of a non-recursive part and a recursive part. Since the two parts form a linear time-invariant network, we can interchange them without affecting the frequency response. We then obtain the structure of Fig. 7.9. However, the two series of delay elements now contain exactly the same signals $v[n]$, $v[n-1]$, ..., $v[n-N]$ and can therefore be combined. We then get the structure of Fig. 7.10, which we call the *direct form 2* (in this example it has been assumed that $M > N$). We now have a total of only M delay elements and this structure is canonic.

With both the direct form 1 and the direct form 2 we can think of the structure as being split into two parts, with one part realizing all the poles and the other part realizing all the zeros. This leads at once to one of the more undesirable features of these two structures: a small variation in (say) one of the coefficients b_i affects the location of all the N zeros, and similarly a small variation in one of the coefficients a_i affects the location of all the M poles. This means that the entire frequency response can change considerably. These filters are therefore said to have a high *parameter sensitivity*.

This effect can be prevented by dividing the system function $H(z)$ into a number of smaller sections $H_1(z)$, $H_2(z)$, ..., $H_K(z)$, each representing only a limited number of poles and zeros of $H(z)$. These smaller sections can then be realized either as a cascade structure or as a parallel structure.

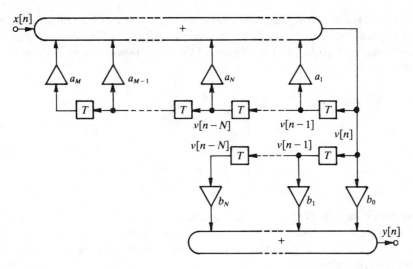

FIGURE 7.9 Interchanging the recursive and non-recursive parts of the circuit of Fig. 7.8.

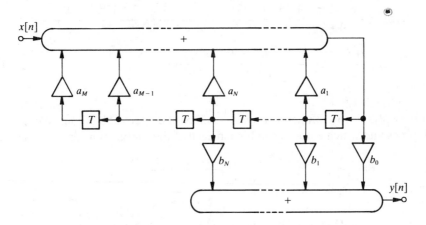

FIGURE 7.10 The direct-form-2 structure.

7.3.3 The cascade structure

To realize a desired system function $H(z)$ as a cascade structure it is rewritten in the form (see Fig. 7.11):

$$H(z) = H_1(z)H_2(z) \ldots H_i(z) \ldots H_K(z) \qquad (7.11)$$

FIGURE 7.11 The cascade structure.

We recall here that complex poles and zeros in a system with a real impulse response $h[n]$ always occur in complex conjugate pairs. This must also apply for each of the subfunctions $H_1(z)$ to $H_K(z)$. Usually one of the two following forms is chosen for such a subfunction $H_i(z)$:

$$H_i(z) = \frac{1 + c_i z^{-1}}{1 + d_i z^{-1}} \tag{7.12}$$

or

$$H_i(z) = \frac{1 + c_i z^{-1} + d_i z^{-2}}{1 + e_i z^{-1} + f_i z^{-2}} \tag{7.13}$$

The system function $H_i(z)$ of (7.12) represents one real zero and one real pole (a 'first-order section'[†]); while the $H_i(z)$ of (7.13) represents two zeros and two poles that may be complex (a 'second-order section'). An example of a cascade realization of a third-order system given by:

$$H(z) = \frac{23 + 40z^{-1} + 36z^{-2} + 19z^{-3}}{10 + 9z^{-1} + 8z^{-2} + 3z^{-3}} = \frac{(1 + z^{-1})(23 + 17z^{-1} + 19z^{-2})}{(2 + z^{-1})(5 + 2z^{-1} + 3z^{-2})}$$

$$= \frac{0.5 + 0.5z^{-1}}{1 + 0.5z^{-1}} \times \frac{4.6 + 3.4z^{-1} + 3.8z^{-2}}{1 + 0.4z^{-1} + 0.6z^{-2}} \tag{7.14}$$

is given in Fig. 7.12.

Splitting $H(z)$ into subfunctions gives a number of degrees of freedom. We can choose which poles and which zeros to combine in each of the subfunctions, and the sequential

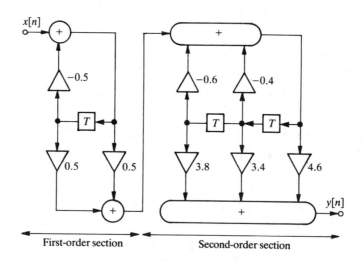

FIGURE 7.12 A cascade realization of the system function given in (7.14).

[†]For the notion of 'order' see (4.112).

order (the 'ordering') of the subfunctions can also be chosen freely. If we are not concerned with quantization effects these choices will not affect the total resultant system function. However, as soon as we begin to take the effects of quantization into account, the particular method of splitting into subfunctions becomes very important, and generally requires very careful consideration.

We hardly need to add that the poles and the zeros of the subfunctions taken together exactly represent the poles and the zeros of $H(z)$. This is not so for the structure we shall discuss in the next section.

7.3.4 The parallel structure

We can also think of $H(z)$ as the *sum* of K subfunctions:

$$H(z) = H_0 + H_1(z) + \ldots + H_i(z) + \ldots + H_K(z) \tag{7.15}$$

where H_0 is a constant and each of the subfunctions $H_1(z)$, $H_2(z)$, $\ldots$ represents a first-order or second-order section, such as we saw in the previous section. We then obtain a parallel structure (see Fig. 7.13).

The poles of the subfunctions taken together represent the poles of $H(z)$ exactly. However, this is *not* true for the zeros. We can see this by realizing the $H(z)$ of (7.14) as a parallel structure:

$$H(z) = \frac{23 + 40z^{-1} + 36z^{-2} + 19z^{-3}}{(2 + z^{-1})(5 + 2z^{-1} + 3z^{-2})} = \frac{19}{3} - \frac{5}{(2 + z^{-1})} - \frac{23 - z^{-1}}{3(5 + 2z^{-1} + 3z^{-2})}$$

$$= H_0 + H_1(z) + H_2(z) \tag{7.16a}$$

with

$$H_0 = \frac{19}{3}, \quad H_1(z) = \frac{-2.5}{1 + 0.5z^{-1}} \quad \text{and} \quad H_2(z) = \frac{-23/15 + z^{-1}/15}{1 + 0.4z^{-1} + 0.6z^{-2}} \tag{7.16b}$$

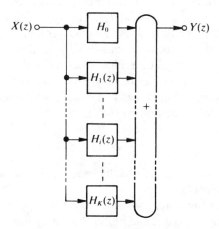

FIGURE 7.13 The parallel structure.

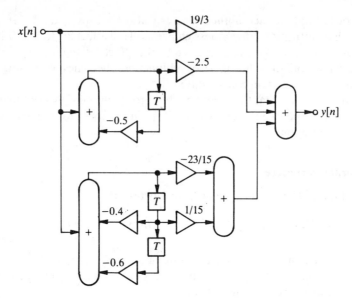

FIGURE 7.14 A parallel structure with the system function given by (7.16).

The poles of $H(z)$ are at $z = -\frac{1}{2}$ and $z = (-2 \pm j\sqrt{56})/10$; $H_1(z)$ has one pole at $z = -\frac{1}{2}$ and $H_2(z)$ has two poles at $z = (-2 \pm j\sqrt{56})/10$. The zeros of $H(z)$ are at $z = -1$ and $z = (-17 \pm j\sqrt{1459})/46$; while only $H_2(z)$ has a zero, at $z = 1/23$. This system function can be realized with the circuit of Fig. 7.14.

7.4 SOME SPECIAL FILTER STRUCTURES

7.4.1 Comb filters

An interesting type of filter, known as a comb filter, is obtained from a given arbitrary discrete filter with the system function $H(z)$ by replacing *each* delay element by N cascaded delay elements. We then obtain a new filter with the system function $G(z) = H(z^N)$. This means that the frequency response in the fundamental interval $(-\pi \le \theta < \pi)$ is periodically repeated N times. This is illustrated in Fig. 7.15 for $N = 3$ and $N = 4$.

We obtain a simple example of a non-recursive comb filter by starting from $H(z) = 1 - z^{-1}$. The resulting comb filter consists of N delay elements, one multiplier and one adder (Fig. 7.16(a)). The system function of this filter is:

$$G(z) = Y(z)/X(z) = 1 - z^{-N} \qquad (7.17)$$

It is easily seen that this filter has N equally spaced zeros on the unit circle, with $z = 1$ representing a zero in any case.

Figure 7.16(b) shows the poles-and-zeros plot for $N = 20$ and Figs 7.16(c) and (d) show the corresponding amplitude and phase characteristics. We see that the frequency response is always zero for the frequencies $\theta_k = 2\pi k/N$.

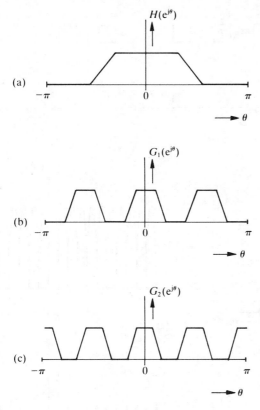

FIGURE 7.15 Frequency responses of comb filters obtained by replacing *each* delay element in the
filter of frequency response $H(e^{j\theta})$ by three and four delay elements.

7.4.2 The frequency-sampling filter

If the comb filter of Fig. 7.16 is followed by a recursive network that has a number of poles
that coincide exactly with the same number of zeros of the comb filter, we obtain a
frequency-sampling filter. This recursive part usually consists of the parallel connection
of a number of second-order sections (possibly with a first-order section to realize a pole
at $z = 1$ or $z = -1$).

This structure gives us a simple way of choosing the value of the frequency response
exactly at N frequencies. These N frequencies are equally spaced throughout the interval
$-\pi \leq \theta < \pi$, i.e. at:

$$\theta = 0, \pm\frac{2\pi}{N}, \pm\frac{4\pi}{N}, \ldots, \pm\left(\frac{N-2}{2}\right)\frac{2\pi}{N}, \pi \quad \text{if } N = \text{even}$$

and at:

$$\theta = 0, \pm\frac{2\pi}{N}, \pm\frac{4\pi}{N}, \ldots, \pm\left(\frac{N-3}{2}\right)\frac{2\pi}{N}, \pm\left(\frac{N-1}{2}\right)\frac{2\pi}{N} \quad \text{if } N = \text{odd}$$

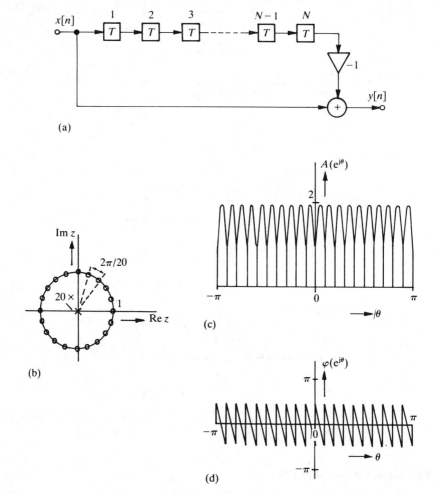

FIGURE 7.16 (a) A simple Nth-order comb filter. (b) The poles-and-zeros plot for $N = 20$. (c) The corresponding amplitude characteristic $A(e^{j\theta})$. (d) The corresponding phase characteristic $\varphi(e^{j\theta})$.

For a realizable filter (i.e. with a real impulse response) we must of course have (see 4.34a):

$$H(e^{j2\pi i/N}) = H^*(e^{-j2\pi i/N}) \tag{7.18}$$

For simplicity we call the chosen values of the frequency response:

$$H(e^{j2\pi i/N}) = H_i \tag{7.19}$$

and

$$H(e^{-j2\pi i/N}) = H_{-i} = H_i^* \tag{7.20}$$

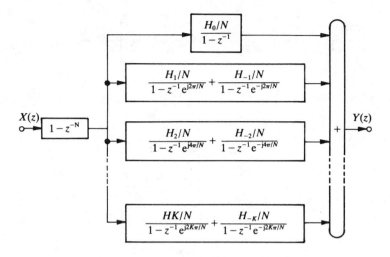

FIGURE 7.17 The frequency-sampling filter.

Figure 7.17 shows the general block diagram for N = odd, in which one or more of the H_i's may be zero (in this figure $K = (N-1)/2$). Figure 7.18(a) gives an example for $N = 5$ with $H_{-2} = H_0 = H_2 = 0$ and $H_{-1} = H_1 = 3$. The total system function here is:

$$H(z) = (1-z^{-5})\left[\frac{3/5}{1-z^{-1}e^{j2\pi/5}} + \frac{3/5}{1-z^{-1}e^{-j2\pi/5}}\right] \quad (7.21a)$$

$$= \frac{(1-z^{-5})[6-6z^{-1}\cos(2\pi/5)]}{5(1-z^{-1}e^{j2\pi/5})(1-z^{-1}e^{-j2\pi/5})} \quad (7.21b)$$

$$= \frac{(z^5-1)(6z-1.854)}{5(z-e^{j2\pi/5})(z-e^{-j2\pi/5})z^4} \quad (7.21c)$$

This system therefore has six zeros z_i and six poles p_i:

$$z_1 = 1; \quad z_{2,3} = e^{\pm j2\pi/5}; \quad z_{4,5} = e^{\pm j4\pi/5} \quad \text{and} \quad z_6 = 0.309$$

$$p_{1,2} = e^{\pm j2\pi/5} \quad \text{and} \quad p_3 = p_4 = p_5 = p_6 = 0 \quad (7.22)$$

The corresponding poles-and-zeros plot is shown in Fig. 7.18(b) and the amplitude and phase characteristics are shown in Fig. 7.18(c) and 7.18(d). Since two of the poles coincide with two of the zeros, effectively four poles remain – all at $z = 0$ in this case; this system is therefore of order four. The total frequency response is still zero for frequencies that correspond to zeros on the unit circle that are not compensated by a pole; hence for $\theta = 0$ and $\theta = \pm 4\pi/5$. At the frequencies where there is compensation (i.e. at $\theta = \pm 2\pi/5$), the frequency response is exactly equal to $H_1 = H_{-1} = 3$. (By giving H_1 a

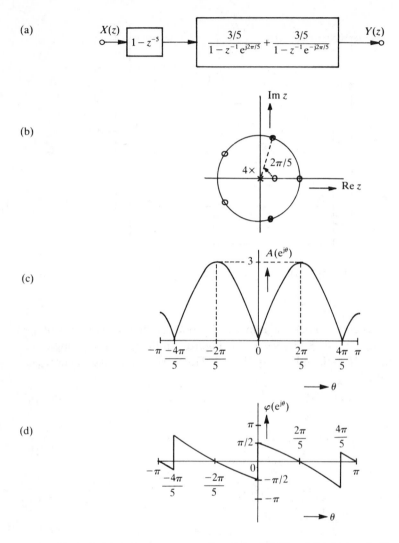

FIGURE 7.18 (a) A frequency-sampling filter with $N = 5$ in which only H_1 and H_{-1} are not zero. (b) Poles-and-zeros plot. (c) Amplitude characteristic. (d) Phase characteristic.

complex value, $H_1 = a_1 + jb_1$, so that $H_{-1} = a_1 - jb_1$, and making a correct choice of a_1 and b_1 we can obtain any desired amplitude and phase at the frequency $\theta = \pm 2\pi/5$.)

The frequency-sampling filter can be a particularly attractive solution if we want to make a narrow-band filter; then N is large, while the number of recursive sections can be very small. This means that the filter need contain only very few multipliers and adders.

For practical applications we have to keep an eye on a number of things. In the theory we start by assuming that a number of poles and zeros coincide exactly on the unit circle in the z-plane. This requires that we should be able to realize the filter coefficients to an accuracy of 100%. However, apart from a few obvious exceptions such as -1, 0 and $+1$, this is never the case.[58] This means that while we can locate the zeros of the comb-filter section in exactly the right positions, we cannot do so for the corresponding poles. The best we can do is to get them 'in the vicinity'. This can lead to a very erratic local variation of the frequency response, or even to an unstable system if one of the poles lies just *outside* the unit circle. In practice, both zeros and poles are therefore deliberately located just *inside* the unit circle. For the comb filter the system function is then chosen not from (7.17), but from:

$$H(z) = 1 - (\alpha z^{-1})^N \qquad (7.23)$$

where α is made slightly less than 1.

We also have to remember that any initial conditions for the delay elements or any interference in them can have an effect for a very long time because of the poles on or close to the unit circle.

7.4.3 Ladder and lattice filters

In recent years a great deal of interest has arisen in a few types of filters that cannot be identified with the structures we have already discussed. These are the *ladder filters* and *lattice filters*. These filters are found in so many variations that it is not possible to give an exhaustive treatment of them here. To give an idea of these types, however, a few typical examples are shown in Fig. 7.19.

The common feature of all the filters of Fig. 7.19 is that we can identify basic units (indicated by a dashed line) that each have *two* inputs and *two* outputs. In principle, an arbitrary number of these basic units can be connected in cascade. Lattice filters are typified by a crossover structure within the basic unit. These filters are interesting because of one or more of the following features:[30]

1. These structures give good correspondence with certain models that have been formulated for the human speech organs. This can offer considerable advantages in the discrete processing of speech signals.
2. The requirement that a filter should be stable can easily be expressed in terms of specifications for each separate coefficient (this is not at all the case for the filters in a direct-form-1 or direct-form-2 realization, for example).
3. The frequency response is not very sensitive to variations in the precise values of the coefficients ('low parameter sensitivity'); other filter structures are far more sensitive.
4. Properties (2) and (3) make these structures attractive for realizing 'adaptive filters', i.e. filters whose coefficients are automatically adjusted to satisfy a certain criterion (see also section 7.6).

7.5 TRANSPOSITION

A general method of deriving from any given arbitrary discrete filter another filter with exactly the same frequency response is based on the 'transposition theorem'.[31] This

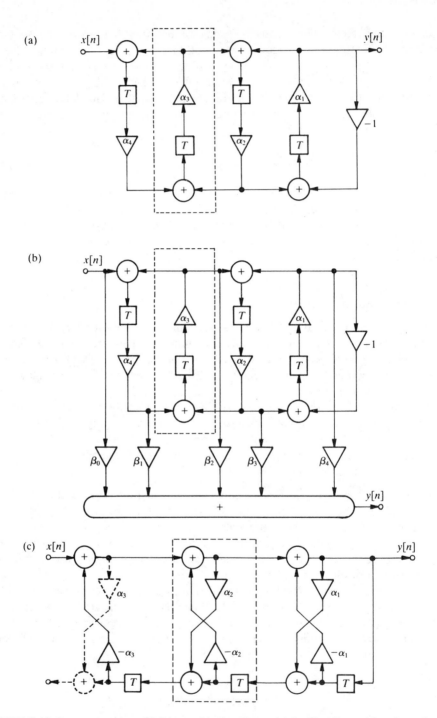

FIGURE 7.19 Some examples of ladder and lattice filters: (a) Ladder filter with poles only. (b) Ladder filter with poles and zeros. (c) Lattice filter with poles only. (d) Lattice filter with poles and zeros. (e) Lattice filter with zeros only.

(d)

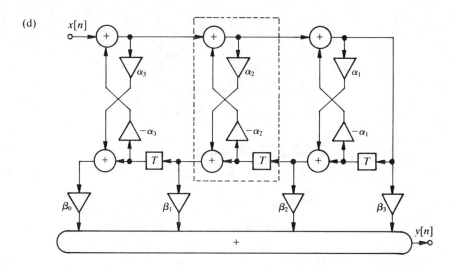

(e)

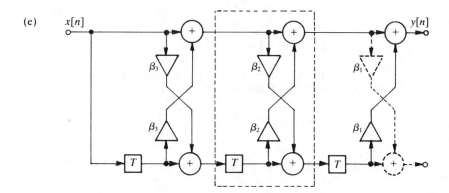

FIGURE 7.19 *contd*

theorem states that the frequency response of an LTD system is unchanged on transposition, i.e. if:

(a) the signal flows reverse direction (implying that the input is made the output and vice versa), *and*

(b) adders are replaced by nodes, and nodes by adders.

This is illustrated in Fig. 7.20, where we derive the transposed version of a transversal filter. (Since input and output are interchanged the positions of $x[n]$ and $y[n]$ are also interchanged.) As a second example of the transposition theorem, Fig. 7.21 shows the transposed form of the lattice filter of Fig. 7.19(c).

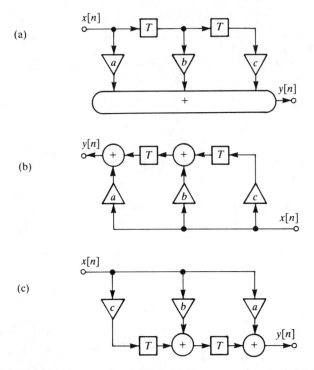

FIGURE 7.20 (a) Simple transversal filter. (b) Transposed version of this transversal filter. (c) Transposed form with the input shown on the left and the output on the right.

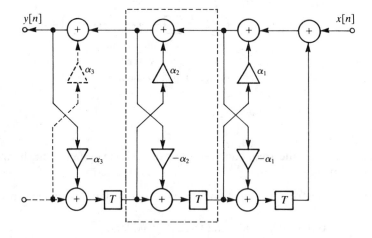

FIGURE 7.21 The transposed version of the lattice filter of Fig. 7.19(c).

7.6 ADAPTIVE FILTERS

So far we have always been concerned with 'fixed' filters, i.e. filters whose coefficients are calculated once and for all in the design and realized as fixed parameters in the filter. There are however a large number of cases in which we cannot say beforehand what the best values of the coefficients are, or in which the best values of the coefficients vary as a function of time. We can then make use of *adaptive filters*, which determine the values of the filter coefficients themselves and automatically modify them if the circumstances require it.

An adaptive filter consists of two distinct parts: the filter proper, which in principle can have any of the structures described earlier and is characterized at any discrete instant n by a set of filter coefficients $c_0[n], c_1[n], \ldots, c_N[n]$ and a control unit. The values of the coefficients are automatically calculated in the control unit in accordance with a prearranged control criterion (usually based on minimizing the difference $\epsilon[n]$ between the actual output signal $y[n]$ from the filter and a reference signal $g[n]$). This principle is illustrated in Fig. 7.22.

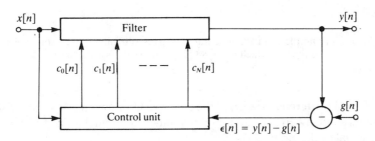

FIGURE 7.22 Principle of an adaptive filter.

An adaptive filter of this type can be used in television, for example, to derive an improved version $y[n]$ from a received signal $x[n]$ that is degraded by echoes ('ghosts') due to reflections. Here we require a reference signal $g[n]$ that is unaffected by the echoes (the signal during the flyback times could be used as the reference signal, since it is known exactly what this should look like when there are no echoes). Now we do not have to know exactly what the reflections are in advance, and the adaptive filter itself determines the best settings for the coefficients.

A comparable application (and also the oldest) of adaptive digital filters is in the automatic equalization of cables in data transmission, where the cable characteristics again are not precisely known beforehand and can also vary slowly as a result of temperature fluctuations.

In another category of applications of adaptive filters (noise cancellation) we are not in the first place interested in the signal $y[n]$, but in the signal $\epsilon[n]$. Examples of this are:

1. The improvement of the characteristics of a microphone with output signal $g[n]$ in a noisy environment. The signal $g[n]$ contains a considerable amount of unwanted information as well as the desired information. A separate monitoring microphone

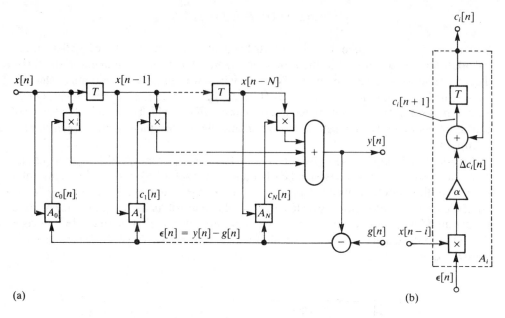

FIGURE 7.23 (a) Example of an adaptive transversal discrete filter. (b) Example of an adaptation
unit A_i in which the coefficient value $c_i[n]$ is calculated.

picks up the interference $x[n]$, and the adaptive filter now readjusts itself so that $\epsilon[n]$
represents a much better ratio of useful to unwanted signal than $g[n]$; $\epsilon[n]$ is now the
output signal of the circuit.

2. A similar application is found in the medical field, where we want to measure the
   heartbeat of an unborn baby although the mother's heartbeat acts as a strong
   interfering signal. By measuring the mother's heartbeat at a position where the baby's
   heartbeat is virtually undetectable, we obtain $x[n]$. The output signal $\epsilon[n]$ now
   represents the baby's heartbeat with greatly reduced interference from the mother.

3. In the same way, in telephony and data transmission, for example, two-way transmission can be obtained on a two-wire circuit: echo cancellation.[19]

In yet another and quite different category of applications, neither the signal $y[n]$ nor the
signal $\epsilon[n]$ are the significant quantities; what we are interested in here are the actual
values of the filter coefficients to which the adaptive filter sets itself. This is called
'modeling'. Here we start from a signal $x[n]$ and a signal $g[n]$ that has been produced from
$x[n]$ by filtering it in an unknown way. Now we would like to know what that unknown
filter is like. After the adaptive filter has set itself the coefficients $c_0[n]$ to $c_N[n]$ give an
approximation to the unknown filter. An example of this application is found in speech
processing (linear predictive coding, or LPC).

Of all the possible filter structures, the ones most widely encountered in adaptive filters
are the (non-recursive) transversal structure and the (recursive) ladder and lattice
structures. While the application of other recursive structures has been investigated, the
associated problems of possible instabilities and the absence of any certainty that the

filter will automatically set itself to the optimum setting make their application less attractive.

To give an impression of the complexity of an adaptive filter, the block diagram of an adaptive filter of the transversal type[19] is shown in Fig. 7.23; each coefficient $c_i[n]$ from Fig. 7.23(a) is calculated in an adaptation unit A_i (Fig. 7.23(b)).

In Fig. 7.23 we can see a number of details that were not entirely clear in the general circuit of Fig. 7.22:

1. The coefficients $c_i[n]$ are calculated by an iterative method, i.e. the coefficient value $c_i[n]$ is continually changed by a small amount $\Delta c_i[n]$, to give a new coefficient value $c_i[n + 1]$:

$$c_i[n + 1] = c_i[n] + \Delta c_i[n] \tag{7.24}$$

Each coefficient is gradually adjusted in this way to its optimum value. When this has been reached, each coefficient then fluctuates about that value.
2. In the calculation of all the coefficients the same error signal $\epsilon[n]$ is used, but different delayed versions of the input signal $x[n]$.
3. In the adaptation unit A_i there is a multiplication by a constant α. This determines the dynamic behavior of the adaptive filter, i.e. the rate of control of the adaptive filter and the magnitude of the fluctuations in the coefficients in the steady state.
4. The precise operations performed in the adaptation unit determine the control algorithm of the adaptive filter. For the adaptation unit of Fig. 7.23 we have:

$$\Delta c_i[n] = \alpha \epsilon[n] x[n - i] \tag{7.25}$$

The combination of (7.24) and (7.25) represents the widely used 'stochastic-iteration algorithm' or Widrow's Least Mean Squares (LMS) algorithm.[32] This is just one of the much larger class of gradient algorithms.

7.7 EXERCISES

Exercises to section 7.2

7.1 A transversal filter has eight filter coefficients b_0 to b_7, none of them equal to zero, with $b_0 = 1$, $b_1 = 2$, $b_2 = 3$ and $b_3 = 4$.
   (a) Determine b_4 to b_7 such that this filter has a linear phase characteristic (two solutions).
   (b) Give the resulting amplitude characteristics.
7.2 A transversal filter has seven filter coefficients b_0 to b_6, with $b_0 = 1$, $b_1 = 2$, $b_2 = 3$ and $b_3 = 4$.
   (a) Determine b_4 to b_6 such that this filter has a linear phase characteristic. Are there one or two solutions?
   (b) Give the resulting amplitude characteristic(s).
7.3 Draw a canonic structure for the circuit of Fig. 7.2(b).
7.4 Give a transversal filter that has the same $h[n]$ as the circuit of Fig. 7.2(a).

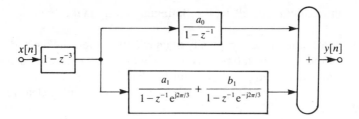

FIGURE 7.24 Exercise 7.5.

Exercises to sections 7.3 and 7.4

7.5 A frequency-sampling filter as shown in Fig. 7.24 has $N = 3$.
  (a) Determine a_0, a_1 and b_1 such that this filter has a real impulse response $h[n]$ and such that for the FTD of $h[n]$ we have $H(1) = 3$ and $H(e^{j2\pi/3}) = 6 + 3j\sqrt{3}$.
  (b) Give a detailed diagram of this circuit, with delay elements, multipliers and adders.
  (c) Give a general expression for $H(e^{j\theta})$.
  (d) Give a filter that has the same frequency response $H(e^{j\theta})$, but is realized as a transversal filter.

7.6 (a) Show the complete block diagram of the frequency-sampling filter of Fig. 7.18(a) with adders, multipliers and delay elements.
  (b) This frequency-sampling filter has an amplitude characteristic $A(e^{j\theta})$ that has a value of 3 at $\theta = \pm 2\pi/5$ (see Fig. 7.18(c)). The amplitude characteristic of the comb-filter part of this filter, however, is exactly zero at this frequency. So how can there nevertheless be an output signal at $\theta = \pm 2\pi/5$? (Hint: find out how the comb filter reacts to the input signal $x[n] = \sin(2\pi n/5)u[n]$.)

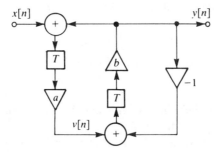

FIGURE 7.25 Exercise 7.7.

7.7 Determine the frequency response $H(e^{j\theta})$ of the ladder filter of Fig. 7.25.
7.8 Determine the frequency response $H(e^{j\theta})$ of the lattice filter of Fig. 7.26.

Exercise to section 7.5

7.9 Use the transposition theorem to determine an alternative circuit for the second-order direct-form-2 filter of Fig. 7.27.

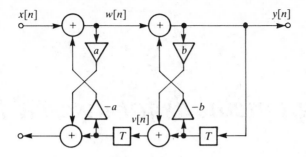

FIGURE 7.26 Exercise 7.8.

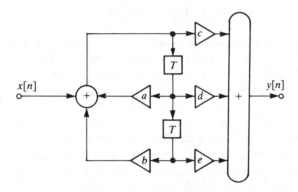

FIGURE 7.27 Exercise 7.9.

8
Design methods for discrete filters

8.1 INTRODUCTION

The design of a discrete filter usually starts with a specification of the frequency behavior required. As a rule the specification states the limits for the required amplitude and phase characteristics. The shape of the phase characteristic is sometimes left completely unspecified. In other cases the phase characteristic may be required to be linear, for example.

The specifications usually take the form of a *tolerance diagram*, as shown in Fig. 8.1 for the amplitude characteristic of a low-pass filter. The desired characteristic must not pass through the hatched areas. Since we are almost invariably concerned with real impulse responses, the specifications need only be indicated in the interval $0 \le \theta < \pi$; see (4.34).

We see three frequency ranges, which are called the *pass band*, the *stop band* and the *transition band*. In this example the maximum error for the amplitude in the pass band $(0 \le \theta < \theta_1)$ must be no more than δ_1; in the stop band $(\theta_h \le \theta < \pi)$ it must be no more than δ_2. In the transition band $(\theta_1 \le \theta < \theta_h)$ there is a gradual unspecified transition in the amplitude characteristic. The curve A in Fig. 8.1 represents an amplitude characteristic that exactly meets the specification.

If we start from given filter specifications we find that the design process contains the following stages:

1. We decide whether we want to approximate the desired frequency characteristic with an FIR filter or an IIR filter.
2. We choose the order of the filter and we try to calculate the coefficients of the system function of the filter in the best possible way (this is the main subject of this chapter).
3. We choose a filter structure and take into account any quantization effects for the input signal, output signal and filter coefficients (this is of particular importance in digital filters).
4. We check whether the resulting filter meets the original specification; if not, we repeat the design procedure, adopting a different type of filter or filter structure or a different order or a different form of quantization, or combinations of these alternatives.

In the design of discrete filters certain steps in the design process are usually repeated several times. The process may therefore be described as *iterative*.

Just as with continuous filters, we can classify discrete filters by the type of frequency

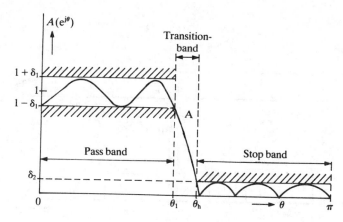

FIGURE 8.1 Example of a tolerance diagram for the amplitude characteristic $A(e^{j\theta})$ of a low-pass filter.

characteristic. The filters in which the amplitude characteristic is the most important feature can be subdivided into:

- Low-pass filters;
- High-pass filters;
- Band-pass filters;
- Band-stop filters.

Examples of ideal filters are shown in Fig. 8.2. (To emphasize the periodic nature of the frequency characteristics we show here a frequency interval larger than the fundamental interval $-\pi \leq \theta < \pi$.)

If we want to design a filter of one of the last three types, we can follow the design process described previously, starting from the original filter specifications. It is however also possible to obtain the desired result from a known *low-pass* filter by using transformation equations (see for example in Table 5.1 refs [9] and [60]).

The subdivision into filter types given above is by no means complete. Examples of filters that cannot be classified under these headings, yet sometimes play an important part in discrete signal processing, are:

- the differentiator: $H_D(e^{j\theta}) = j\theta$ for $|\theta| \leq \pi$;
- the integrator: $H_I(e^{j\theta}) = 1/j\theta$ for $|\theta| \leq \pi$;

- the Hilbert transformer: $H_H(e^{j\theta}) = \begin{cases} j \text{ for } 0 \leq \theta < \pi \\ -j \text{ for } -\pi \leq \theta < 0; \end{cases}$

- the phase-shifter (all-pass filter): $|H_F(e^{j\theta})| = 1$ for $|\theta| \leq \pi$.

The ideal frequency characteristics for the first three of these filters are shown in Fig. 8.3. The phase-shifter is not one particular filter, but is the collective name for all the filters with a constant amplitude characteristic. It is often used in combination with one of the filter types from Fig. 8.2 to obtain a specified phase characteristic in addition to the

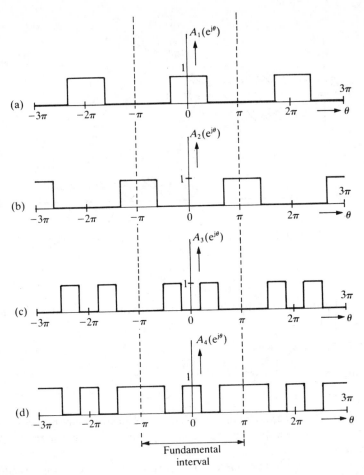

FIGURE 8.2 Examples of ideal discrete filters of (a) a low-pass nature; (b) a high-pass nature; (c) a band-pass nature; (d) a band-stop nature.

desired amplitude characteristic. In the following sections we shall describe a number of design methods for discrete filters in more detail; first for FIR filters and then for IIR filters. Finally, we shall list as clearly as possible all the differences we have encountered between FIR and IIR filters, since the choice between FIR and IIR filters is an important first step in the design process.

8.2 DESIGN METHODS FOR FIR FILTERS

The most essential feature of FIR filters is – by definition – the finite length of the impulse response. As soon as this is known, moreover, at least one form of realization (namely as a transversal filter) is also directly available. For these reasons the impulse response plays a very central part in various design methods for FIR filters, as we shall see in the

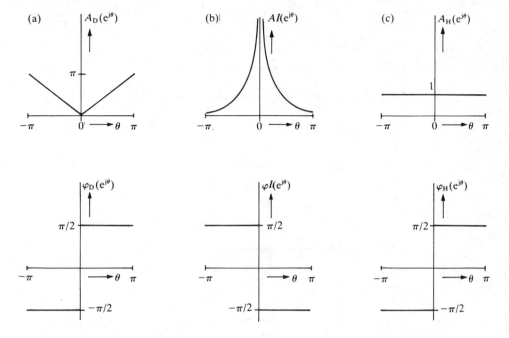

FIGURE 8.3 Ideal frequency characteristics of three important types of filter. (a) Differentiator, (b) Integrator, (c) Hilbert transformer.

following sections. Another important point is that it can be seen directly from the impulse response of an FIR filter whether we have a linear phase characteristic or not. As we know from section 7.2 there are four different cases here. In this chapter we shall look at this more closely in a separate section.

8.2.1 Design based on the FTD and windowing

The starting point in this design method is the frequency response $H_d(e^{j\theta})$ that we want to realize as closely as possible with an FIR filter. With the aid of the IFTD we find the ideal impulse response $h_d[n]$ directly from this. In general, however, we cannot immediately make a practical filter on the basis of $h_d[n]$, because:

(a) $h_d[n]$ is of very large or even infinite duration ('length'), and
(b) $h_d[n]$ is non-causal (i.e. $h_d[n] \neq 0$ for $n < 0$).

We shall therefore have to:

(a) limit the length of $h_d[n]$ to an acceptable number of L samples, and
(b) introduce sufficient shift ('delay') to obtain a causal impulse response.

In Fig. 8.4 we show how we can take these steps to arrive at a realizable impulse response $h[n]$, which is an approximation to $h_d[n]$. The impulse response $h[n]$ gives the coefficients directly for a realization as a transversal filter, for example.

So just how good is our approximation of the desired filter? Truncation of the infinitely

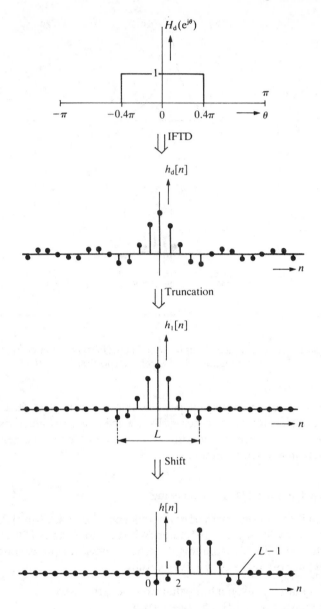

FIGURE 8.4 Transformation in stages from a filter specification in the frequency domain to a realizable impulse response by the method based on the FTD.

long impulse response $h_d[n]$ leads to errors in the frequency response of the discrete filter and the introduction of the shift gives extra linear phase shift, which we shall not consider further here. To give an impression of the amplitude errors, the amplitude characteristic that we obtain if we choose $L = 11, 21$ or 31 is shown in Fig. 8.5. We see that there is an oscillatory variation in both the pass band and the stop band and that the greatest

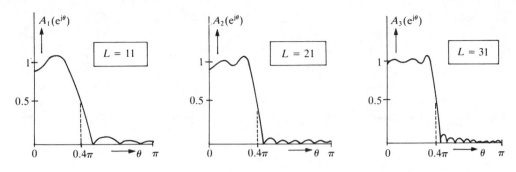

FIGURE 8.5 Approximation of an ideal low-pass filter with impulse responses of length $L = 11$, $L = 21$ and $L = 31$.

departure from the desired characteristic occurs in the vicinity of the steep transitions of the desired function $H_d(e^{j\theta})$. These can be explained in the following way. In fact the truncation of the infinitely long impulse response $h_d[n]$ corresponds to a multiplication in the time domain by a rectangular 'window' function $w[n]$ of length L:

$$h[n] = w[n]h_d[n] \qquad (8.1a)$$

In the frequency domain this corresponds to the convolution of $H_d(e^{j\theta})$ and $W(e^{j\theta})$, where $W(e^{j\theta})$ is the FTD of $w[n]$:

$$H(e^{j\theta}) = W(e^{j\theta}) * H_d(e^{j\theta}) \qquad (8.1b)$$

The function $W(e^{j\theta})$ has already been given in Fig. 4.5(d) and the processing with the window function is shown graphically in Fig. 8.6 for both the time domain and the frequency domain.

However large we make L, there will always be oscillations in the function $H(e^{j\theta})$ in the vicinity of steep transitions in $H_d(e^{j\theta})$. The striking feature here is that although the *number* of oscillations increases with L, the *maximum value* of the oscillations remains practically constant. However large we make L, the oscillations always remain. This is known as the *Gibbs phenomenon*.

If we want to reduce the oscillations in $H(e^{j\theta})$ we must choose another window function $w[n]$ to derive the finite-duration impulse response $h[n]$ from $h_d[n]$. We have to choose $w[n]$ such that $W(e^{j\theta})$:

(a) has as narrow a 'main lobe' as possible, and
(b) has 'side lobes' that contain as little energy as possible.

We do not obtain the reduction in the oscillations for nothing, however; it is always accompanied by a broadening of the transition band (this can be affected, however, by increasing the number of samples L).

Extended studies on window functions can be found in the literature.[55] A number of widely used functions of length L are specified by the following functions:

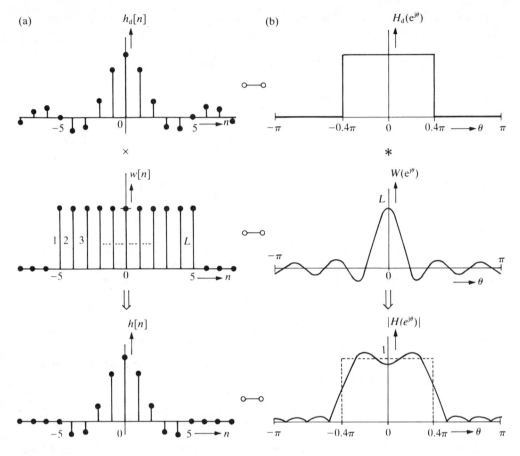

FIGURE 8.6 The effect of the window function $w[n]$ of length $L = 11$: (a) multiplication in the time domain, and (b) convolution in the frequency domain.

Rectangular window:

$$w[n] = \begin{cases} 1 & \text{for } 0 \le n \le L - 1 \\ 0 & \text{elsewhere} \end{cases} \tag{8.2a}$$

Bartlett window:

$$w[n] = \begin{cases} \dfrac{2n}{L-1} & \text{for } \quad 0 \le n \le \dfrac{L-1}{2} \\[2ex] 2 - \dfrac{2n}{L-1} & \text{for } \quad \dfrac{L-1}{2} < n \le L - 1 \\[2ex] 0 & \text{elsewhere} \end{cases} \tag{8.2b}$$

Hanning window:

$$w[n] = \begin{cases} \dfrac{1}{2}\left\{1-\cos\left(\dfrac{2\pi n}{L-1}\right)\right\}, & \text{for } 0 \le n \le L-1 \\[2ex] 0 & \text{elsewhere} \end{cases} \tag{8.2c}$$

Hamming window:

$$w[n] = \begin{cases} 0.54 - 0.46\cos\left(\dfrac{2\pi n}{L-1}\right) & \text{for } 0 \le n \le L-1 \\[2ex] 0 & \text{elsewhere} \end{cases} \tag{8.2d}$$

Blackman window:

$$w[n] = \begin{cases} 0.42 - 0.5\cos\left(\dfrac{2\pi n}{L-1}\right) + 0.08\cos\left(\dfrac{4\pi n}{L-1}\right) & \text{for } 0 \le n \le L-1 \\[2ex] 0 & \text{elsewhere} \end{cases} \tag{8.2e}$$

These windows are shown in Fig. 8.7(a) (for clarity they are shown as continuous functions $w(n)$ of n).

The *Kaiser windows* form a rather special family. They are defined as follows:

$$w[n] = \begin{cases} \dfrac{I_0\left(2\beta\sqrt{\dfrac{n}{L-1}-\left(\dfrac{n}{L-1}\right)^2}\right)}{I_0(\beta)} & \text{for } 0 \le n \le L-1 \\[2ex] 0 & \text{elsewhere} \end{cases} \tag{8.2f}$$

where $I_0(x)$ is a Bessel function, which is defined as follows:

$$I_0(x) = 1 + \sum_{k=1}^{\infty} \left[\frac{(x/2)^k}{k!}\right]^2 \tag{8.2g}$$

We can determine the exact shape of the Kaiser window by our choice of β: as β is made larger the width of the transition band increases, but the ripple becomes smaller. The Kaiser window is shown for various values of β in Fig. 8.7(b) (again, for clarity, as a continuous function $w(n)$).

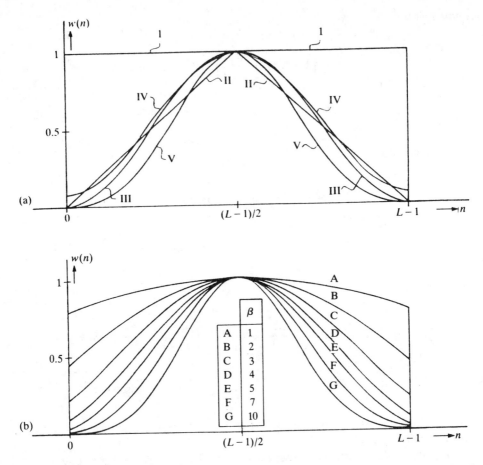

FIGURE 8.7 (a) Five widely used window functions: I. Rectangular, II. Bartlett, III. Hanning, IV. Hamming, V. Blackman. (b) The Kaiser windows for various values of β.

For the calculation of the Bessel function $I_0(x)$ it is not necessary to use an infinite number of terms from the series expansion in (8.2g). A computer program (in ALGOL 60) that calculates $I_0(x)$ to sufficient accuracy in a simple way is included in ref. [9].

To give an impression of the results that can be obtained by the application of a window function, a number of amplitude characteristics $A(e^{j\theta})$ that have been obtained as an approximation to $H_d(e^{j\theta})$ from Fig. 8.4 with $L = 31$ are shown (on a dB scale) in Fig. 8.8. Figure 8.8(a) gives the results for the use of the rectangular window, the Bartlett window, the Hanning window, the Hamming window and the Blackman window. Figure 8.8(b) gives the results for the use of the Kaiser window with $\beta = 1$, $\beta = 6$ and $\beta = 10$.

The window functions that have been described in this section can also be used in the application of the DFT to periodic signals (see section 5.4.1). Depending on the application we find slightly different definitions in the literature for one and the same window. Thus, for example, we find three different versions for the Hanning window:

- in [26]: $w_1[n] = [1 - \cos(2\pi n/L)]/2$ for $0 \le n \le L - 1$
- in [8]: $w_2[n] = [1 + \cos(2\pi n/L)]/2$ for $-(L-1)/2 \le n \le (L-1)/2$
- in [9]: $w_3[n] = [1 - \cos\{2\pi n/(L-1)\}]/2$ for $0 \le n \le L - 1$

These three windows are shown in Fig. 8.9 for $L = 13$. Although the windows have much in common (especially for high values of L), $w_1[n]$, $w_2[n]$ and $w_3[n]$ each give different numbers of samples differing from zero. These are $(L-1)$, L and $(L-2)$ respectively.

One window may sometimes offer advantages over another; it depends on the application. In general the window $w_1[n]$ is used in conjunction with the DFT (see section 5.4.1).

8.2.2 Design based on the DFT (frequency-sampling design)

In this method we start from the observation that the L values of the impulse response of an FIR filter we want to design by means of the L-point DFT can be unambiguously converted into L values of the frequency response $H_d(e^{j\theta})$. We can also turn the argument around, and, starting from L frequency samples of $H_d(e^{j\theta})$, try to determine L values of the impulse response by means of the L-point IDFT. This method is shown in Fig. 8.10 for an ideal low-pass filter $H_d(e^{j\theta})$ and $L = 33$. We calculate a 33-point IDFT starting from the $H_p[k]$ whose amplitude values are shown in Fig. 8.10(b). This gives the $h_p[n]$ of Fig. 8.10(c). For the desired impulse response $h[n]$ we then take:

$$h[n] = \begin{cases} h_p[n] & \text{for } 0 \le n \le 32 \\ 0 & \text{elsewhere} \end{cases} \tag{8.3}$$

How well can we approximate the desired $H_d(e^{j\theta})$ in this way? At the frequencies that coincide exactly with a sample of $H_p[k]$ the approximation is exact, but at intermediate frequencies there are errors that we have no direct control over. We can illustrate this by applying the FTD (*not* the DFT!) to the $h[n]$ from (8.3). We then obtain the frequency response $H(e^{j\theta})$ whose amplitude characteristic $A(e^{j\theta})$ and phase characteristic $\varphi(e^{j\theta})$ are shown in Fig. 8.10(d) and (e). At the 33 positions indicated by a 'dot', $A(e^{j\theta})$ corresponds exactly to $|H_p[k]|$ and $|H_d(e^{j\theta})|$, as would be expected. In calculating the IDFT, however, we have to specify not only the amplitude values of $H_p[k]$, but also the corresponding phase values. In this case we have made use of:

$$H_p[k] = \begin{cases} e^{-j16(2\pi k/33)} & \text{for } 0 \le k \le 8 \quad \text{and} \quad 25 \le k \le 32 \\ 0 & \text{for } 9 \le k \le 24 \end{cases} \tag{8.4a}$$

This gives a filter with a perfectly linear phase characteristic, as we can also see from the symmetry of the impulse response.

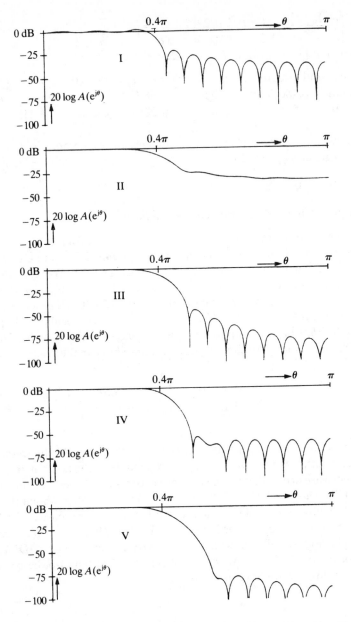

(a)

FIGURE 8.8 (a) Effects of five windows on the example of Fig. 8.4 when $L = 31$. I. Rectangular window, II. Bartlett window, III. Hanning window, IV. Hamming window, V. Blackman window. (b) Effects of the Kaiser window (with $\beta = 1, \beta = 6$ and $\beta = 10$) on the example of Fig. 8.4 when $L = 31$.

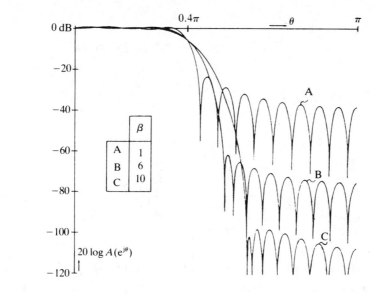

FIGURE 8.8 *contd*

We can see from Fig. 8.11 how important the choice of the correct phase values is. The figure shows the impulse response $h_p[n]$, the amplitude characteristic $A(e^{j\theta})$ and the phase characteristic $\varphi(e^{j\theta})$ that we obtain if in calculating the IDFT we do not start from $H_p[k]$ as given by (8.4a), but from

$$H_p[k] = \begin{cases} 1 & \text{for } 0 \le k \le 8 \quad \text{and} \quad 25 \le k \le 32 \\ 0 & \text{for } 9 \le k \le 24 \end{cases} \tag{8.4b}$$

(Both (8.4a) and (8.4b) comply with the $|H_p[k]|$ of Fig. 8.10(b); verify this!)

At 33 points $A(e^{j\theta})$ from Fig. 8.11 is again exactly equal to the desired $|H_d(e^{j\theta})|$; in between, however, the errors are considerable! With some experience, and possibly after trying out several alternatives, choosing the optimum phase values for an arbitrary value of L is no great problem.

The design method described above can be refined by trying to obtain an even better approximation to $|H_d(e^{j\theta})|$ by varying one or more of the values of $|H_p[k]|$. Here we usually take the values of $|H_p[k]|$ in or near the transition band, since the exact value there is generally of less interest to us. We are then dealing with a simple optimization procedure, in which we can use *linear programming*, for example.

In the example of Fig. 8.10b we have varied the values of the samples $|H_p[8]|$ and $|H_p[25]|$. The resulting amplitude characteristics $A(e^{j\theta})$ for $|H_p[8]| = |H_p[25]| = 1$ and $|H_p[8]| = |H_p[25]| = 0.3904$ are shown (on a dB scale) in Fig. 8.12.

8.2.3 Equiripple design

The design method described above resulted in discrete filters in which the departures from the ideal filter characteristics were mainly to be found near the transition band (or

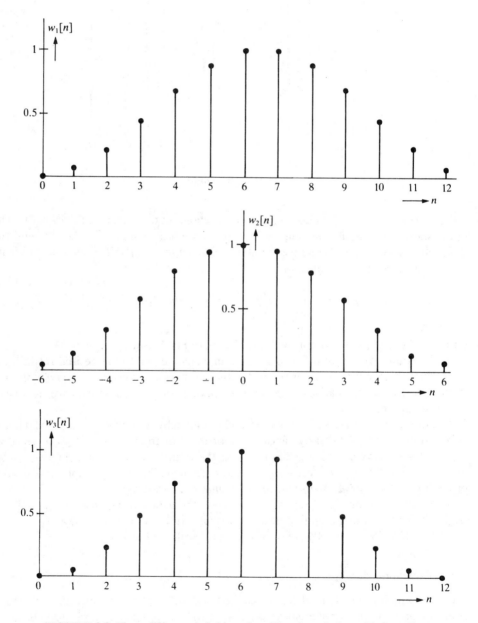

FIGURE 8.9 Three different definitions for a Hanning window with $L = 13$.

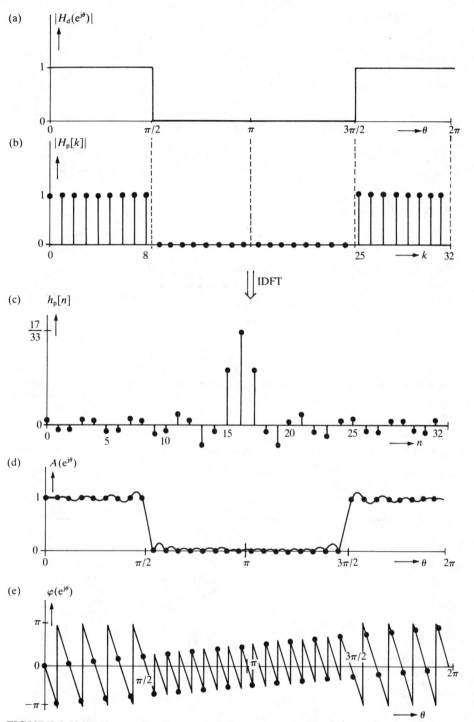

FIGURE 8.10 Design method based on the L-point DFT (with $L = 33$). (a) Desired amplitude characteristic $|H_d(e^{j\theta})|$; (b) 'sampled' amplitude characteristic $|H_p[k]|$; (c) $h_p[n]$ obtained by means of the IDFT; (d) final amplitude characteristic $A(e^{j\theta})$; (e) final phase characteristic $\varphi(e^{j\theta})$.

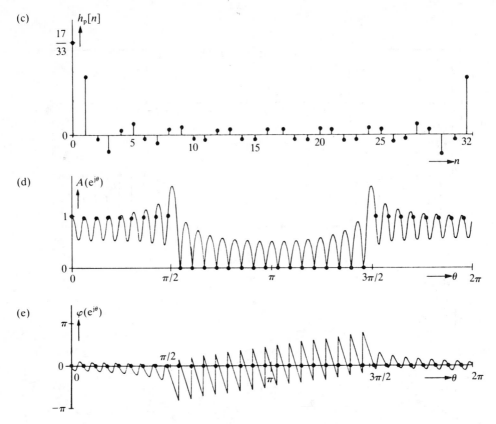

(c)

$h_p[n]$

$\dfrac{17}{33}$

0

0   5   10   15   20   25   32

$\longrightarrow n$

(d)

$|A(e^{j\theta})|$

1

0

0   $\pi/2$ π $3\pi/2$ $\longrightarrow \theta$ 2π

(e)

$\varphi(e^{j\theta})$

π

$\pi/2$

0

0   π $3\pi/2$ $\longrightarrow \theta$ 2π

$-\pi$

FIGURE 8.11 Alternatives to Fig. 8.10(c), (d) and (e) for another choice of the phase characteristic of $H_p[k]$, i.e. $\arg\{H_p[k]\} = 0$.

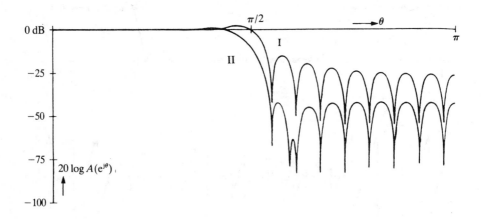

0 dB

$\pi/2$ $\longrightarrow \theta$

π

I

II

-25

-50

-75

$20 \log A(e^{j\theta})$

-100

FIGURE 8.12 Effect of varying $|H_p[8]|$ and $|H_p[25]|$ in Fig. 8.10(b) from the value 1 to the value 0.3904. I. original amplitude characteristic of Fig. 8.10(d) (on a dB scale); II. new amplitude characteristic.

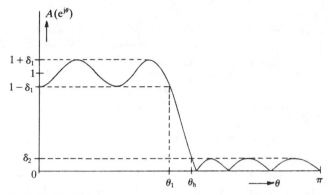

FIGURE 8.13 General example of a low-pass equiripple filter.

bands). Away from the transition band such errors were in general much smaller. It seems much more logical, however, to try to distribute these approximation errors as evenly as possible over *all* the frequencies. It is in fact found that this approximation does give the smallest maximum error, which then occurs not once but many times. We call such filters *equiripple filters*. The maximum error in the pass band (δ_1) can also be different from the maximum error in the stop band (δ_2). An example of a low-pass equiripple filter is shown in Fig. 8.13. The calculation of the filter coefficients for an equiripple filter is not easily performed without a computer; it usually requires an iterative optimization procedure.

Various programs have been developed for equiripple design of linear-phase filters, and these are described in the literature.[8,9] We shall return to linear-phase filters in section 8.2.4. It goes beyond the scope of this book to explain the operation of the programs in detail; they can be used for designing not only low-pass filters, but also circuits such as band-pass and band-stop filters (with one or more pass bands or stop bands) and differentiators.

However, we should like to say a little more about this here in general terms, but confining ourselves to the case of a filter with a low-pass characteristic (see Fig. 8.13). In such a filter we have five parameters: the number of coefficients L, the maximum error in the pass band δ_1, the maximum error in the stop band δ_2, the lowest frequency θ_1 and the highest frequency θ_h of the transition band. If we choose four of these parameters, the fifth one is then in principle fixed. It depends on the program we use which four should be specified and which one we obtain as the result. Parks and McLellan have written a widely used program, in which we have to specify L, θ_1, θ_h and the ratio of δ_1 to δ_2. The results thus obtained are the minimum values for δ_1 and δ_2 and the filter coefficients with which these can be realized. In the use of programs of this type it is useful to have some idea beforehand of the values of the various parameters. Various rules of thumb that can help here are given in the literature.[33,34] A rough estimate of the number of filter coefficients necessary to make a low-pass filter as in Fig. 8.13 can for example be obtained from the relation:[34]

$$L \approx -\frac{10 \log_{10}(\delta_1 \delta_2) + 15}{14 \Delta f} + 1 \tag{8.5}$$

where $\Delta f = (\theta_h - \theta_1)/2\pi$

Example

The Parks and McLellan program has been used to design a low-pass filter with the parameter values $L = 17$, $\theta_1 = 0.3\pi$, $\theta_h = 0.38\pi$ and $\delta_1/\delta_2 = 1$. The resulting optimum impulse response and amplitude characteristic are shown in Fig. 8.14; the value obtained for δ_1 and δ_2 is 0.13.

8.2.4 Linear-phase filters

Since by far the most important FIR filters are the ones with a linear phase characteristic, we should just like to look in a little more detail at a number of their properties. We shall look in turn at the four possible types of FIR filters with linear phase that we have already encountered in Fig. 7.5, and again we designate the samples of the impulse responses by a_i, b_i, c_i and d_i.

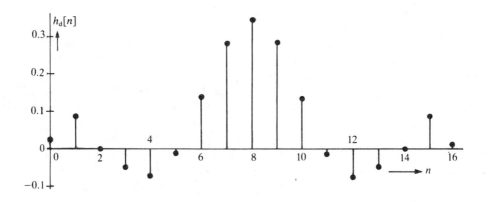

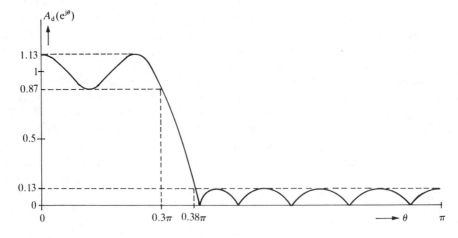

FIGURE 8.14 The impulse response and amplitude characteristic of the equiripple low-pass filter from the example of section 8.2.3.

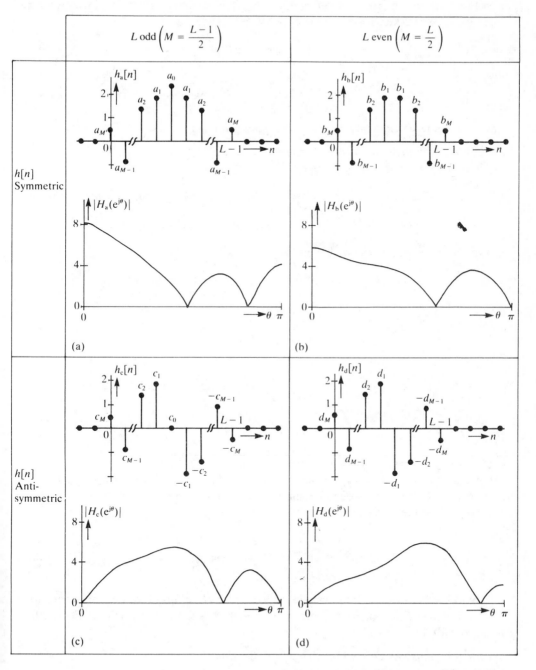

FIGURE 8.15 The four basic types of filter with linear phase and impulse response $h[n]$ of length L. We always have $H_b(e^{j\pi}) = 0$; $H_c(e^{j0}) = H_c(e^{j\pi}) = 0$ and $H_d(e^{j0}) = 0$.

In Fig. 8.15 the impulse responses $h_a[n]$, $h_b[n]$, $h_c[n]$ and $h_d[n]$ are again shown for arbitrary L. We make use here of the quantity M, for which:

$$M = (L-1)/2 \qquad \text{if } L \text{ is odd} \tag{8.6a}$$

and

$$M = L/2 \qquad \text{if } L \text{ is even} \tag{8.6b}$$

We shall now calculate the frequency responses for all four cases:

- *Case (a): Symmetric impulse response, L is odd (Fig. 8.15(a))*

The impulse response $h_a[n]$ is given by:

$$h_a[n] = a_M \delta[n] + a_{M-1} \delta[n-1] + \ldots + a_1 \delta[n-M+1] + a_0 \delta[n-M]$$
$$+ a_1 \delta[n-M-1] + \ldots + a_{M-1} \delta[n-L+2] + a_M \delta[n-L+1] \tag{8.7a}$$

For the frequency response we have:

$$H_a(e^{j\theta}) = a_M + a_{M-1} e^{-j\theta} + \ldots + a_1 e^{-j(M-1)\theta} + a_0 e^{-jM\theta}$$
$$+ a_1 e^{-j(M+1)\theta} + \ldots + a_{M-1} e^{-j(L-2)\theta} + a_M e^{-j(L-1)\theta}$$
$$= e^{-jM\theta} \left[a_0 + \sum_{i=1}^{M} a_i (e^{ji\theta} + e^{-ji\theta}) \right] \tag{8.7b}$$

or

$$H_a(e^{j\theta}) = e^{-j\theta(L-1)/2} \left[a_0 + 2 \sum_{i=1}^{(L-1)/2} a_i \cos(\theta i) \right] \tag{8.7c}$$

- *Case (b): Symmetric impulse response, L is even (Fig. 8.15(b))*

We are now dealing with an impulse response $h_b[n]$ that is symmetric with respect to a point exactly half-way between two samples. In the same way as in case (a) we now find for the total frequency response:

$$H_b(e^{j\theta}) = 2e^{-j\theta(L-1)/2} \sum_{i=1}^{L/2} b_i \cos\{\theta(i-\tfrac{1}{2})\} \tag{8.8}$$

A striking feature of this frequency response is that $H_b(e^{j\theta})$ is always equal to zero for $\theta = \pi$, independently of b_i; this means that high-pass filter characteristics cannot be realized with this type of filter.

- *Case (c): Antisymmetric impulse response, L is odd (Fig. 8.15(c))*

For the frequency response we now have:

$$H_c(e^{j\theta}) = 2j e^{-j\theta(L-1)/2} \sum_{i=1}^{(L-1)/2} c_i \sin(\theta i) \tag{8.9}$$

The notable feature of this frequency response is the constant factor j. This means that, except for the term $e^{-j\theta(L-1)/2}$ (which corresponds to a constant delay), we are dealing with a purely imaginary frequency response. Also, independently of the choice of c_i, $H_c(e^{j\theta}) = 0$ for $\theta = 0$ and $\theta = \pi$. We can therefore only realize filters of a band-pass nature.

● *Case (d): Antisymmetric impulse response, L is even (Fig. 8.15(d))*
In the now familiar way we find:

$$H_d(e^{j\theta}) = 2je^{-j\theta(L-1)/2} \sum_{i=1}^{L/2} d_i \sin\{\theta(i - \tfrac{1}{2})\} \tag{8.10}$$

In this case again we find a constant factor j in the frequency response. Moreover, independently of the choice of d_i, $H_d(e^{j\theta}) = 0$ for $\theta = 0$. In this case, therefore, we cannot make low-pass filters.

● *Comment 1.* There is a useful way of remembering the characteristic location of the zeros in the frequency responses of the four different types of linear-phase filters. From the definition (4.18) of the FTD we always have:

$$H(e^{j0}) = \sum_{n=-\infty}^{\infty} h[n] \quad \text{and} \quad H(e^{\pm j\pi}) = \sum_{n=-\infty}^{\infty} (-1)^n h[n] \tag{8.11}$$

Thus by summing all the samples of an impulse response we find $H(e^{j0})$; by first multiplying them alternately by +1 and −1 and then making the summation we find $H(e^{\pm j\pi})$. From this it follows that in cases (c) and (d) $H(e^{j0}) = 0$ and that in cases (b) and (c) $H(e^{\pm j\pi}) = 0$.

● *Comment 2.* The property $H(e^{j0}) = 0$ means that the corresponding system function $H(z)$ has a zero at $z = 1$, and similarly $H(e^{\pm j\pi}) = 0$ means that $H(z)$ has a zero at $z = -1$. Since linear-phase filters are always of the FIR type, their system functions have only zeros and no poles (except at the origin). From the symmetry properties of the impulse responses it can however be shown that these zeros cannot just occur at any arbitrary position. It can be proved fairly easily that in linear-phase filters – apart from any zeros at $z = \pm 1$ – these zeros will only occur:
   (a) in pairs on the real axis, reflected in the unit circle, e.g. at $z = a$ and $z = 1/a$;
   (b) as a complex conjugate pair on the unit circle, e.g. at $z = e^{j\theta_1}$ and $z = e^{-j\theta_1}$;
   (c) as a group of four at $z = z_1$, $z = 1/z_1$, $z = z_1^*$ and $z = 1/z_1^*$ (* = complex conjugate).
This is shown schematically in Fig. 8.16.

8.3 DESIGN METHODS FOR IIR FILTERS

The great distinction between IIR filters and FIR filters is that in IIR filters there are poles as well as zeros in the system function $H(z)$. In certain respects this gives a direct correspondence with conventional continuous filters, whose design also often starts from

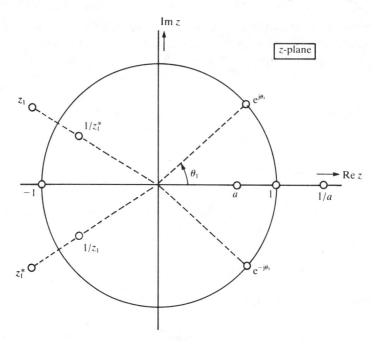

FIGURE 8.16 In the system function of a linear-phase filter zeros can only occur at $z = \pm 1$, in pairs $(a, 1/a)$ on the real axis, in pairs $(e^{j\theta_1}, e^{-j\theta_1})$ on the unit circle and as groups of four $(z_1, 1/z_1, z_1^*$ and $1/z_1^*)$.

a description in terms of poles and zeros. Consequently there are a number of techniques for designing IIR filters that are based on existing continuous filters (such as Butterworth, Bessel, Chebyshev and elliptic or Cauer filters[39]). In the following sections we shall consider the most widely used versions of these techniques.

In addition there is the possibility of designing IIR filters without starting from a known continuous filter, but basing the design entirely on an analysis in the z-plane. Here use is made of optimization procedures that are performed with the aid of a computer. We shall say something in general terms about these optimization methods, but a detailed description is beyond the scope of this book.

Because of the presence of poles in the system function of IIR filters we can encounter filters that are not stable since the poles occur outside the unit circle $|z| = 1$. In various design procedures there is no guarantee in advance that the resulting filter is stable. We therefore always have to check this afterwards.

With IIR filters it is not in general possible to give conditions for the filter coefficients in advance such that the phase characteristic satisfies certain conditions (as was possible for example in the design of linear-phase FIR filters). If we want to design an IIR filter that meets certain specifications for the amplitude and the phase, we set to work as follows:

1. First we design an IIR filter that has the desired amplitude characteristic.
2. Then we design a phase-shifter that gives the desired phase characteristic when combined with the first filter.
3. Cascading the two filters from (1) and (2) gives the desired total filter.

8.3.1 Impulse-response invariance

In this method we start from an analog filter of impulse response $h_a(t)$ and the corresponding frequency response $H_a(\omega)$ and system function $H_a(p)$; see section 2.10. The objective of our design is to realize an IIR filter with an impulse response $h_d[n]$, which satisfies:

$$h_d[n] = h_a(nT) \tag{8.12}$$

where $1/T$ is the sampling frequency of the discrete system. Now what is the relation between the frequency responses $H_d(e^{j\theta})$ of the discrete filter and $H_a(\omega)$ of the analog filter? With the aid of Chapter 3 (in particular Fig. 3.6) we can easily show that:

$$H_d(e^{j\theta}) = \frac{1}{T} \sum_{k=-\infty}^{\infty} H_a(\omega - 2\pi k/T) \text{ with } \theta = \omega T \tag{8.13}$$

This is illustrated in Fig. 8.17. We see that with this method there are problems to a greater or lesser extent, depending on the choice of T, due to aliasing in the frequency response. This means that apparent similarity in the time domain may correspond to a poor approximation in the frequency domain.

So how do we find the filter coefficients of the IIR filter in this design method? Let us

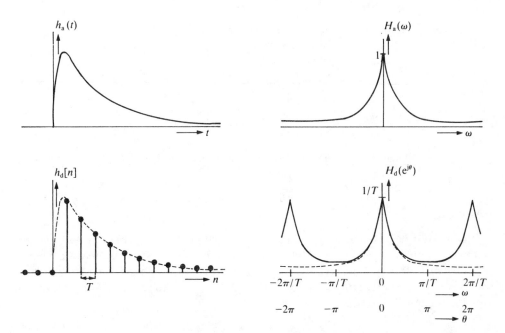

FIGURE 8.17 Schematic presentation of the relationship between the impulse responses $h_a(t)$ and $h_d[n] = h_a(nT)$ and the corresponding frequency responses $H_a(\omega)$ and $H_d(e^{j\theta})$.

take a simple example. We consider an analog filter whose frequency response $H_a(\omega)$ and system function $H_a(p)$ are given by:

$$H_a(\omega) = \frac{A}{j\omega - B} \tag{8.14}$$

and hence:

$$H_a(p) = \frac{A}{p - B} \tag{8.15}$$

From (8.14) and using (2.5) and (2.6) we can find the impulse response $h_a(t)$:

$$h_a(t) = \begin{cases} A e^{Bt} & \text{for } t \geq 0 \\ 0 & \text{for } t < 0 \end{cases} \tag{8.16}$$

We thus find for the discrete impulse response $h_d[n]$:

$$h_d[n] = h_a(nT) = A e^{BnT} u[n] \tag{8.17}$$

and for its z-transform:

$$H_d(z) = \sum_{n=-\infty}^{\infty} h_d[n]z^{-n} = A \sum_{n=0}^{\infty} (e^{BT} z^{-1})^n$$

$$= \frac{A}{1 - e^{BT} z^{-1}} \tag{8.18}$$

This system function can be realized with the circuit of Fig. 8.18.

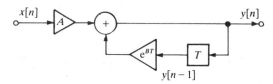

FIGURE 8.18 Discrete approximation of the system function

$$H_a(p) = \frac{A}{p - B} \text{ by } H_d(z) = \frac{A}{1 - e^{BT} z^{-1}}.$$

It is interesting just to compare the system functions $H_a(p)$ and $H_d(z)$. We see that $H_a(p)$ has a pole for $p = B$ and $H_d(z)$ has a pole for $z = e^{BT}$. This represents a general property of this design method: every (simple) pole $p = p_k$ of the analog filter that we started with is converted into a (simple) pole $z = z_k = e^{p_k T}$. In general, if we start from a system function $H_a(p)$ given by:

$$H_a(p) = \sum_{k=1}^{N} \frac{A_k}{p - B_k} \tag{8.19}$$

then with this design method we find a discrete filter with a system function $H_d(z)$ given by:

$$H_d(z) = \sum_{k=1}^{N} \frac{A_k}{1 - e^{B_k T} z^{-1}} \tag{8.20}$$

The way in which in the transition from (8.19) to (8.20) the poles of the analog filter are converted into poles of the discrete filter guarantees that a stable continuous filter is converted into a stable discrete filter. This can be explained as follows. Poles of a stable continuous filter lie in the left-hand half of the p-plane; i.e. $\text{Re}\{B_k\} \leq 0$. Therefore $|e^{B_k T}| \leq 1$, which corresponds to a pole inside or on the unit circle in the z-plane.

For the conversion of zeros of $H_a(p)$ to zeros of $H_d(z)$ in this design method no such simple relation exists, since this conversion depends not only on the actual zeros but also on the poles.

We shall illustrate what we have said above by two examples:

Example 1
We have a second-order system function $H_a(p)$ given by:

$$H_a(p) = \frac{2(p+A)}{(p+A)^2 + B^2} = \frac{1}{p+A+jB} + \frac{1}{p+A-jB} \tag{8.21}$$

Using the impulse-invariance method we find for $H_d(z)$:

$$H_d(z) = \frac{1}{1 - e^{-AT} e^{-jBT} z^{-1}} + \frac{1}{1 - e^{-AT} e^{jBT} z^{-1}}$$

$$= \frac{2 - e^{-AT} z^{-1}(e^{-jBT} + e^{jBT})}{(1 - e^{-AT} e^{-jBT} z^{-1})(1 - e^{-AT} e^{jBT} z^{-1})}$$

$$= \frac{2z[z - e^{-AT}\cos(BT)]}{(z - e^{-AT} e^{-jBT})(z - e^{-AT} e^{jBT})} \tag{8.22}$$

The system function $H_a(p)$ has poles at $p_{1,2} = -A \pm jB$ and a zero at $p_3 = -A$; the system function $H_d(z)$ has, as would be expected, poles at $z_{1,2} = e^{(-A \pm jB)T}$. There are two zeros: at $z_3 = 0$ and at $z_4 = e^{-AT}\cos(BT)$; the position of *zero* z_4 of $H_d(z)$ is thus partly determined (through the value of B) by the position of the *poles* of $H_a(p)$.

Example 2
We have a system function given by:

$$H_a(p) = \frac{2p+22}{(p+1)(p^2 + 4p + 13)} = \frac{2}{p+1} - \frac{2p+4}{p^2 + 4p + 13}$$

$$= \frac{2}{p+1} - \frac{1}{p+2+3j} - \frac{1}{p+2-3j} \tag{8.23}$$

Using the impulse-invariance method we find:

$$H_d(z) = \frac{2}{1 - e^{-T}z^{-1}} - \frac{1}{1 - e^{-(2+3j)T}z^{-1}} - \frac{1}{1 - e^{-(2-3j)T}z^{-1}}$$

$$= \frac{2z\{[e^{-T} - e^{-2T}\cos(3T)]z + e^{-4T} - e^{-3T}\cos(3T)\}}{(z - e^{-T})[z^2 - 2ze^{-2T}\cos(3T) + e^{-4T}]} \tag{8.24}$$

$H_a(p)$ has poles at $p_1 = -1$ and $p_{2,3} = -2 \pm 3j$ and one zero at $p_4 = -11$. $H_d(z)$ has poles at $z_1 = e^{-T}$ and $z_{2,3} = e^{-(2\pm 3j)T}$ and zeros at $z_4 = 0$ and $z_5 = [-e^{-4T} + e^{-3T}\cos(3T)]/[e^{-T} - e^{-2T}\cos(3T)]$. To permit a true comparison between $H_a(\omega)$ and $H_d(e^{j\theta})$ we must make a choice of T. In Fig. 8.19 two cases are shown: $H_{d1}(e^{j\theta})$ for $T = 1$ and $H_{d2}(e^{j\theta})$ for $T = \frac{1}{2}$. The functions $h_a(t)$ and $H_a(\omega)$ are shown in dashed lines. *Note*: Since $H_d(e^{j\theta})$ is inversely proportional to T, as can be seen from (8.13), $\frac{1}{2}H_{d2}$ is shown for $T = \frac{1}{2}$ instead of H_{d2}; this facilitates the comparison.

- *Comments*
1. For the application of this design method it is essential that a given continuous system function $H_a(p)$ is first written in *exactly* the form given in (8.19).
2. For the application of the simple transition from (8.19) to (8.20) $H_a(p)$ should have only *single* poles. This means that the quantities B_k can be either real or complex, but that no two or more quantities B_k may have the same value. It is certainly possible to extend this method to include the case of multiple poles, but the transition from (8.19) to (8.20) is then clearly more complicated;[29,35] see also Appendix II.4.
3. For completeness we should also add that another design method exists, known as the 'matched z-transform'. In this method not only is each pole at $p = p_k$ replaced by a pole at $z = z_k = e^{p_k T}$, but the zeros are also replaced in accordance with the same rule. Because of certain disadvantages this method is not often used.

8.3.2 Replacing differentials by differences

In this method we make use of the fact that a continuous system with a system function $H_a(p)$ given by:

$$H_a(p) = \frac{\displaystyle\sum_{i=0}^{N} b_i p^i}{1 - \displaystyle\sum_{i=1}^{M} a_i p^i} \tag{8.25}$$

can also be described by a differential equation containing the input signal $x_a(t)$ and the output signal $y_a(t)$ (see section 2.6):

$$y_a(t) = \sum_{i=0}^{N} b_i \frac{d^i x_a(t)}{dt^i} + \sum_{i=1}^{M} a_i \frac{d^i y_a(t)}{dt^i} \tag{8.26}$$

We shall now replace the derivatives by an approximation:

$$\left.\frac{dy_a(t)}{dt}\right|_{t=nT} \approx \frac{y_a(nT) - y_a(nT - T)}{T} = \frac{y_d[n] - y_d[n-1]}{T} \tag{8.27}$$

where $y_d[n]$ represents the output signal of a discrete system. This approximation of a differential by a difference is also used in numerical analysis, for example. Intuitively we expect that the accuracy of the approximation will improve as we make T smaller. This is indeed the case.

Now what is the relationship between the system function $H_d(z)$ of a discrete filter that we ultimately obtain in this way and the system function $H_a(p)$ that we started with? Here we have to remember two things:

1. if
$$y_a(t) \ \circ\!\!-\!\!-\!\!\circ \ Y_a(p)$$

then
$$\frac{dy_a(t)}{dt} \ \circ\!\!-\!\!-\!\!\circ \ pY_a(p) \qquad (8.28a)$$

2. if
$$y_d[n] \ \circ\!\!-\!\!-\!\!\circ \ Y_d(z)$$

then
$$\frac{y_d[n] - y_d[n-1]}{T} \ \circ\!\!-\!\!-\!\!\circ \ \frac{Y_d(z) - z^{-1}Y_d(z)}{T} = \left(\frac{1 - z^{-1}}{T}\right)Y_d(z) \qquad (8.28b)$$

With this method we therefore realize an $H_d(z)$ that we can obtain from $H_a(p)$ by replacing p by $(1 - z^{-1})/T$. Using (8.25) we then obtain:[†]

$$H_d(z) = \frac{\displaystyle\sum_{i=0}^{N} b_i \left(\frac{1 - z^{-1}}{T}\right)^i}{1 - \displaystyle\sum_{i=1}^{M} a_i \left(\frac{1 - z^{-1}}{T}\right)^i} \qquad (8.29)$$

[†]In (8.27) and (8.28) we have actually restricted our attention to the first-order derivative of $y_a(t)$. However, by repeatedly applying the approximation rule of (8.27) the discrete approximations for the higher-order derivatives of $y_a(t)$ can easily be found:

$$\frac{d^2 y_a(t)}{dt^2}\bigg|_{t=nT} = \frac{d}{dt}\left(\frac{dy_a(t)}{dt}\right)\bigg|_{t=nT} \approx \frac{\dfrac{y_d[n] - y_d[n-1]}{T} - \dfrac{y_d[n-1] - y_d[n-2]}{T}}{T}$$

so
$$\frac{d^2 y_a(t)}{dt^2}\bigg|_{t=nT} \approx \frac{y_d[n] - 2y_d[n-1] + y_d[n+2]}{T^2} = y_{d_2}[n]$$

and, similarly,
$$\frac{d^3 y_a(t)}{dt^3}\bigg|_{t=nT} \approx \frac{y_d[n] - 3y_d[n-1] + 3y_d[n-2] - y_d[n-3]}{T^3} = y_{d_3}[n].$$

The z-transforms $Y_{d2}(z)$ and $Y_{d3}(z)$ of $y_{d2}[n]$ and $y_{d3}[n]$ are given by

$$Y_{d_2}(z) = \frac{Y_d(z) - 2Y_d(z)z^{-1} + Y_d(z)z^{-2}}{T^2} = \left(\frac{1 - z^{-1}}{T}\right)^2 Y_d(z)$$

and

$$Y_{d_3}(z) = \left(\frac{1 - z^{-1}}{T}\right)^3 Y_d(z).$$

In general, for the ith-order derivative of $y_a(t)$ which has Laplace-transform $p^i Y_a(p)$ we find the discrete approximation $y_{di}[n]$ with z-transform $Y_{di}(z)$:

$$Y_{d_i}(z) = \left(\frac{1 - z^{-1}}{T}\right)^i Y_d(z).$$

From this we can conclude more firmly that p^i in $H_c(p)$ is replaced by $(1 - z^{-1})^i/T^i$ in $H_d(z)$.

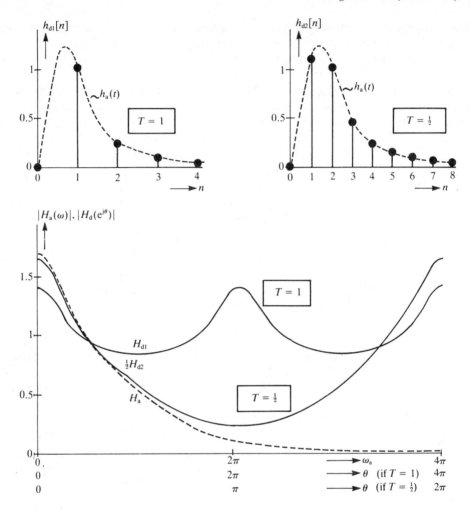

FIGURE 8.19 Results of two designs based on the invariance of the impulse response $h_a(t)$ at different values of T.

What happens to the poles and zeros because of this substitution? Instead of:

$$p = \frac{1 - z^{-1}}{T} \tag{8.30}$$

we can also write:

$$z = \frac{1}{1 - pT} \tag{8.31}$$

Each pole (or zero) at $p = p_k$ is thus replaced by a pole (or zero) at $z = 1/(1 - p_k T)$. This applies not just for the poles and zeros, of course, but also for every arbitrary value of p.

We say that (8.31) represents a mapping of the p-plane on to the z-plane. We can show this graphically as in Fig. 8.20. The imaginary axis of the p-plane is mapped into the small circle of radius 0.5 in the z-plane. The origin of the p-plane ($p = 0$) maps to the point $z = 1$; the points $p = j\infty$ and $p = -j\infty$ map to the origin $z = 0$ of the z-plane. The hatched region to the left of the vertical axis in the p-plane maps into the *inside* of the hatched circle in the z-plane; the region to the right of that axis maps into the *outside* of the hatched circle. This therefore means that the effect of this mapping is to replace a stable continuous system (with all its poles to the left of the vertical axis in the p-plane) by a stable discrete system (in which all the poles are inside the small circle and hence certainly inside the unit circle).

However, this mapping is not ideal in the sense that the discrete system has exactly the same frequency behavior in the fundamental interval as the analog system. This would require that the vertical axis in the p-plane maps exactly into the unit circle in the z-plane. It is only in the region in which the small circle comes close to the unit circle in the z-plane (i.e. for low relative frequencies: $|\theta| \ll \pi$) that there is good agreement between the frequency behavior of the continuous and discrete systems.

In the case just described we have approximated the continuous derivative $dy_a(t)/dt$ by the *backward difference*: $[y_a(nT) - y_a(nT - T)]/T$; (8.27). We could also have used the *forward difference*:

$$\left.\frac{dy_a(t)}{dt}\right|_{t=nT} \approx \frac{y_a(nT + T) - y_a(nT)}{T} \tag{8.32}$$

In the same way as above we would then have found:

$$p = \frac{z - 1}{T} \tag{8.33}$$

or

$$z = 1 + pT \tag{8.34}$$

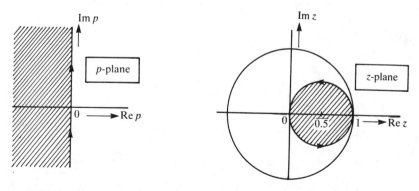

FIGURE 8.20 Mapping of the p-plane on the z-plane by $z = 1/(1 - pT)$.

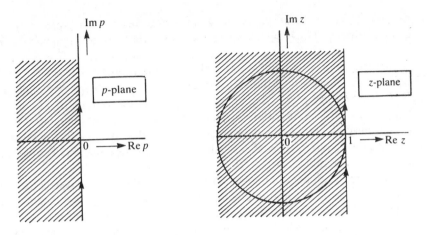

FIGURE 8.21 Mapping of the p-plane on the z-plane by $z = 1 + pT$.

This again represents a mapping of the p-plane on to the z-plane, but in which the vertical axis of the p-plane now maps into a vertical line in the z-plane through the point $z = 1$; Fig. 8.21. The origin $p = 0$ of the p-plane again maps to the point $z = 1$ and the entire region to the left of the vertical axis of the p-plane maps to the left of the vertical line in the z-plane; the region to the right of the vertical axis of the p-plane maps to the right of the vertical line in the z-plane. A stable continuous system can be converted into an unstable discrete system by this mapping. And if this does not happen, the frequency behavior of the discrete system will again only be the same as that of the continuous system at low relative frequencies.

Because of the imperfections associated with the two methods so far described, they are little used in designing digital systems. They are used, however, in the design of circuits such as switched-capacitor filters (see also section 8.3.4). This is because filters of this type more or less automatically contain subcircuits whose system function is of the type $(1 - z^{-1})/T$ or $(z - 1)/T$.[2]

To illustrate what we have said above we shall again start from the continuous system with the system function of (8.23) in the following two examples.

Example 1
We have the system function given by (8.23):

$$H_a(p) = \frac{2p + 22}{(p + 1)(p^2 + 4p + 13)} \tag{8.35}$$

With the choice $T = \frac{1}{2}$ the substitution required for the method of backward differences becomes:

$$p = \frac{1 - z^{-1}}{T} = 2 - 2z^{-1} \tag{8.36}$$

This gives:

$$H_{d3}(z) = \frac{4 - 4z^{-1} + 22}{(3 - 2z^{-1})(4 - 8z^{-1} + 4z^{-2} + 8 - 8z^{-1} + 13)}$$

$$= \frac{2z^2(13z - 2)}{(3z - 2)(5z - 1.6 + 1.2j)(5z - 1.6 - 1.2j)} \qquad (8.37)$$

$H_{d3}(z)$ therefore has zeros at $z_{1,2} = 0$ and $z_3 = 2/13$ and poles at $z_4 = 2/3$ and $z_{5,6} = 0.32 \pm 0.24j$. The corresponding amplitude characteristic $|H_{d3}(e^{j\theta})|$ is shown in Fig. 8.22.

Example 2
With the choice $T = \frac{1}{2}$ the substitution required for the method of forward differences becomes:

$$p = \frac{z - 1}{T} = 2z - 2 \qquad (8.38)$$

The system function of (8.35) then gives:

$$H_{d4}(z) = \frac{4z + 18}{(2z - 1)(4z^2 - 8z + 4 + 8z - 8 + 13)}$$

$$= \frac{2(2z + 9)}{(2z - 1)(2z + 3j)(2z - 3j)} \qquad (8.39)$$

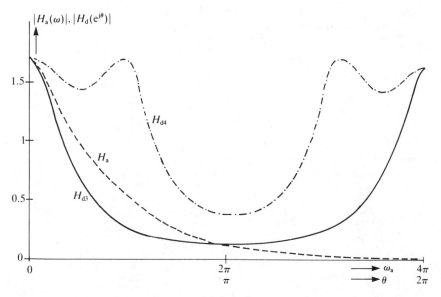

FIGURE 8.22 Results H_{d3} and H_{d4} of two designs (with $T = \frac{1}{2}$) based on backward differences and forward differences respectively.

$H_{d4}(z)$ therefore has a zero at $z_1 = -9/2$ and poles at $z_2 = \frac{1}{2}$ and $z_{3,4} = \pm 3j/2$. If we calculate $|H_{d4}(z)|$ on the unit circle (i.e. for $z = e^{j\theta}$), then we find the curve designated H_{d4} in Fig. 8.22. Now we shall be under a considerable misapprehension here if we imagine that the method of forward differences gives a usable system with $H_{d4}(e^{j\theta})$ as the frequency response, for since the poles $z_{3,4} = \pm 3j/2$ are *outside* the unit circle, $H_{d4}(z)$ is an *unstable* system! The unit circle in the z-plane is then outside the region of convergence of $H_{d4}(z)$ and expression (8.39) is only valid *in* the region of convergence. (The region of convergence of $H_{d4}(z)$ occupies the entire z-plane *outside* the circle $|z| = 3/2$.)

8.3.3 Bilinear transformation

The most widely used method of designing discrete filters, starting from continuous filters, is known as the *bilinear transformation*. In this transformation we again map the p-plane on to the z-plane. Here we use the substitution:

$$p = \frac{2}{T} \cdot \frac{1 - z^{-1}}{1 + z^{-1}} \tag{8.40}$$

and hence:

$$z = \frac{2 + pT}{2 - pT} \tag{8.41}$$

We can show this mapping graphically, as in Fig. 8.23.

The vertical axis of the p-plane is now mapped exactly into the unit circle of the z-plane, with the point $p = 0$ transforming to the point $z = 1$ and the points $p = +j\infty$ and $p = -j\infty$ transforming to $z = -1$. The left-hand half of the p-plane transforms to the region inside the unit circle; the right-hand half-plane to the region outside it. In this method we therefore obtain stable discrete systems if we start from stable continuous systems.

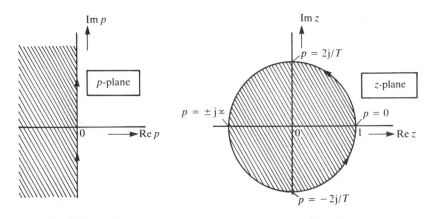

FIGURE 8.23 Mapping of the p-plane on to the z-plane by $z = (2 + pT)/(2 - pT)$.

Example

We have a system function $H_a(p)$ such that:

$$H_a(p) = \frac{A}{p+A} \tag{8.42}$$

Applying the bilinear transformation to this gives a system function $H_d(z)$:

$$H_d(z) = \frac{A}{\dfrac{2}{T} \cdot \dfrac{1-z^{-1}}{1+z^{-1}} + A} = \frac{AT(1+z^{-1})}{AT+2+(AT-2)z^{-1}} \tag{8.43}$$

The modulus of the frequency responses $H_a(\omega)$ and $H_d(e^{j\theta})$ for $A = 1000$ and $T = 1/1000$ is shown in Fig. 8.24.

In the example above we are dealing with two different angular frequencies ω; one from the continuous filter we started with, and one from the discrete filter we have calculated. So that we can clearly differentiate between the two frequencies we have called them ω_a (a for 'analog') and $\omega_d = \theta/T$ (d for 'discrete'). We shall use this same notation in the rest of our discussion at places where confusion is possible (with $f_a = \omega_a/2\pi$ and $f_d = \omega_d/2\pi$, of course).

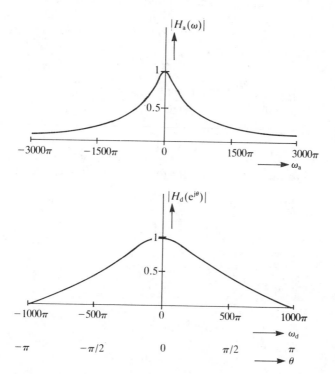

FIGURE 8.24 The modulus of $H_a(\omega) = A/(j\omega + A)$ and of $H_d(e^{j\theta})$ obtained from $H_a(\omega)$ by the bilinear transformation.

There is a very important property of the bilinear transformation that can be seen in the above example. The entire frequency range $(-\infty \leq \omega_a \leq \infty)$ of the continuous system maps into the fundamental interval $(-\pi \leq \theta \leq \pi)$ of the discrete system, where $\omega_a = 0$ corresponds to $\theta = 0$, $\omega_a = +\infty$ to $\theta = \pi$ and $\omega_a = -\infty$ to $\theta = -\pi$. We can easily demonstrate this from (8.40). An arbitrary continuous frequency ω_a corresponds to a discrete frequency θ through the relation (substitute $p = j\omega_a$ and $z = e^{j\theta}$ in (8.40)):

$$j\omega_a = \frac{2}{T} \cdot \frac{1 - e^{-j\theta}}{1 + e^{-j\theta}} = \frac{2}{T} \cdot \frac{e^{-j\theta/2}(e^{j\theta/2} - e^{-j\theta/2})}{e^{-j\theta/2}(e^{j\theta/2} + e^{-j\theta/2})} = \frac{2}{T} \cdot \frac{2j\sin(\theta/2)}{2\cos(\theta/2)}$$

so that
$$\omega_a = \frac{2}{T}\tan(\theta/2) \tag{8.44}$$

or
$$\theta = 2\arctan(\omega_a T/2) \tag{8.45a}$$

and
$$\omega_d = \frac{2}{T}\arctan(\omega_a T/2) \tag{8.45b}$$

We see that a nonlinear relation exists between ω_d and ω_a. This effect is called 'warping', and is shown in Fig. 8.25.

From (8.45b) we also see why there was a factor of $2/T$ in (8.40); this factor is included to make $\omega_d \approx \omega_a$ at low frequencies (since arctan $(x) \approx x$ if x is very small, of course).

The great advantage of warping is that no aliasing of the frequency characteristic can occur in the transformation of a continuous filter to a discrete filter, such as we encountered in the impulse-invariance method. We must however check carefully just how the various characteristic frequencies of the continuous filter convert to characteristic frequencies of the discrete filter. We can illustrate this with the aid of Fig. 8.26 for a band-pass filter. Using (8.45) we see immediately that:

$$\theta_i = 2\arctan(\omega_i T/2) \quad \text{for } i = 1, 2 \text{ and } 3 \tag{8.46}$$

In designing a discrete filter by this method we must first 'prewarp' the given filter specifications to find the continuous filter to which we are going to apply the bilinear

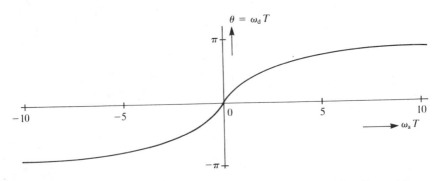

FIGURE 8.25 The relation between the 'continuous' frequency ω_a and the 'discrete' frequency ω_d in the bilinear transformation.

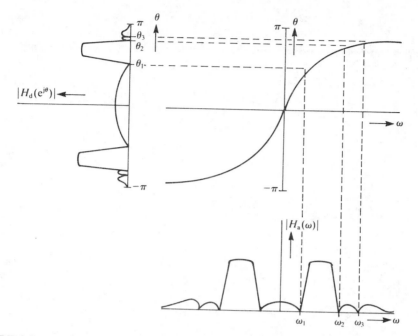

FIGURE 8.26 The effect of 'warping' in the conversion of a continuous filter of frequency response $H_a(\omega)$ to a discrete filter of frequency response $H_d(e^{j\theta})$.

transformation. This is illustrated in Fig. 8.27, where the specification of a desired discrete low-pass filter is shown with a sampling frequency of 8 kHz (so that $T = 125\,\mu s$), a pass band up to 2.6 kHz ($\theta_1 = 0.65\pi$) and a stop band above 3 kHz ($\theta_h = 0.75\pi$).

Using (8.44) we find that we must start from a continuous filter with:

$$\omega_1 = 2\pi f_1 = \frac{2}{T}\tan(\theta_1/2) = 2\pi \cdot 4155 \text{ rad/s}$$

$$\omega_h = 2\pi f_h = \frac{2}{T}\tan(\theta_h/2) = 2\pi \cdot 6148 \text{ rad/s} \tag{8.47}$$

We have to remember here that the warping of the frequency axis applies not only to the amplitude characteristic but also to the phase characteristic. Furthermore, we should also like to mention that the bilinear transformation does *not* give us a direct relation between the impulse responses of the discrete and continuous filters.

Finally, we should like to apply the bilinear transformation to the system function of (8.23) as an example.

Example

With the choice $T = \frac{1}{2}$ the substitution required for the bilinear transformation becomes:

$$p = \frac{2}{T}\cdot\frac{z-1}{z+1} = 4\frac{z-1}{z+1} \tag{8.48}$$

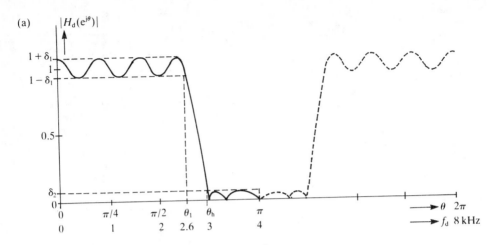

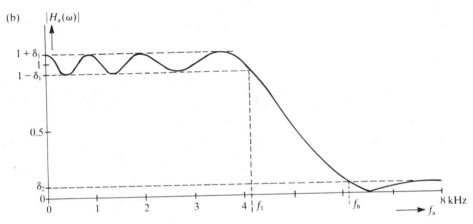

FIGURE 8.27 Taking account of warping of the frequency axis in designing a discrete filter (a) on the basis of a continuous filter (b) with the aid of the bilinear transformation. $\theta_l = 0.65\pi$, $\theta_h = 0.75\pi$, $f_l = \omega_l/2\pi = 4.155$ kHz and $f_h = \omega_h/2\pi = 6.148$ kHz.

The system function of (8.23) now becomes:

$$H_{d5}(z) = \cfrac{\dfrac{8z-8}{z+1} + \dfrac{22(z+1)}{z+1}}{\left(\dfrac{4z-4}{z+1} + \dfrac{z+1}{z+1}\right)\left[\dfrac{16z^2-32z+16}{(z+1)^2} + \dfrac{16(z-1)(z+1)}{(z+1)(z+1)} + \dfrac{13(z+1)^2}{(z+1)^2}\right]}$$

$$= \frac{(30z+14)(z+1)^2}{(5z-3)(45z^2-6z+13)} \tag{8.49}$$

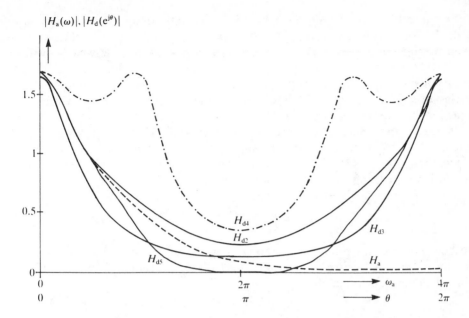

FIGURE 8.28 The results of four different design methods for discrete IIR filters (with $T = \frac{1}{2}$), starting from the same continuous filter H_a. H_{d2} = impulse invariance; H_{d3} = backward differences; H_{d4} = forward differences (*note*: unstable); H_{d5} = bilinear transformation.

$H_{d5}(z)$ therefore has zeros at $z_{1,2} = -1$ and $z_3 = -7/15$ and poles at $z_4 = 0.6$ and $z_{5,6} = 0.067 \pm 0.533\,j$. The corresponding amplitude characteristic $|H_{d5}(e^{j\theta})|$ is shown in Fig. 8.28. The results of the design methods discussed earlier are also shown here for comparison. We see from the curves that each design method gives a different approximation to the continuous filter H_a that we started with in each case.

8.3.4 Transformation of elements

A completely different approach to the design of discrete filters starts out from the structure (i.e. the exact composition in terms of R, L and C elements) of a continuous filter, instead of the impulse response, the differential equation or the system function, as discussed in the previous sections. In this method each element from the continuous circuit is separately converted into a discrete circuit, and in principle these are connected together in the same way as in the original structure. The main reason for using this method is that some of the desirable features of LC filters (in particular), such as zero loss and above all the low sensitivity to coefficient variations, are retained in the conversion of a continuous filter to a discrete filter.

One application of this method is in designing special digital ladder filters.[36] The method has also given rise to a completely separate class of digital filters, which we have not so far mentioned: *wave digital filters*.[37] These filters are characterized by the fact that the separate elements obtained after conversion of the continuous domain to the discrete domain are connected to one another via 'adaptors'.

Until now this method has been most widely used in the design of switched-capacitor filters. We should like to illustrate this with the example of Fig. 8.29.[2,38] Here we have taken a third-order low-pass filter as the starting point and show what a switched-capacitor filter designed by this method will look like. It can clearly be seen how each of the reactive elements (C_1, L and C_2) has been converted into a discrete element (a 'lossless discrete integrator' or LDI), consisting of an operational amplifier with one fixed capacitor and one switched capacitor.

There are various ways of converting a continuous filter element into a discrete filter element. For example, differentials can be replaced by differences (both forward and backward; section 8.3.2). The bilinear transformation can also be used; section 8.3.3. Another transformation used in switched-capacitor filters is:

$$p = \frac{1}{T}(z^{1/2} - z^{-1/2}) \tag{8.50}$$

We have not encountered this transformation previously, and it cannot be translated directly into a mapping of the p-plane on to the z-plane (in fact the conversion of Fig. 8.29 is based on this transformation).

8.3.5 Optimization methods

Sometimes we want to design a discrete filter for which no continuous filter is available as a starting point. We can then base the design on an optimization procedure performed by

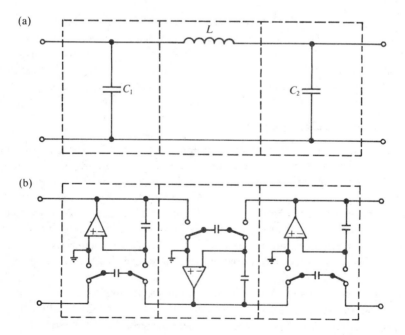

FIGURE 8.29 (a) Third-order continuous *LC* filter and (b) switched-capacitor filter obtained from it using the design method based on element transformation.

computer. The starting point here is the definition of a permitted error between the desired frequency response $H_d(e^{j\theta})$ and the approximated version of this $H(e^{j\theta})$, which is characterized by the system function $H(z)$:

$$H(z) = \frac{\displaystyle\sum_{i=0}^{N} b_i z^{-i}}{1 - \displaystyle\sum_{i=1}^{M} a_i z^{-i}} \tag{8.51}$$

We start from roughly estimated values of a_i and b_i and determine an error ϵ by comparing the desired and approximated functions at particular frequencies θ_i. (These frequencies can be chosen arbitrarily; they do not have to be evenly spaced, for example.) The error criterion usually takes the form:

$$\epsilon = \sum_{i=1}^{K} W(e^{j\theta_i})\{|H(e^{j\theta_i})| - |H_d(e^{j\theta_i})|\}^{2P} \tag{8.52}$$

where $W(e^{j\theta_i})$ is a frequency-dependent weighting function, which can be used to weight the error more heavily at certain frequencies than at others; P is a constant whose value can be chosen. Special computational methods (such as the Fletcher-Powell algorithm) are used on the computer to vary the constants a_i and b_i in steps in such a way that the value of the error criterion ϵ decreases until the minimum value of ϵ is reached. In (8.52) we can make various choices for the function $W(e^{j\theta_i})$ and the value of P. A frequently made choice is:

$$W(e^{j\theta_i}) = 1 \quad \text{for all } \theta_i \quad \text{and} \quad P = 1 \tag{8.53}$$

We then have the 'minimum mean-square error method'.

In using (8.52) only the error in the amplitude characteristic is considered. The equation can be modified so that errors in the phase characteristic (especially those relating to the group delay) are taken into account.[53]

8.4 COMPARISON OF FIR AND IIR FILTERS

The first question we encountered in the design procedure for a discrete filter related to the choice between an FIR filter and an IIR filter. Many factors come into play here, so that it is not always clear beforehand what the final choice will be.

Sometimes it helps in reaching a well-considered decision if two complete designs are made: one an FIR filter and the other an IIR filter. Then the two are assessed to see which gives the best solution for a particular application. Here very practical factors such as complexity, power consumption, speed of computation, ease of integration and availability of certain circuit modules can sway the balance.

In this section we should now like to collect together a number of more or less theoretical factors that are also of interest in our choice and which we have already encountered at various points in our discussion.

FIR filters	IIR filters
1. *System function*	
Contains only zeros.	Contains both poles and zeros.
2. *Frequency response*	
The normal design methods are suitable for arbitrary frequency responses; e.g. filters with several pass bands, differentiators and filters with a specified frequency characteristic in the transition band.	The design methods are mainly suitable for designing low-pass, high-pass, band-pass and band-stop filters.
3. *Phase characteristic*	
• Exactly linear phase possible.	• Linear phase can only be approximated; if a separate phase equalizer is required for this it can greatly increase the complexity of the filter. The filter specification often relates only to the amplitude characteristic.
• Phase-shifters (all-pass filters) not possible.	• All-pass filters possible.
4. *Stability*	
Filter is always stable.	Filter is unstable if there are poles outside the unit circle.
5. *Design aids*	
A medium-sized computer is usually required for iterative filter-design procedures.	It is not necessary to use a larger computer if the existing 'ready-made' design formulae for continuous filters are used, and the bilinear transformation, for example; a pocket calculator is then often sufficient.
6. *Complexity*	
Proportional to the length of the impulse response.	No direct relation between the complexity and the length of the impulse response (which is infinite by definition); filters with a high selectivity can be realized with hardware of relatively low complexity.
7. *Structure*	
A recursive structure (infrequent) or a non-recursive structure are both possible; the best known is the (non-recursive) transversal structure.	Only the recursive structure is possible; the most widely used form is the cascade connection of first-order and second-order sections. The distribution of the poles and zeros over the different sections is an important part of the design procedure.
8. *Sensitivity to interference*	
The initial state of the memory elements and any brief interfering signals (e.g. via the supply) can affect the output signal only for the length of	In principle, the initial state of the memory elements and any brief interfering signals can affect the output signal for an infinite length of time.

the impulse response (this only applies for non-recursive realization!).

9. *Quantization*

Quantization effects, e.g. in the realization as a digital filter, play a subordinate part. An exception occurs in the recursive structure in which exact compensation of poles and zeros is also required *after* quantization (see Chapter 10).

Because of quantization of the filter coefficients, a pole can in principle move from a position inside the unit circle to a position outside the unit circle and hence cause instability. Quantization effects can also lead to unwanted oscillations such as limit cycles and overflow oscillations (see Chapter 10).

10. *Adaptive filters*

The transversal structure is very suitable for making adaptive filters.

Adaptive filters are mainly based on lattice and ladder structures.

8.5 EXERCISES

Exercises to section 8.2

8.1 The frequency response $H(e^{j\theta})$ is defined by:

$$H(e^{j\theta}) = \begin{cases} 1 & \text{for } |\theta| < \pi/3 \\ 0 & \text{for } \pi/3 < |\theta| < \pi \end{cases}$$

(a) Use the IFTD to determine the function $h[n]$ that is the Fourier transform of $H(e^{j\theta})$.

(b) Determine from $h[n]$ an impulse response $h_1[n]$ of length $L = 9$ of a causal filter with a linear phase characteristic, which approximates the modulus of the given frequency response; use a rectangular window function here.

(c) Do the same as in (b), making use of a Hanning window.

8.2 Calculate the impulse response $h[n]$, the system function $H(z)$ and the frequency response $H(e^{j\theta})$ of the five filters with a linear phase characteristic that are shown in Fig. 8.30. Draw the amplitude and phase characteristics of these filters.

Exercises to section 8.3

8.3 Calculate the system function $H(z)$ of a discrete filter by the method of impulse-response invariance $(T = \frac{1}{4})$, starting from continuous filters with the system functions:

(a) $H_1(p) = \dfrac{1}{p+1} - \dfrac{2}{p+2} + \dfrac{1}{p+3}$

(b) $H_2(p) = \dfrac{2}{(p+1)(p+2)(p+3)}$

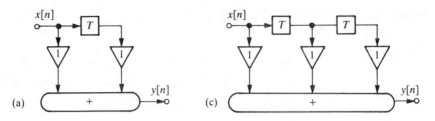

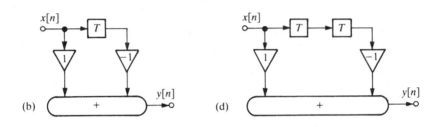

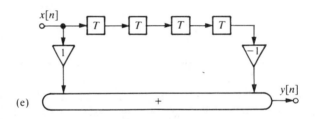

FIGURE 8.30 Exercise 8.2.

8.4 A continuous system has the frequency response $H_a(\omega)$:

$$H_a(\omega) = \frac{2}{2 + 3j\omega - \omega^2}$$

(a) Determine the system function $H_a(p)$ of this system and calculate the positions of the poles and zeros. Indicate whether this system is stable or not.

(b) Determine the impulse response $h_a(t)$ of this system and calculate the values of the samples $h_a(nT)$ for $T = \frac{1}{2}$.

(c) Determine the system function $H_1(z)$ of a discrete filter with an impulse response $h_1[n]$ for which $h_1[n] = h_a(nT)$ with $T = \frac{1}{2}$.

(d) Determine the poles and zeros of $H_1(z)$ and give a verdict on the stability of this discrete system.

(e) Use the method in which differentials are replaced by 'backward differences' to determine the system function $H_2(z)$ that follows from $H_a(p)$ (take the value $T = \frac{1}{2}$ for the constant T).

(f) Determine the poles and zeros of $H_2(z)$ and give a verdict on the stability.

(g) Determine the impulse response $h_2[n]$ that corresponds to the discrete system with system function $H_2(z)$. Compare $h_2[n]$ with $h_a(nT)$ for $T = \frac{1}{2}$.

(h) Now use 'forward differences' (with $T = \frac{1}{2}$) to derive a discrete system whose system function is $H_3(z)$ from $H_a(p)$. Answer for $H_3(z)$ and $h_3[n]$ the questions that applied to $H_2(z)$ and $h_2[n]$ in (f) and (g).

(i) Determine the values of T for which the method of forward differences (see (h)) gives a stable discrete system.

(j) Use the bilinear-transformation method (with $T = \frac{1}{2}$) to derive $H_4(z)$ from $H_a(p)$. Answer for $H_4(z)$ and $h_4[n]$ the questions that applied to $H_2(z)$ and $h_2[n]$ in (f) and (g).

8.5 For a continuous fifth-order filter:

$$H_a(p) = \frac{1}{(p+1)(p^2+p+1)(p^2+\sqrt{3}p+1)}$$

(a) Determine the poles of $H_a(p)$.

(b) Determine the poles of $H_d(z)$ that are obtained from $H_a(p)$ by the bilinear transformation with $T = \frac{1}{2}$.

8.6 We have determined the filter of Fig. 8.31 with the aid of reference [39], page 177. The system function of this filter is:

$$H_a(p) = \frac{0.43639p^2 + 1.45265}{(p+1.03213)(p^2+0.59572p+1.40743)}$$

and is characterized by the following parameters:

$$\delta_1 = 0.02 \qquad \omega_3 = 1.00 \text{ rad/s}$$
$$\delta_2 = 0.10 \qquad \omega_4 = 1.62 \text{ rad/s}$$
$$\omega_1 = 0.56 \text{ rad/s} \qquad \omega_5 = 1.82 \text{ rad/s}$$
$$\omega_2 = 0.89 \text{ rad/s} \qquad \omega_6 = 3.00 \text{ rad/s}$$

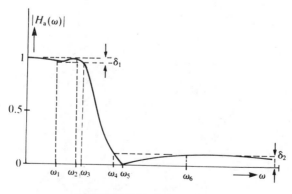

FIGURE 8.31 Exercise 8.6.

(a) Use the bilinear transformation (with $T = \frac{1}{2}$) to determine the system function $H_d(z)$ from $H_a(p)$.

(b) For $1 \leq i \leq 6$ determine the relative angular frequencies θ_i and the absolute frequencies f_i (in Hz) of the discrete filter that correspond to the frequencies ω_i of the continuous filter.

9

Multirate systems

9.1 INTRODUCTION

So far we have always been concerned with discrete systems in which only one sampling rate $1/T$ was used; input signal, output signal and all the other signals in a system always had the same sampling rate. In the conversion of continuous signals to discrete signals (Chapter 3) we have seen that T is a parameter that can be freely chosen, provided we make sure that $1/T$ is larger than twice the highest frequency present in the continuous signal. If we satisfy this condition, the discrete signal is a unique representation of the continuous signal. This means that the discrete signal contains all the information necessary for a complete and undistorted reconstruction of the continuous signal.

Figure 9.1 shows how an arbitrary continuous signal $x(t)$ with a highest frequency f_h can be converted into various discrete signals $x_1[n]$, $x_2[n]$ and $x_3[n]$ with sampling rates (or sampling frequencies) $f_1 = 1/T_1 = 6f_h$, $f_2 = 1/T_2 = 4f_h$ and $f_3 = 1/T_3 = 2f_h$. The corresponding spectra $X_1(e^{j\omega T_1})$, $X_2(e^{j\omega T_2})$ and $X_3(e^{j\omega T_3})$ are also shown.

Each of the three signals $x_1[n]$, $x_2[n]$ and $x_3[n]$ represents the same information, and in principle we can convert *any* of these signals into the others,[40] even if it is just by first recovering the continuous signal and sampling it again. (Soon we shall see how we can also do this in a completely discrete manner.) But why should we ever want to make this kind of conversion? The most important reason is that the signal $x_3[n]$ is represented by fewer samples per second than $x_2[n]$ and $x_1[n]$. This also means that fewer operations per second are required in the processing of $x_3[n]$, so that economies can be made in, say, the costs or the power consumption of a discrete system. In a discrete system that has signals with different highest frequencies at different points, there may therefore be advantages in working at different sampling rates (a 'multirate system'). We then try to keep the sampling rate at each point as low as possible (i.e. close to twice the highest signal frequency in the fundamental interval).

We have a simple example of this in a discrete low-pass filter (Fig. 9.2(a)) with input signal $x[n]$ and output signal $y[n]$. The signal $y[n]$ has a narrower frequency band than $x[n]$ and we can convert $y[n]$ into another discrete signal with a *lower* sampling rate before we perform any further discrete operations. We shall see later not only that this provides advantages in any further operations performed on $y[n]$, but also that fewer operations are required in the low-pass filter itself!

The converse situation occurs in a discrete modulator, for example (Fig. 9.2(b)). The modulator must operate with a sampling rate $1/T$ higher than twice the highest frequency

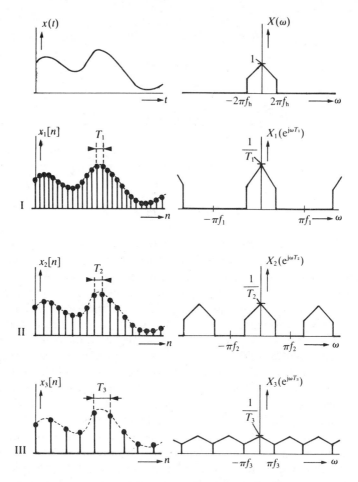

FIGURE 9.1 Conversion of a continuous signal into a discrete signal with various sampling rates:
(I) $f_1 = 1/T_1 = 6f_h$; (II) $f_2 = 1/T_2 = 4f_h$ and (III) $f_3 = 1/T_3 = 2f_h$.

in the fundamental interval of the *output signal* $y[n]$, so that $1/T \geq (\omega_c + \omega_1)/\pi$. The input signal $x[n]$ must also have this sampling rate, which is actually much higher than is strictly necessary on the basis of the narrow spectrum of $x[n]$. Often $x[n]$ may not be directly available, but instead a signal that contains the same information and has a lower sampling rate. We can derive $x[n]$ from this by an *increase* in the sampling rate.

Besides the two examples given here there are countless other situations, perhaps not so immediately obvious, in which a change in the sampling rate can offer advantages. This is the case in very narrow-band discrete filters in which input and output signals do have to have the same sampling rate. Sometimes one or more lower sampling rates are then used internally to keep the total number of discrete operations per unit time as low as possible.[41]–[45]

Yet another application can be found in converters from continuous time to discrete time (and vice versa). By working with more than one sampling rate continuous filtering

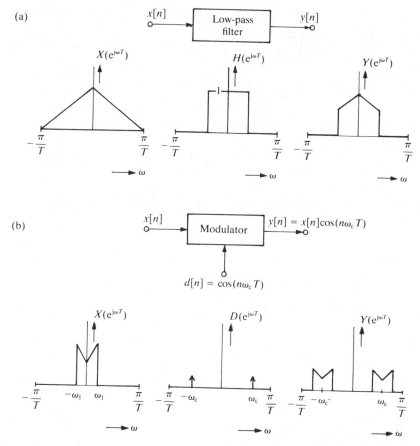

FIGURE 9.2 (a) Situation suitable for a decrease in the sampling rate. (b) Situation suitable for an increase in the sampling rate.

can be exchanged for discrete filtering. We shall return to this later in this chapter (section 9.8).

Before we look at multirate operation in more detail we ought to think about our notation for a moment. Since we must always be fully aware of which sampling rate we are dealing with for a particular signal, if there is more than one sampling rate we shall from now on write:

- for the signals: $x[nT_1]$, $y[nT_2]$ instead of $x[n]$, $y[n]$, and
- for the spectra: $X(e^{j\omega T_1})$, $Y(e^{j\omega T_2})$ instead of $X(e^{j\theta})$, $Y(e^{j\theta})$.

9.2 DECREASING THE SAMPLING RATE

We shall first look to see how we can decrease the sampling rate of a given signal $x[nT_1]$; this is sometimes called *downsampling*. We shall first confine our attention to a decrease

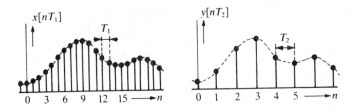

FIGURE 9.3 Decreasing the sampling rate by a factor of $R = 3$.

by an integer factor (the *decimation factor*) R, e.g. $R = 3$. In Fig. 9.3 a signal $x[nT_1]$ is shown. For the case where $R = 3$ it seems obvious that we should just take every third sample of $x[nT_1]$ to form the desired signal $y[nT_2]$. We can express the relation between $x[nT_1]$ and $y[nT_2]$ formally (for $R = 3$) by:

$$y[nT_2] = x[3nT_1] \quad \text{with} \quad n = 0, \pm 1, \pm 2, \pm 3, \ldots \tag{9.1}$$

Or, more generally, for a decrease by an integer factor R:

$$y[nT_2] = x[RnT_1] \quad \text{with} \quad n = 0, \pm 1, \pm 2, \pm 3, \ldots \tag{9.2}$$

The conversion of $x[nT_1]$ to $y[nT_2]$ takes place in a sampling rate decreaser (SRD), which we represent symbolically as in Fig. 9.4.

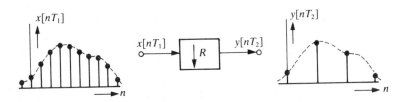

FIGURE 9.4 Symbol for a sampling rate decreaser (SRD) by a factor of $R = T_2/T_1$.

We can *not* characterize the SRD by an impulse response, a frequency response or a system function, because we are dealing with a time-varying system here (a unit pulse at $n = 0$ at the input also appears at the output, but a unit pulse at $n = 1$ does not!). But we *can* give a relation between the spectrum $Y(e^{j\omega T_2})$ of $y[nT_2]$ and the spectrum $X(e^{j\omega T_1})$ of $x[nT_1]$. We could give a strict mathematical derivation of this here, but the expressions obtained are rather opaque, so we shall merely quote the results here and show that they are in fact reasonable. We make use here of a hypothetical experiment, in which we proceed *as if* we can derive $y[nT_2]$ from $x[nT_1]$ in the following way:

1. We derive from $x[nT_1]$ a continuous signal $x_a(t)$ for which $x[nT_1]$ is a unique representation. We know that the spectrum $X_a(\omega)$ of $x_a(t)$ is limited to the interval $|\omega| \leq \pi/T_1$.
2. By sampling the analog signal $x_a(t)$ at a sampling rate $1/T_2$ we obtain the desired signal $y[nT_2]$.

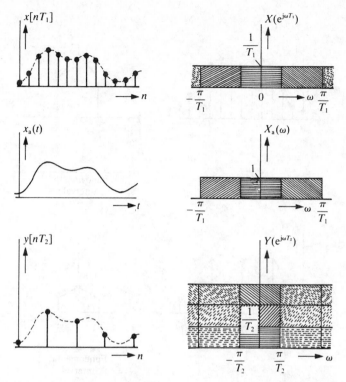

FIGURE 9.5 The relation between $X(e^{j\omega T_1})$ and $Y(e^{j\omega T_2})$ for decreasing the sampling rate by a factor $R = 3$.

The spectrum $Y(e^{j\omega T_2})$ has a fundamental interval $|\omega| \leq \omega/T_2$. This means that spectral components from $X_a(\omega)$ for which $\pi/T_2 \leq |\omega| \leq \pi/T_1$ give rise to aliasing in $Y(e^{j\omega T_2})$. This is shown graphically in Fig. 9.5 for $R = 3$, with the spectrum $X_a(\omega)$ only shown schematically. Frequency components in the horizontally hatched part of the spectrum $X(e^{j\omega T_1})$ do not give rise to aliasing in $Y(e^{j\omega T_2})$; frequency components in the obliquely hatched parts do.

We see that we can also find the fundamental interval of $Y(e^{j\omega T_2})$ directly from the fundamental interval of $X(e^{j\omega T_1})$ by multiplying this by a factor $T_1/T_2 = 1/R$, dividing it up into R equal parts and putting them together in the way indicated. We should realize that if aliasing occurs (i.e. if the obliquely hatched parts of the spectrum of $X(e^{j\omega T_1})$ are not zero) it will never be possible to recover $x[nT_1]$ uniquely from $y[nT_2]$. Formally, for $Y(e^{j\omega T_2})$ we can write:

$$Y(e^{j\omega T_2}) = \frac{1}{R} \sum_{i=0}^{R-1} X(e^{j(\omega - i2\pi/T_2)T_1)}) \quad \text{with} \quad T_2 = RT_1 \tag{9.3}$$

If $X(e^{j\omega T_1}) = 0$ for $|\omega| \geq \pi/T_2$ in the fundamental interval of $X(e^{j\omega T_1})$, then in the fundamental interval of $Y(e^{j\omega T_2})$:

$$Y(e^{j\omega T_2}) = \frac{1}{R} X(e^{j\omega T_1}) \quad \text{for} \quad |\omega| \leq \pi/T_2 \tag{9.4}$$

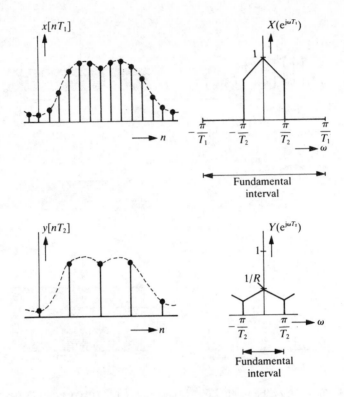

FIGURE 9.6 There is no aliasing on decreasing the sampling rate if $X(e^{j\omega T_1})$ is sufficiently band-limited.

which means that no aliasing is introduced. This is shown in Fig. 9.6 (again for $R = T_2/T_1 = 3$).

The spectral properties of $x[nT_1]$ therefore determine whether the sampling rate decreaser (SRD) of Fig. 9.4 introduces aliasing or not. This has led to the definition of a *decimator*; this consists of an ideal low-pass discrete filter (with a sampling rate $1/T_1$ and a cut-off frequency $\omega_c = \pi/T_2 = \pi/RT_1$) cascaded with an SRD (see Fig. 9.7).

A decimator therefore gives an output signal with a sampling rate R times lower than for the input signal, after all the spectral components that could cause aliasing have been removed. The signals from Fig. 9.7 are shown in Fig. 9.8.

The combination of a discrete filter with an arbitrary frequency response and an SRD is called a *decimating filter*. This is a filter with a high input sampling rate and a low output sampling rate. A general representation of a decimating filter is shown in Fig. 9.9(a). Because the discrete filter of frequency response $H(e^{j\omega T_1})$ cannot have an infinitely large attenuation in the stop band, unlike the ideal low-pass filter, there is always a certain amount of aliasing in $Y(e^{j\omega T_2})$. Also, if $x[nT_1]$ represents a useful signal whose spectrum is in fact limited to $|\omega| < \pi/T_2$, then in a decimating filter we must be prepared for interfering signals (such as noise) that do not have these frequency limitations to appear in the fundamental interval of $Y(e^{j\omega T_2})$ because of aliasing. This effect can be particularly important at large values of R (why?).

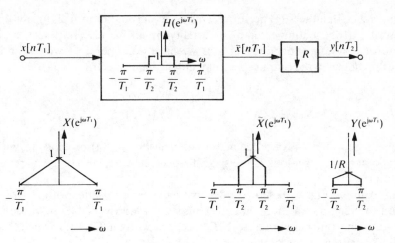

FIGURE 9.7 A decimator.

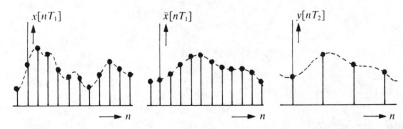

FIGURE 9.8 The signals $x[nT_1]$, $\tilde{x}[nT_1]$ and $y[nT_2]$ from Fig. 9.7.

(a)

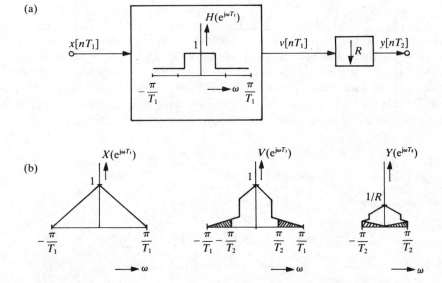

(b)

FIGURE 9.9 (a) A decimating filter. (b) The spectra at various points for $R = 2$.

We have seen that the SRD performs a time-varying operation. This means that we can *not* describe a decimating filter in its totality by an impulse response, a frequency response or a system function. But – as we have done before – we *can* indicate the relationship between the output spectrum and the input spectrum.

9.3 REALIZATION OF DECIMATING FILTERS

The generalized diagram of a decimating filter in Fig. 9.9(a) provides us with a good starting point for a closer look at the special aspects that arise in a practical realization of such a filter.

As an example Fig. 9.10 shows a decimating transversal filter with $R = 2$, which is based on the diagram of Fig. 9.9(a). It consists of a transversal filter whose input and output sampling rates are equal to $1/T_1$ cascaded with an SRD for which $R = 2$. The following equations apply for this filter:

$$v[nT_1] = \sum_{i=0}^{3} b_i x[nT_1 - iT_1] \tag{9.5a}$$

$$y[nT_2] = v[2nT_1] \tag{9.5b}$$

Combining (9.5a) and (9.5b) gives:

$$y[nT_2] = v[2nT_1]$$
$$= b_0 x[2nT_1] + b_1 x[2nT_1 - T_1] + b_2 x[2nT_1 - 2T_1] + b_3 x[2nT_1 - 3T_1] \tag{9.6}$$

The circuit of Fig. 9.10 calculates $v[nT_1]$ for all values of n. In the SRD $y[nT_2]$ is formed by discarding $v[T_1], v[3T_1], v[5T_1], \ldots$. In fact, it is not very sensible to calculate these values first and then just ignore them. From (9.6) we see that for the calculation of $v[2T_1]$, $v[4T_1], v[6T_1], \ldots$ it is only necessary to multiply the coefficients b_0 and b_2 by the *even* samples of $x[nT_1]$ and the coefficients b_1 and b_3 by the *odd* samples of $x[nT_1]$. We can do this if instead of having a single SRD at the output of the filter, we put a number of SRDs directly in front of the multipliers. This gives the structure of Fig. 9.11.

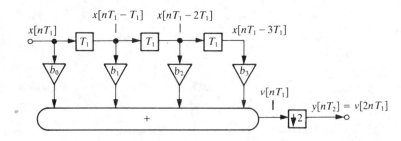

FIGURE 9.10 Decimating transversal filter with $R = 2$.

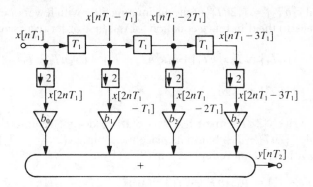

FIGURE 9.11 Decimating transversal filter ($R = 2$) with a reduced number of multiplications per unit time.

The great advantage of the circuit of Fig. 9.11 over the circuit of Fig. 9.10 is that the multipliers b_0, b_1, b_2 and b_3 and the adder only have to work at half the speed. The larger number of SRDs of Fig. 9.11 compared with Fig. 9.10 is not a disadvantage. We should appreciate that an SRD represents an operation that requires no extra hardware; in practice the samples are discarded by taking appropriate measures in the timing circuits of the discrete filter.

Decimating versions of filters with a recursive structure can aɩsɔ be derived. Here we have to be extra careful about the way in which we reduce the number of multiplications per unit time. We should like to make this clear from a simple example with a first-order recursive network. In Fig. 9.12 we have derived a decimating version of such a filter, by putting an SRD with $R = 2$ after a standard filter.

In this circuit the decimation has not yet led to the reduction of the number of operations required per unit time. Each sample $v[nT_1]$ is calculated. Even though half of them are ignored by the SRD, we require all of the samples of $v[nT_1]$ so as to be able to calculate future output samples via the feedback. It is however still possible to calculate

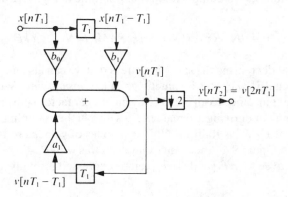

FIGURE 9.12 Decimating recursive filter with $R = 2$.

the output signal $y[nT_2] = v[2nT_1]$ of the decimating filter with fewer operations per unit time, if we go back to the difference equation for this first-order recursive section:

$$v[nT_1] = b_0 x[nT_1] + b_1 x[nT_1 - T_1] + a_1 v[nT_1 - T_1] \tag{9.7a}$$

$$y[nT_2] = v[2nT_1] \tag{9.7b}$$

Now we try to write $v[2nT_1]$ in such a way that we do not require the immediately preceding sample $v[2nT_1 - T_1]$, but in its place the sample $v[2nT_1 - 2T_1]$. We do this in the following way. First of all we replace n in (9.7a) by $n - 1$:

$$v[nT_1 - T_1] = b_0 x[nT_1 - T_1] + b_1 x[nT_1 - 2T_1] + a_1 v[nT_1 - 2T_1] \tag{9.8a}$$

Substituting (9.8a) in (9.7a) then gives:

$$v[nT_1] = b_0 x[nT_1] + (b_1 + a_1 b_0) x[nT_1 - T_1] + a_1 b_1 x[nT_1 - 2T_1] + a_1^2 v[nT_1 - 2T_1] \tag{9.8b}$$

Finally, combining (9.8b) and (9.7b):

$$\begin{aligned}
y[nT_2] &= v[2nT_1] \\
&= b_0 x[2nT_1] + (b_1 + a_1 b_0) x[2nT_1 - T_1] + a_1 b_1 x[2nT_1 - 2T_1] \\
&\quad + a_1^2 v[2nT_1 - 2T_1]
\end{aligned} \tag{9.9}$$

We therefore see that the immediately preceding sample $v[2nT_1 - T_1]$ is indeed not now required for the calculation of $v[2nT_1]$. We also see from (9.9) that in the calculation of $v[2nT_1]$ the coefficients b_0 and $a_1 b_1$ are multiplied by *even* samples of $x[nT_1]$ and only the coefficient $(b_1 + a_1 b_0)$ is multiplied by *odd* samples of $x[nT_1]$. We can now draw a circuit in which the SRD operation is brought as far forward as possible. This is done in Fig. 9.13.

In Fig. 9.12 three multiplications per sampling period T_1 were required. In Fig. 9.13, however, four multiplications are required in the time $T_2 = 2T_1$, so that in Fig. 9.13 only two-thirds of the number of multiplications per unit time in Fig. 9.12 are required.

9.4 INCREASING THE SAMPLING RATE

The counterpart of decreasing the sampling rate is that we can also increase the sampling rate of a given signal $x[nT_1]$; this is sometimes called *upsampling*. We shall confine our attention in the first instance to an increase by an integer factor (the *interpolation factor*) R. The simplest way of deriving a signal $y[nT_2]$ with a higher sampling rate $1/T_2 = R/T_1$ from a given signal $x[nT_1]$ is to insert $(R - 1)$ samples of value zero between every two samples of $x[nT_1]$. Figure 9.14 shows an example of this with $R = 3$.

The relation between $y[nT_2]$ and $x[nT_1]$ can be expressed formally as:

$$y[nT_2] = \begin{cases} x[nT_1/R] & \text{for } n = 0, \pm R, \pm 2R, \ldots \\ 0 & \text{elsewhere} \end{cases} \tag{9.10}$$

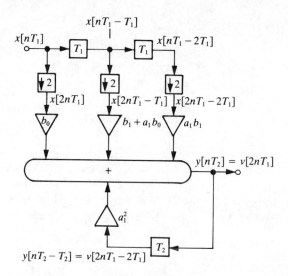

FIGURE 9.13 Decimating recursive filter ($R = 2$) with a reduced number of multiplications per unit time.

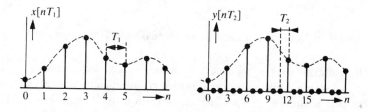

FIGURE 9.14 Increasing the sampling rate by a factor of $R = 3$.

The conversion of $x[nT_1]$ to $y[nT_2]$ takes place in a sampling rate increaser (SRI), which we represent symbolically as in Fig. 9.15.

Like the SRD, the SRI is a time-varying system, since shifting the input signal through one (input) sampling interval results in an output signal shifted through R (output) sampling intervals. An SRI, like an SRD, cannot therefore be described in terms of an impulse response, frequency response or system function. We *can* however indicate a relation between the spectra $X(e^{j\omega T_1})$ and $Y(e^{j\omega T_2})$. This is easily derived. In general, from (4.14) we have:

$$Y(e^{j\omega T_2}) = \sum_{n=-\infty}^{\infty} y[nT_2]e^{-jn\omega T_2} \tag{9.11}$$

Substituting (9.10) in this equation gives:

$$Y(e^{j\omega T_2}) = \sum_{\substack{n=0,\pm R, \\ \pm 2R, \, \ldots}}^{\infty} x[nT_1/R]e^{-jn\omega T_2} \tag{9.12}$$

Substituting $n = iR$ in (9.12) gives:

$$Y(e^{j\omega T_2}) = \sum_{i=-\infty}^{\infty} x[iT_1]e^{-ji\omega T_2 R}$$

$$= \sum_{i=-\infty}^{\infty} x[iT_1]e^{-ji\omega T_1} = X(e^{j\omega T_1}) \qquad (9.13)$$

We therefore see that the insertion of zeros has no effect on the spectrum of the signal, apart from the fact that the fundamental interval of $Y(e^{j\omega T_2})$ is larger by a factor of R than the fundamental interval of $X(e^{j\omega T_1})$. We show this in Fig. 9.16 for $R = 3$. In the fundamental interval of $Y(e^{j\omega T_2})$ we see a periodic spectrum, in which the original

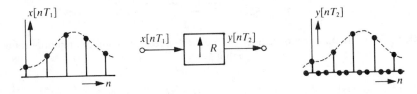

FIGURE 9.15 Symbol for a sampling rate increaser (SRI) by a factor of $R = T_1/T_2$.

frequency information of $X(e^{j\omega T_1})$ occurs R times. We wonder what happens if we use an ideal low-pass discrete filter (with sampling rate $1/T_2$ and gain factor R) to remove all the frequencies in the interval $\pi/T_1 \le |\omega| \le \pi/T_2$ from $Y(e^{j\omega T_2})$ (see Fig. 9.17). We then obtain a signal $\tilde{y}[nT_2]$.

What does the signal $\tilde{y}[nT_2]$ look like as a function of time? We can easily find out by comparing the spectra $X(e^{j\omega T_1})$ and $\tilde{Y}(e^{j\omega T_2})$ with the spectra from Fig. 9.1. We then see that we can consider $x[nT_1]$ and $\tilde{y}[nT_2]$ as being derived from a continuous signal $x(t)$ from which $x[nT_1]$ has been obtained by sampling at a rate $1/T_1$ and $\tilde{y}[nT_2]$ by sampling at a rate $1/T_2 = 3/T_1$.

The signal $\tilde{y}[nT_2]$ is an *interpolated* version of $x[nT_1]$; we therefore call the combination of an SRI with interpolation factor R and an ideal low-pass filter with sampling rate $1/T_2$, gain factor R and cut-off frequency $|\omega| = \pi/T_1 = \pi/RT_2$ an *interpolator*. The relationship between $x[nT_1]$, $y[nT_2]$ and $\tilde{y}[nT_2]$ is illustrated in Fig. 9.18.

The combination of an SRI and a discrete filter with an arbitrary frequency response is called an *interpolating filter*. This is therefore a filter with a low input sampling rate and a high output sampling rate. A general representation of an interpolating filter is shown in Fig. 9.19.

Since the discrete filter of frequency response $H(e^{j\omega T_2})$ (unlike the ideal low-pass filter in the interpolator) cannot introduce infinitely large attenuation in the stop band, the periodic spectral nature of $V(e^{j\omega T_2})$ always shows up to a certain extent in the fundamental interval of the output spectrum $Y(e^{j\omega T_2})$ (the hatched regions in Fig. 9.19(b)). Since the SRI performs a time-varying operation, we cannot describe an interpolating filter in its totality by an impulse response, frequency response or system function. As before,

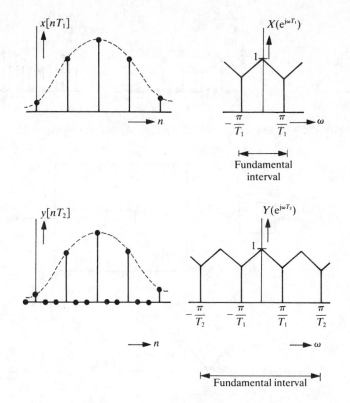

FIGURE 9.16 The relationship between $X(e^{j\omega T_1})$ and $Y(e^{j\omega T_2})$ when the sampling rate is increased by a factor of $R = 3$.

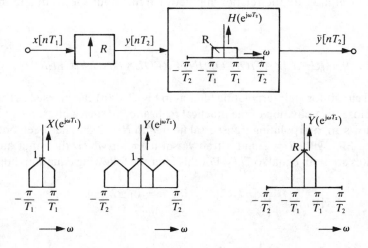

FIGURE 9.17 An interpolator.

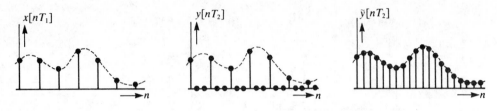

FIGURE 9.18 The signals $x[nT_1]$, $y[nT_2]$ and $\bar{y}[nT_2]$ from Fig. 9.17.

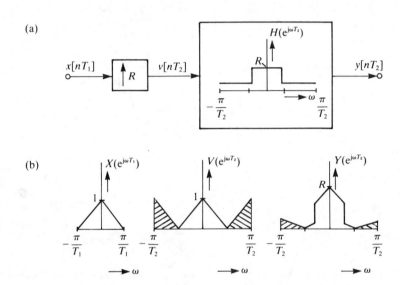

FIGURE 9.19 (a) An interpolating filter. (b) The spectra at the various points for $R = 2$.

however, we *can* indicate the relationship between the input spectrum and the output spectrum.

9.5 REALIZATION OF INTERPOLATING FILTERS

The general circuit for an interpolating filter as in Fig. 9.19(a) can be used as the starting point for a closer examination of the practical realization of such a filter. As an example, Fig. 9.20 shows an interpolating transversal filter with $R = 2$ that has been obtained by cascading an SRI with $R = 2$ and a transversal filter for which the input and output sampling rates are both equal to $1/T_2$. For this filter the following equations apply:

$$v[nT_2] = \begin{cases} x[nT_1/2] & \text{for } n = 0, \pm2, \pm4, \ldots \\ 0 & \text{elsewhere} \end{cases} \tag{9.14a}$$

$$y[nT_2] = \sum_{i=0}^{3} b_i v[nT_2 - iT_2] \tag{9.14b}$$

Now let us just write out a few values of $y[nT_2]$, making use of the fact that $v[nT_2] = 0$ if n is odd:

$$
\begin{aligned}
y[0] &= b_0 v[0] &&+ b_2 v[-2T_2] = b_0 x[0] &&+ b_2 x[-T_1] \\
y[T_2] &= b_1 v[0] &&+ b_3 v[-2T_2] = b_1 x[0] &&+ b_3 x[-T_1] \\
y[2T_2] &= b_0 v[2T_2] + b_2 v[0] &&= b_0 x[T_1] &&+ b_2 x[0] \\
y[3T_2] &= b_1 v[2T_2] + b_3 v[0] &&= b_1 x[T_1] &&+ b_3 x[0] \\
y[4T_2] &= b_0 v[4T_2] + b_2 v[2T_2] &&= b_0 x[2T_1] &&+ b_2 x[T_1]
\end{aligned}
\tag{9.15}
$$

We see that only two multiplications are required for each output sample $y[nT_2]$, and that if n is even the two filter coefficients used (b_0 and b_2) are not the same ones used if n is odd (b_1 and b_3). We can modify the circuit of Fig. 9.20 for these operations, so that we can again obtain a reduction in the number of multiplications required per unit time. One such circuit is shown in Fig. 9.21. We can see that it contains two separate filters; one with the coefficients b_0 and b_2, which provides the output samples of $y[nT_2]$ for even n; the other with the coefficients b_1 and b_3, which provides the output samples of $y[nT_2]$ for odd n. It is possible to think of an arbitrary interpolating transversal filter with interpolation

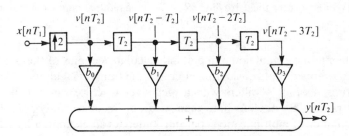

FIGURE 9.20 Interpolating transversal filter with $R = 2$.

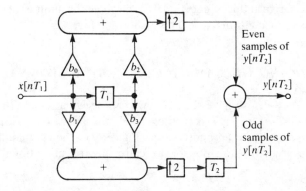

FIGURE 9.21 Interpolating transversal filter ($R = 2$) with a reduced number of multiplications per unit time.

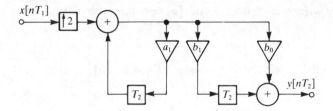

FIGURE 9.22 Interpolating recursive filter with $R = 2$.

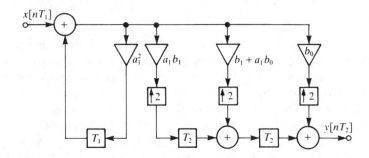

FIGURE 9.23 Interpolating recursive filter ($R = 2$) with a reduced number of multiplications per unit time.

factor R as being like R individual filters that calculate samples of the output signal in turn. This is a useful way of considering interpolating transversal filters.

Interpolating versions of filters with a recursive structure can also be derived. In general it is rather less obvious here than with non-recursive filters just how we can reduce the number of multiplications per unit time. As an example Fig. 9.22 shows an interpolating first-order recursive filter with $R = 2$ in which no reduction in the number of multiplications has as yet been made.

This same filter can also be realized in another way (for the moment we shall not consider how we arrive at this; see section 9.7), in which the number of multiplications in the interval T_1 is reduced from 6 to 4. This is shown in Fig. 9.23.

9.6 *CHANGING THE SAMPLING RATE BY A RATIONAL FACTOR*

In the previous sections we have seen how we can derive a signal $y[nT_2]$ from a given signal $x[nT_1]$ when T_2/T_1 or T_1/T_2 is an integer R. In the first case we use a decimator (Fig. 9.7) and in the second case an interpolator (Fig. 9.17). The combination of these two principles in one circuit makes it possible to realize an arbitrary rational ratio of T_1 to T_2, i.e.:

$$T_2 = \frac{R_2}{R_1} T_1 \qquad\qquad (9.16)$$

where R_2 and R_1 are integers. If we put the interpolator in front of the decimator we can make do with only one ideal low-pass filter with gain factor R_1 (Fig. 9.24).

We have two separate cases here:

1. $R_2 < R_1$; the sampling rate of $y[nT_2]$ is higher than that of $x[nT_1]$ and therefore $\pi/T_1 < \pi/T_2$. The ideal low-pass filter now has a passband for $|\omega| < \pi/T_1$. For the signal spectrum $Y(e^{j\omega T_2})$ of $y[nT_2]$ in the fundamental interval we have:

$$Y(e^{j\omega T_2}) = \begin{cases} \dfrac{R_1}{R_2} X(e^{j\omega T_1}) & \text{for } |\omega| \leq \pi/T_1 \\[2ex] 0 & \text{for } \pi/T_1 \leq |\omega| \leq \pi/T_2 \end{cases} \tag{9.17}$$

2. $R_2 > R_1$; the sampling rate of $y[nT_2]$ is lower than that of $x[nT_1]$ and therefore $\pi/T_2 < \pi/T_1$. The ideal low-pass filter now has a passband for $|\omega| < \pi/T_2$. Any frequency components from $X(e^{j\omega T_1})$ with $\pi/T_2 < |\omega| < \pi/T_1$ are suppressed by this filter. For the fundamental interval of $Y(e^{j\omega T_2})$ we therefore have:

$$Y(e^{j\omega T_2}) = \frac{R_1}{R_2} X(e^{j\omega T_1}) \qquad \text{for } |\omega| \leq \pi/T_2 \tag{9.18}$$

If we replace the ideal low-pass filter in Fig. 9.24 by a filter with an arbitrary frequency response we obtain an interpolating or decimating filter with a rational ratio of input to output sampling rates. However, because of the finite attenuation in the stop band(s) aliasing can now occur, and this must be carefully taken into account.

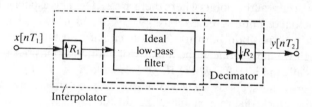

FIGURE 9.24 Changing the sampling rate by a rational factor $T_2/T_1 = R_2/R_1$.

9.7 TRANSPOSITION OF INTERPOLATING AND DECIMATING FILTERS

In section 7.5 we introduced the transposition theorem for linear time-invariant discrete systems (LTD systems). Applying transposition to such a system gives a new LTD system with exactly the same frequency response, but a different structure. Interpolating and decimating filters are not classed as LTD systems, since, as we have seen, they are time-varying. The transposition theorem can be extended so that it also applies to filters of this type. To the transposition rules stated in section 7.5 we must now add:

- SRIs are replaced by SRDs, and
- SRDs are replaced by SRIs.

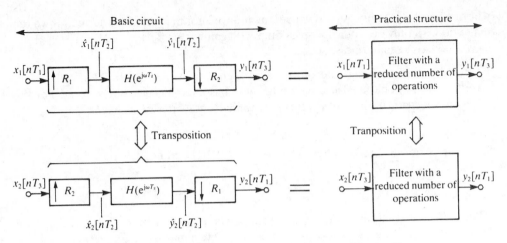

FIGURE 9.25 Transposition of interpolating and decimating filters.

If we apply these extended transposition rules to a filter with input sampling rate $1/T_x$ and output sampling rate $1/T_y$, we obtain a new filter with input sampling rate $1/T_y$ and output sampling rate $1/T_x$; in this way a decimating filter becomes an interpolating filter and vice versa. Since both before and after transposition we are dealing with a time-varying system here, we cannot talk about the frequency responses; the filter properties depend to some extent, as we saw earlier (see section 9.2), on the spectrum of the input signal. There is of course a very close relationship between the system properties before and after transposition. However, it is most important that we should consider carefully whether aliasing occurs at the points where there are SRDs. This aspect is best made clear with the aid of actual examples; see Exercises 9.9, 9.10 and 9.11.

We can apply transposition both to the basic circuit of a filter and to the practical structure in which we have reduced the number of multiplications per unit time. This is shown schematically in Fig. 9.25, where $R_1 = T_1/T_2$ and $R_2 = T_3/T_2$.

In one of the previous sections we used transposition to convert the block diagram of Fig. 9.12 into the block diagram of Fig. 9.22. In the same way we have derived the practical structure of Fig. 9.23 from the structure of Fig. 9.13 (verify this!). A more extensive treatment of the transposition theorem for time-varying systems can be found in the literature.[31]

9.8 APPLICATIONS

A very interesting application of decimating filters is found in the conversion of analog signals into digital signals (A/D conversion). Let us assume we have an analog signal $x_a(t)$ with a wide spectrum, and that we want to convert the signal components with frequencies $|\omega| < \pi/T_1$ into a digital signal $x[nT_1]$. This case occurs in telephony, for example. Here (for speech) we are only interested in frequencies below about 4 kHz, while the microphone provides a much broader spectrum. In principle, we should be able to represent the speech signal in discrete form with a sampling rate of 8 kHz. We must therefore first limit the bandwidth of the analog signal to 4 kHz in an analog low-pass prefilter to prevent aliasing in the A/D conversion (Fig. 9.26).

Since this low-pass filter must have a narrow transition band, it will have to meet a fairly difficult specification. The requirements can be eased considerably, however, if the A/D conversion is performed at a higher sampling rate than is strictly necessary, and we use a decimator to get back to the desired sampling rate. In this way we exchange analog filtering for digital filtering. This is shown in Fig. 9.27, where we have the A/D converter operating at twice the sampling rate, $2/T_1$. The stop band of the analog filter now does not have to start at π/T_1, but only at $3\pi/T_1$. At the output of the A/D converter we then have

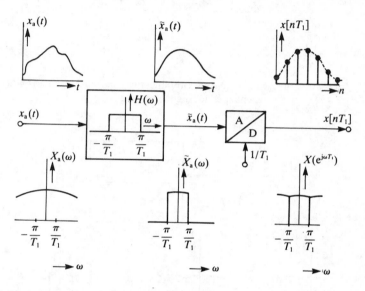

FIGURE 9.26 Analog/digital conversion with minimum sampling rate and consequently a complicated prefilter $H(\omega)$.

to content with aliasing in the range $\pi/T_1 \le |\omega| \le 2\pi/T_1$, but this part of the spectrum is completely removed by the decimator.

A situation closely related to the situation described above is found in the conversion of digital signals to analog signals (D/A conversion). Here we require a 'sharp' analog filter to ensure that no frequency components in the analog signal occur outside the interval $|\omega| \le \pi/T_1$ (Fig. 9.28). By introducing an interpolator (see Fig. 9.29) we can first increase the fundamental interval of the digital signal, so that the requirements for the analog filter are considerably eased. Here we have used an interpolator with an interpolation factor of $R = 2$. Once again, the stop band of the analog filter only needs to start at $3\pi/T_1$. This principle (with $R = 4$) is often applied in Compact Disc players and then, less correctly, called *oversampling*.

To simplify the description we have explained the exchange of analog filtering for digital filtering in the preceding examples on the basis of the use of a decimator and an interpolator. They are both idealized systems, but we can approximate them to any required accuracy with realizable decimating and interpolating filters.

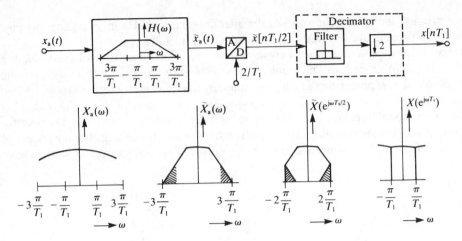

FIGURE 9.27 A/D converter with increased sampling rate, permitting analog filtering to be exchanged for digital filtering.

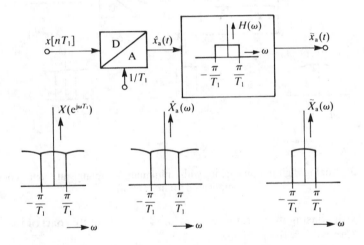

FIGURE 9.28 Digital/analog conversion with minimum sampling rate and consequently a complicated postfilter $H(\omega)$.

9.9 EXERCISES

Exercises to section 9.3

9.1 A decimating second-order recursive discrete filter with $R = 2$ is shown in Fig. 9.30. Derive another decimating filter from it with a reduced number of multiplications. How large is the reduction in the number of multiplications per unit time?

9.2 Take the filter of Fig. 9.12. Derive a decimating filter from it with $R = 3$ (instead of $R = 2$) and with a reduced number of multiplications. How large is the reduction in the number of multiplications per unit time?

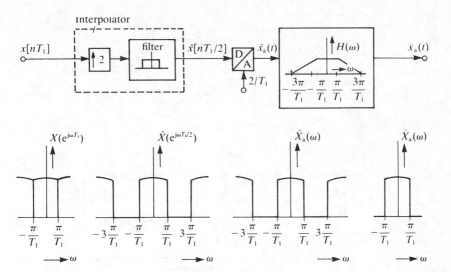

FIGURE 9.29 D/A converter with increased sampling rate, permitting analog filtering to be exchanged for digital filtering.

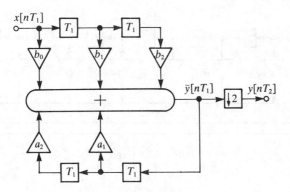

FIGURE 9.30 Exercise 9.1.

9.3 (*For the enthusiast.*) From the decimating recursive ($R = 2$) filter of Fig. 9.31(a) a filter with a reduced number of multiplications has been derived as shown in Fig. 9.31(b). Determine the corresponding coefficients c_0, c_1, c_2, c_3 and c_4. How large is the reduction in the number of multiplications per unit time?

Exercises to section 9.5

9.4 An interpolating transversal filter is shown in Fig. 9.32. Derive an interpolating transversal filter from it in which the number of multiplications per unit time is reduced as far as possible. Draw this filter in such a way that three separate component filters can be clearly identified.

(a)

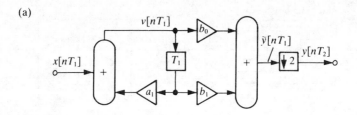

(b)

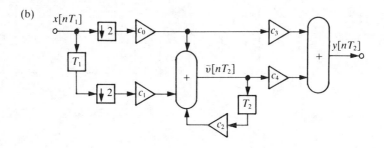

FIGURE 9.31 Exercise 9.3.

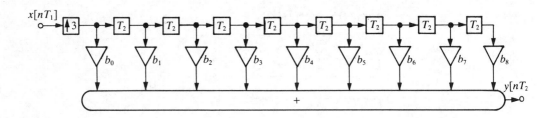

FIGURE 9.32 Exercise 9.4.

9.5 Consider a signal $x[nT_1]$. This signal is interpolated in such a way that the new sampling period $T_2 = T_1/2$, so that $R = 2$. The signal samples $y[nT_2]$ are obtained from the samples $x[nT_1]$ by an interpolating filter in which 'linear interpolation' is applied:

$$y[nT_2] = \begin{cases} x[nT_1/2] & \text{for } n = 0, \pm 2, \pm 4, \ldots \\ \dfrac{1}{2}\left\{ x\left[\dfrac{nT_1 - T_1}{2}\right] + x\left[\dfrac{nT_1 + T_1}{2}\right] \right\} & \text{for } n = \pm 1, \pm 3, \ldots \end{cases}$$

This interpolating filter is shown in Fig. 9.33.
(a) For an arbitrary signal $x[nT_1]$ draw the corresponding $\hat{x}[nT_2]$ and $y[nT_2]$.
(b) Determine $h[nT_2]$ such that the desired $y[nT_2]$ is obtained.

(c) Figure 9.19 shows an interpolating filter as the cascade connection of an SRI and a filter with a frequency response $H(e^{j\omega T_2})$. Determine this frequency response for the interpolating filter of Fig. 9.33.

(d) The spectrum of the input signal is given by:

$$X(e^{j\omega T_1}) = \begin{cases} 1 & \text{for } 0.1\pi/T_1 \le |\omega| \le 0.11\pi/T_1 \\ 0 & \text{elsewhere} \end{cases}$$

Determine the spectrum of the interpolated signal $y[nT_2]$ by making a number of diagrams.

(e) Make a number of diagrams to determine the spectrum of the interpolated signal if the spectrum of the input signal is given by:

$$X(e^{j\omega T_1}) = \begin{cases} 1 & \text{for } 0.25\pi/T_1 \le |\omega| \le 0.75\pi/T_1 \\ 0 & \text{elsewhere} \end{cases}$$

(f) For which of the two input spectra given in (d) or (e) does the operation of the interpolating filter given here approximate best to the function of the ideal interpolator with $R = 2$?

(g) The filter with impulse response $h[nT_2]$ is a non-causal filter. How should the relation shown earlier between $y[nT_2]$ and $x[nT_1]$ be changed to make the filter causal?

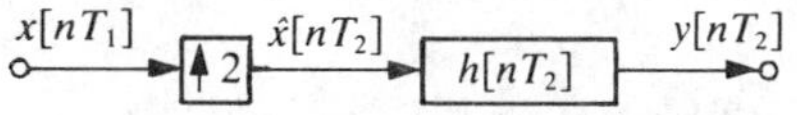

FIGURE 9.33 Exercise 9.5.

Exercise to section 9.6

9.6 A signal $x[nT_x]$ has a spectrum $X(e^{j\omega T_x})$ as shown in Fig. 9.34(a). This signal is applied to the circuit of Fig. 9.34(b). The ideal low-pass filter has a sampling rate $1/T_v = 3/T_x$, a gain factor of 1 in the pass band and a cut-off frequency $|\omega| = \pi/T_x$.

(a) Draw the four signals $x[nT_x]$, $v[nT_v]$, $w[nT_w]$ and $y[nT_y]$.

(b) Draw the four spectra $X(e^{j\omega T_x})$, $V(e^{j\omega T_v})$, $W(e^{j\omega T_w})$ and $Y(e^{j\omega T_y})$ in their fundamental intervals.

(c) Repeat (a) and (b) with the ideal low-pass filter omitted.

Exercises to section 9.7

9.7 (a) Show that the circuit of Fig. 9.22 can be obtained by transposition of the circuit of Fig. 9.12.

(b) Do the same for Figs 9.23 and 9.13.

9.8 Derive an interpolating filter by transposing the filter of Fig. 9.10.

9.9 (a) The time-varying filter shown in Fig. 9.35 has an input signal $x[nT_1]$ with the spectrum $X(e^{j\omega T_1})$ shown. Draw, one above the other, the spectra of $x[nT_1]$, $v[nT_2]$, $w[nT_2]$ and $y[nT_3]$ (in the corresponding fundamental intervals).

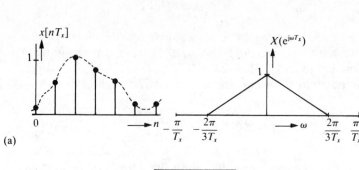

(a)

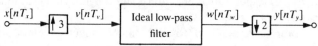

(b)

FIGURE 9.34 (a) and (b) Exercise 9.6.

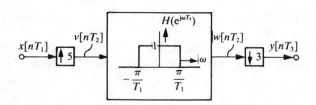

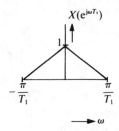

FIGURE 9.35 Exercise 9.9.

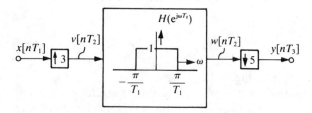

FIGURE 9.36 Exercise 9.10.

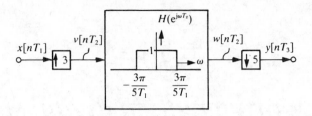

FIGURE 9.37 Exercise 9.11.

(b) Draw the block diagram of the filter obtained by transposing the filter of Fig. 9.35.

(c) Draw, one above the other, the four most important signal spectra for this new filter; use the output signal $y[nT_3]$ from Fig. 9.35 as the input signal here. Compare the output spectrum of this filter with the original input spectrum $X(e^{j\omega T_1})$.

9.10 Repeat Exercise 9.9(a), (b) and (c) for the circuit of Fig. 9.36.

9.11 Repeat Exercise 9.9(a), (b) and (c) for the circuit of Fig. 9.37.

10

Finite word length in digital signals and systems

10.1 INTRODUCTION

In the previous chapters we have dealt at length with various aspects of discrete signals and discrete systems. In doing so we assumed at all times that all signals and other quantities (such as filter coefficients) could assume any value. This assumption does not hold for the very important category of digital signals and digital systems, because every quantity is represented as a combination of a finite number of bits (i.e. a *binary word* or just *word*). A *bit* is a number that can have only two different values (usually 0 and 1).[†] With a word length of B bits we can therefore distinguish between at most 2^B different values. If we are free to choose B, we can make the digital representation as accurate as we like and hence approximate any desired discrete signal or any discrete system to any degree of accuracy.

Actual practice is entirely different, however. For reasons of economy we are often most interested to know just how we can select the lowest possible value of B without introducing *unacceptable* errors. But then we are inevitably confronted with a number of effects of a very diverse nature, which are caused by the *finite word length* we are using. These effects are often very complicated and the only conclusions that can be drawn about them are statistical ones (relating to mean values, root-mean-square values, maximum values, etc.). This is partly because nonlinearities are introduced into the system, and this makes an exact description of the system complicated, if not impossible.

In applying the theory from the previous chapters to digital signal processing, we shall have to pay special attention to the three following situations in which the finite word length is of considerable significance:

1. The conversion of signals with a continuous amplitude into signals with a discrete amplitude; this involves the introduction of noise usually referred to as *A/D conversion noise* (section 10.4).
2. The conversion of the filter coefficients that we obtained from the filter design procedures discussed earlier into *coefficients of finite word length*; this is associated with a change in the frequency response (section 10.5).

[†] *Binary* means 'two-valued'; *bit* is derived from 'binary digit'.

3. The *performance of operations* (such as multiplication and addition) in a way such that the word length does not increase undesirably. This introduces noise and may even cause oscillation (section 10.6).

In all these situations we are concerned with *quantization* or *overflow*, or both. We shall look at these effects more closely in section 10.3.

If we have a word length of B bits we can still choose the number corresponding to each of the 2^B different words: the *number representation* or *numeric code*. There are various ways of doing this, each with its own advantages and disadvantages. We shall look at this last aspect first.

10.2 NUMBER REPRESENTATIONS

The three most common binary number representations are:

1. Sign and magnitude
2. One's complement
3. Two's complement

The difference between these three cases can be illustrated most easily with the aid of Table 10.1, which shows the nature of the relationship between the eight possible 3-bit words ($B = 3$) and the corresponding decimal values.

Positive decimal values are represented in the same way in all three number representations:

- The bit furthest to the left (the 'sign bit') is a 0 if we are dealing with a positive number.
- The bit furthest to the right (the 'least-significant bit' or LSB) represents the value $2^0 = 1$.
- The bit second from the right represents the value $2^1 = 2$.
- If the word length is greater than 3 the bit third from the right represents the value $2^2 = 4$, and so on.

Thus:

$$01101 = + (1 \times 2^3 + 1 \times 2^2 + 0 \times 2^1 + 1 \times 2^0) = + 13$$

TABLE 10.1 Various binary number representations

Decimal value	Sign and magnitude	One's complement	Two's complement
+3	011	011	011
+2	010	010	010
+1	001	001	001
+0	000	000	000
−0	100	111	—
−1	101	110	111
−2	110	101	110
−3	111	100	101
−4	—	—	100

Negative decimal numbers are represented in different ways in each number representation, but the sign bit is always 1. In sign-and-magnitude representation the other bits represent the magnitude of the number. In one's-complement representation, negative numbers are obtained by replacing all the bits in the corresponding positive number by the opposite bit ('bit inversion'). Negative numbers are obtained in two's-complement representation by inverting all the bits of the corresponding positive number and then adding a one in the place corresponding to the least-significant bit.

In Table 10.1 the least-significant bit has been assigned the value 2^0. This means that we can only represent integers. However, we can also assign to this LSB a value corresponding to a negative integer power of 2, e.g. $2^{-3} = 1/8$. We can then represent decimal fractions by binary words. We can illustrate this by giving the representation of the decimal numbers $+3.625$ and -3.625 in each of the three number representations with a word length of 8 bits and an LSB of 2^{-3} (Table 10.2).

TABLE 10.2 Examples of number representations

Decimal value	Sign and magnitude	One's complement	Two's complement
+3.625	00011.101	00011.101	00011.101
−3.625	10011.101	11100.010	11100.011

To make things easier we have placed a point after the bit representing the value $2^0 = 1$ in Table 10.2. This simplifies the intepretation of the binary numbers since we now know the values of all the bits immediately. These number representations are therefore called *fixed-point representations*.

Of the three number representations given so far, the first and the third are the most widely used. Sign-and-magnitude representation has advantages for performing multiplications; two's-complement representation has advantages in addition and subtraction. A nice feature of two's-complement representation is that the intermediate results in a long series of additions can fall *outside* the range of values of the code without causing errors, provided the final result comes *inside* the range of the code. For instance, we can make the following calculation without error in a 3-bit representation:

Decimal		Two's complement
1		001
2		010
—— +		—— +
3	=	011
3		011
—— +		—— +
6	≠	110
−2		110
—— +		—— +
4	≠	(1) 100
−2		110
—— +		—— +
2	=	(1) 010

(The bit in brackets represents a 'carry', which can be disregarded.)

A completely different way of assigning a decimal value to a binary word is used in *floating-point representations*. Here a decimal number A is first written as:

$$A = M \times 2^E \qquad (10.1)$$

where E is a positive or negative integer and $0.5 \le |M| < 1$. M is called the *mantissa* and E the *exponent*. Each is represented as a binary word in one of the fixed-point notations, the mantissa always having just one bit (the sign bit) before the point. The two words taken together represent the decimal number.

Example
The number 3.5 is first rewritten as $(+0.875) \times 2^{(+2)}$. With a 4-bit mantissa and a 3-bit exponent this gives:

$$3.5 \leftrightarrow \underbrace{0.111}_{\text{mantissa}} \underbrace{010}_{\text{exponent}} \qquad (10.2)$$

The great advantage of this type of representation is that we can represent a wide range of numbers; the small numbers are close to one another (e.g. 4/64, 5/64, 6/64 and 7/64) and the large numbers are far apart (e.g. 4, 5, 6 and 7). This means that successive numbers have about the same relative spacing throughout the entire range (verify this for the preceding example of a 4-bit mantissa and a 3-bit exponent).

A disadvantage of floating-point representations is that the operations are more complicated. To multiply two numbers we have to multiply the two mantissas and add the exponents. We then have to make sure that the result satisfies condition (10.1), if necessary by adjusting the mantissa and exponent obtained. When adding two numbers we first have to make sure that the exponents of both numbers are the same. Then we have to add the mantissas and if necessary adjust the result again to satisfy condition (10.1). These operations are clearly more complicated than those occurring in the fixed-point representations.

Practically the only time when we encounter floating-point representations is when we are using a computer for digital signal processing (e.g. for FFT calculations); in other cases (e.g. digital filters in the form of specially designed integrated circuits) fixed-point representations are generally used. We shall therefore give most attention to this last category in the rest of the treatment.

10.3 QUANTIZATION AND OVERFLOW

In working with digital signals and digital systems with finite word length we frequently encounter the idea of *quantization*. This is the process in which a quantity x is converted into a quantity x_Q that is approximately equal to x, but can assume fewer different values than x. For example, substituting the nearest integer to any real number is quantization (4 for 3.67, say). Another example is reducing the word length of a binary quantity x by reducing the number of bits *after* the point (1011.110111 replaced by 1011.110). The

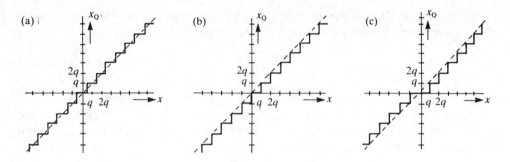

FIGURE 10.1 Three quantization characteristics: (a) Rounding. (b) Value truncation. (c) Magnitude truncation.

relation between x and x_Q is called the *quantization characteristic*. The most widely encountered forms of quantization in digital signal processing are:

1. Rounding
2. Value truncation
3. Magnitude truncation

Figure 10.1 shows the quantization characteristics corresponding to these three cases. The successive possible values of x_Q are separated by a fixed distance q (the *quantization step*). We can see that the conversion of x into x_Q is almost always associated with the introduction of inaccuracy. The magnitude of this inaccuracy, $x_Q - x$, is limited, however, and we can express it in terms of q:

1. For rounding: $\qquad\qquad\quad -q/2 \le x_Q - x < q/2$ $\qquad\qquad\qquad\quad$ (10.3)

2. For value truncation: $\qquad\quad -q \le x_Q - x < 0$ $\qquad\qquad\qquad\qquad$ (10.4)

3. For magnitude truncation: $\quad -q \le x_Q - x < 0 \ \ \text{if} \ x > 0$ $\qquad\qquad\quad$ (10.5)
$$0 \le x_Q - x < q \ \ \text{if} \ x < 0$$

Quantization is essentially a nonlinear operation since in general $(x + y)_Q \ne x_Q + y_Q$.

Another form of nonlinearity that we may encounter in digital signals and digital systems is known as overflow. *Overflow* is what happens when a quantity x seeks a value outside the limits ($-X$ and $+X$) that we must observe. In formal terms we can describe this as a conversion of x into x_P, where

$$x_P \begin{cases} = x & \text{if } |x| \le X \\ \le X & \text{if } |x| > X \end{cases} \qquad\qquad (10.6)$$

The relation between x and x_P is called the *overflow characteristic*. Three examples are given in Fig. 10.2:

1. Saturation
2. Zeroing and
3. 'Sawtooth' overflow

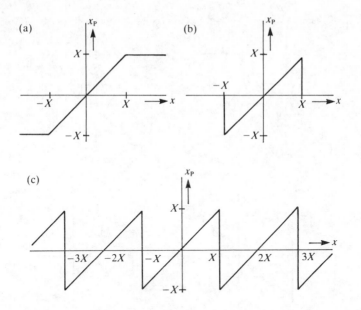

FIGURE 10.2 Three overflow characteristics: (a) Saturation. (b) Zeroing. (c) 'Sawtooth'.

(The third type of overflow is attractive when working in two's complement; why?) From (10.6) and Fig. 10.2 we see that no errors are introduced provided $|x| \leq X$. But for other values of x there is an error $x_P - x$, which in principle is unlimited. For example, we are dealing with an overflow characteristic as shown in Fig. 10.2(a) if for any *real* number we substitute the nearest *real* number between +Ngg and −Ngg.

In digital signal processing we may encounter overflow if we want to reduce the word length of a binary quantity x by reducing the number of bits *before* the point (e.g. for 1011.110111 we would substitute 11.110111).

In certain situations there is a combination of quantization and overflow. This is the case, for example, if we want to substitute, for any *real* number, the nearest *integer* between +Ngg and −Ngg. For numbers whose absolute value is less than Ngg we are dealing with quantization (rounding) and for the other values we are dealing with overflow (saturation). These two effects can be represented in a single characteristic x_F. This has been done in Fig. 10.3 for quantization in the form of rounding and overflow as shown in each of the three cases of Fig. 10.2.

10.4 A/D-CONVERSION NOISE

The first point in a digital system where we encounter quantization is usually in the conversion of an analog input signal into a digital input signal. Besides the changeover from continuous time to discrete time (by sampling), we also have to make the change-over from continuous amplitude to discrete amplitude (by quantization). Here we often make use of rounding (Fig. 10.4).

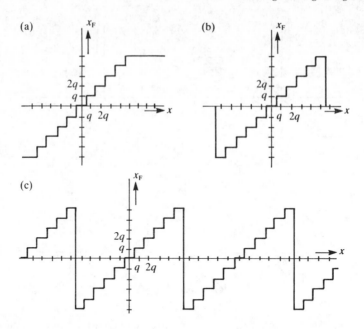

FIGURE 10.3 Combining quantization (rounding) with different types of overflow.

Each input sample $x[n]$ is converted into an output sample $x_Q[n]$. In doing this we introduce an error $e[n]$, given by:

$$e[n] = x_Q[n] - x[n] \qquad (10.7)$$

It can be seen immediately – see also (10.3) – that for Fig. 10.4 we have:

$$|e[n]| \leq q/2 \qquad (10.8)$$

where q is the magnitude of the quantization step. This equation only applies, however, if there is no overflow. If the samples of $x_Q[n]$ have a word length of B bits, then for $x_Q[n]$:

$$-q \cdot 2^{B-1} \leq x_Q[n] \leq q \cdot 2^{B-1} \qquad (10.9)$$

To avoid serious errors due to overflow, we have to make sure that $|x[n]|$ never exceeds – or hardly ever exceeds – the maximum values of $|x_Q[n]|$. We can do this by choosing values of q and B for a given signal $x[n]$ of power[†] P_x (and hence an r.m.s. value $\sqrt{P_x}$) such that:

$$K \cdot q \cdot 2^{B-1} = \sqrt{P_x} \qquad (10.10)$$

†The power P_x of a discrete signal $x[n]$ is defined as:

$$P_x = \lim_{N \to \infty} \left\{ \frac{1}{2N+1} \sum_{n=-N}^{N} x^2[n] \right\}.$$

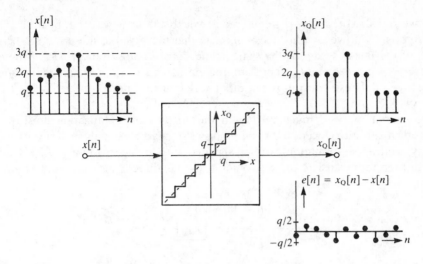

FIGURE 10.4 Quantization in A/D conversion.

where K is a certain factor less than 1. In practice K is often given a value of 1/4. The probability of overflow is then usually negligible (e.g. less than 0.01%), and in the conversion of $x[n]$ into $x_Q[n]$ we only have to take account of the effects resulting from quantization. We now rewrite (10.7) as:

$$x_Q[n] = x[n] + e[n] \qquad (10.11)$$

The quantization of $x[n]$ can therefore be regarded as the addition of a noise signal $e[n]$. This signal is called the *quantization noise*. We can represent Fig. 10.4 in a different, but completely equivalent way, as has been done in Fig. 10.5.

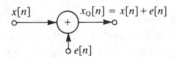

FIGURE 10.5 Quantization represented as the addition of quantization noise $e[n]$.

The quantized signal $x_Q[n]$ is usually applied to a linear system (Fig. 10.6) with an impulse response $h[n]$. The output signal $y[n]$ of this system consists of the filtered version of $x[n]$ plus the filtered version of $e[n]$; we shall call these $v[n]$ and $g[n]$ respectively.

From Fig. 10.6 we see that:

$$\begin{aligned} y[n] &= \{x[n] + e[n]\} * h[n] \\ &= x[n] * h[n] + e[n] * h[n] \\ &= v[n] + g[n] \end{aligned} \qquad (10.12)$$

The two signals $x[n]$ and $e[n]$ are filtered in exactly the same way, i.e. by the same filter characteristic. However, this does *not* mean that the signal-to-noise ratio P_x/P_e at the input of the linear system is the same as the signal-to-noise ratio P_v/P_g at the output (where P_x, P_e, P_v and P_g represent the powers of $x[n]$, $e[n]$, and $v[n]$ and $g[n]$ respectively)! These two ratios can differ widely if the spectra of $x[n]$ and $e[n]$ differ widely.

We cannot proceed much further in the analysis of quantization noise with the theoretical aids introduced so far, since noisy ('stochastic') signals such as $e[n]$ and such as any arbitrary information-carrying signal $x[n]$ (e.g. speech) have *no* FTD, and we cannot therefore translate the convolutions (10.12) directly into a multiplication in the

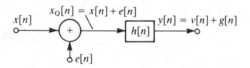

FIGURE 10.6 A linear system with a quantized input signal.

frequency domain. At this point we really need to bring in the theory of stochastic discrete signals, but that would take us beyond the scope of this book. Nevertheless, we shall use some of the results based on this theory, and we shall proceed from a number of common assumptions:

1. The samples of $e[n]$ assume all values between $+q/2$ and $-q/2$ with equal probability.
2. There is no correlation between the individual samples of $e[n]$.
3. The signal $e[n]$ is not correlated with the signal $x[n]$.

(In rather more general terms, assumptions 2 and 3 mean that the value of a certain sample $e[n]$ does not depend on the values of preceding or subsequent samples of $e[n]$ and $x[n]$.)

In practice these assumptions turn out to be justified, especially for fairly random signals (such as speech and music) and where there is a sufficiently large number of quantization steps in the quantization characteristic. From the theory of stochastic signals it can be shown that a signal (continuous or discrete) in which all amplitude values between $+A$ and $-A$ can appear with equal probability has an r.m.s. value of $A/\sqrt{3}$ and hence a (mean) power[46] of $\frac{1}{3}A^2$. Taking note of assumption (1) above, we therefore find for the power P_e of signal $e[n]$:

$$P_e = q^2/12 \qquad (10.13)$$

Assumption (2) means that the power P_e is evenly distributed over all frequencies in the fundamental interval: we say that $e[n]$ has a flat ('white') spectrum. If we apply such a signal to a linear network with a frequency response $H(e^{j\theta})$, we find at the output a signal whose power P_g is given by:

$$P_g = \frac{1}{2\pi} \int_{-\pi}^{\pi} P_e \cdot |H(e^{j\theta})|^2 \, d\theta = \frac{q^2}{12} \cdot \frac{1}{2\pi} \int_{-\pi}^{\pi} |H(e^{j\theta})|^2 \, d\theta \qquad (10.14)$$

(We state this relation here without derivation. We shall use it later; see also Appendix III.3.) On the basis of assumption (3), combining (10.10) and (10.13) gives us the signal-to-noise ratio P_x/P_e:

$$\frac{P_x}{P_e} = \frac{K^2 q^2 2^{2(B-1)}}{q^2/12} \tag{10.15a}$$

With $K = 1/4$ this gives:

$$\frac{P_x}{P_e} = \frac{12}{16} 2^{2(B-1)} = \frac{3}{16} 2^{2B} \tag{10.15b}$$

or expressed in decibels:

$$10\log_{10}(P_x/P_e) = 10\log_{10}\left(\frac{3}{16}2^{2B}\right) = 20B\log_{10}(2) + 10\log_{10}(3/16)$$

$$= (6B - 7.3)\,\text{dB} \tag{10.16}$$

The constant term $(-7.3\,\text{dB})$ in (10.16) depends directly on the choice of K. We also see that the signal-to-noise ratio increases by 6 dB for every bit we add to the word length of $x_Q[n]$.

We should not forget that we assumed above that we were starting from a noise-free signal $x[n]$. If $x[n]$ has been derived from an analog signal, however, it is always accompanied by a certain amount of noise (for an analog telephone signal, for example, a signal-to-noise ratio of 36 dB is not bad). There is then no point in quantizing to a very high accuracy (e.g. $B = 14$), since the least-significant bits only give a precise representation of the analog noise and might just as well be discarded!

10.5 QUANTIZATION OF FILTER COEFFICIENTS

From the design procedures for discrete filters that we described in Chapter 8 'Design methods for discrete filters', we generally find values for the filter coefficients to a very high accuracy. To obtain a realizable digital filter in which the coefficients only have a limited word length, we have to quantize these values. However, when we do so, we alter the corresponding frequency response of the filter, or – in other words – the positions of the poles and zeros of the filter. These changes can be quite considerable. It can happen that after quantization the filter no longer meets the specification on which the calculations of the non-quantized coefficients were based. We should like to illustrate this with the aid of Fig. 10.7, where we show how the amplitude characteristic A of a digital filter can change when we quantize the coefficients.

In extreme cases even a stable filter can change into an unstable filter, if a pole shifts from a position inside the unit circle in the z-plane to a position outside it! Quantizing the coefficients does *not* introduce any nonlinearities *nor* does it introduce any effects that

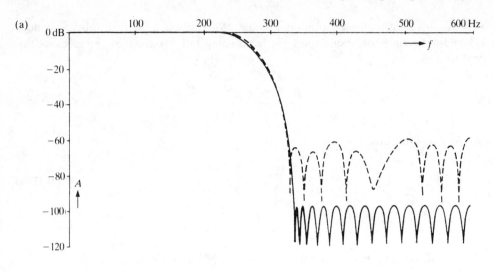

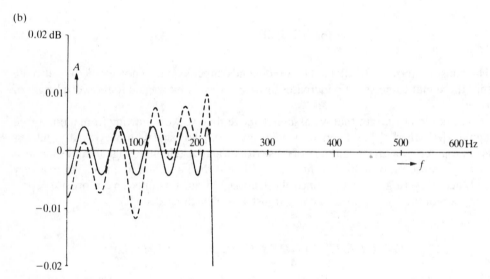

FIGURE 10.7 Amplitude characteristic A of a transversal filter[47] ($L = 49$) with unquantized coefficients (———) and coefficients rounded-off to 12 bits (-----). (a) Coarse dB to show changes in the stop band clearly. (b) Fine dB scale for the pass band.

depend on the input signal or vary with time. It results only in a 'once-only' change in the properties of the filter.

We can analyze this change in detail by calculating the impulse response or the frequency response that results after quantization. From this we can see whether the filter still meets the design specification. It is good practice to make the specification tighter than is strictly necessary for the original calculation of the unquantized coefficients, so that there is a margin for the errors that arise because of the subsequent quantization.

It has been found that one filter structure may be much more sensitive to quantization of the coefficients than another. In general we can say that a filter structure becomes *less* sensitive as the location of each pole and each zero becomes dependent on *fewer* coefficients:

1. In *direct form 1* and *direct form 2* each pole is affected by the values of *all* the coefficients a_i and similarly each zero by the values of *all* the coefficients b_i (section 7.3). Consequently, small changes in a_i and b_i can produce large shifts in the poles and zeros. This is particularly so if the poles or the zeros – or both – are close together, as in narrow-band filters, for example.
2. In the *parallel structure* (section 7.3.4) each of the poles (or pairs of poles) is determined by a small number of coefficients in each of the parallel branches. The zeros, on the other hand, arise because of cancellations between output signals on the different branches and therefore depend on all the coefficients. This means that this kind of filter is fairly insensitive in the pass band (which is mainly determined by the poles), but very sensitive in the stop band (which is mainly determined by the zeros).
3. In the *cascade structure* (section 7.3.3) both the poles and the zeros are always determined by a small number of coefficients. Filters with this structure therefore have low sensitivity both in the pass band and in the stop band. An important property is also found in the cascade connection of first-order sections and second-order sections where both have the direct-form structure. Zeros on the unit circle still remain on the unit circle after quantization, though they may have shifted a little (see Exercises 10.4 and 10.5).
4. The *transversal filter* (section 7.2): the properties of this most popular form of FIR filter are fairly similar to those described above under point (1). Each of the zeros is determined by the value of *all* the coefficients b_i. This filter is therefore also sensitive to quantization of the coefficients. However, it does have the advantage that if a linear phase characteristic is desired, this linearity is retained after quantization, since it is easy to ensure the symmetry of the coefficient values that this requires.
5. *Ladder filters*, *lattice filters* and *wave digital filters* (sections 7.4.3 and 8.3.4): all these filters are based on continuous filters that are characterized by low sensitivity to parameter variations. The transformation from continuous to discrete filter results in this property being retained, so that the sensitivity to quantization of the coefficients is also low.

We can obtain some idea of the effect of quantization on the positions of poles and zeros by taking a closer look at a fairly simple filter. Let us consider a purely recursive second-order filter section with a direct-form structure (Fig. 10.8). The system function $H(z)$ of this filter is:

$$H(z) = \frac{Y(z)}{X(z)} = \frac{1}{z^2 - a_1 z - a_2} \tag{10.17}$$

This function has two poles $z_{1,2}$. If these poles are complex conjugates, then we must have:

$$z_{1,2} = Re^{\pm j\varphi} = R\{\cos(\varphi) \pm j\sin(\varphi)\} \tag{10.18}$$

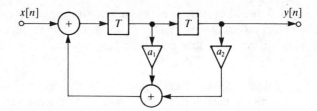

FIGURE 10.8 Purely recursive second-order filter (direct form).

Therefore:

$$H(z) = \frac{1}{z^2 - a_1 z - a_2} = \frac{1}{(z - Re^{j\varphi})(z - Re^{-j\varphi})}$$

$$= \frac{1}{z^2 - 2Rz\cos(\varphi) + R^2} \qquad (10.19)$$

or

$$a_2 = -R^2 \quad \text{and} \quad a_1 = 2R\cos(\varphi) \qquad (10.20)$$

If a_1 and a_2 are unquantized, we can obtain any desired pair of complex conjugate poles by a correct choice of these two parameters. After quantization of a_1 and a_2, however, only a limited number of pole pairs can be realized, because a_1 and a_2 (and hence R and φ also) can only assume a finite number of different values. Figure 10.9 shows an example of the poles that can be obtained inside the unit circle if a_1 and a_2 are quantized to 4 bits (with one bit for the sign). For reasons of symmetry only one quadrant of the z-plane is shown. We see from Fig. 10.9 that the possible positions for the poles are not distributed evenly over the z-plane. The largest errors in the frequency response will most probably

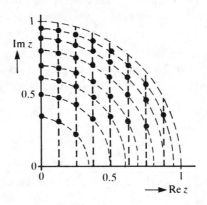

FIGURE 10.9 Possible positions for the poles of the filter of Fig. 10.8, if a_1 and a_2 are quantized to 4 bits.

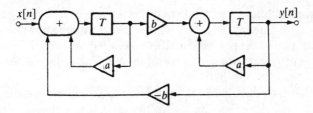

FIGURE 10.10 Second-order section in the 'coupled-form' structure.

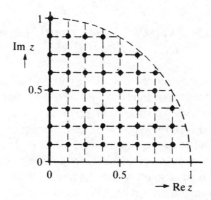

FIGURE 10.11 Possible positions for the poles of the network of Fig. 10.10, if both *a* and *b* are quantized to 4 bits.

occur when the desired filter has poles in a region where the realizable poles are far apart. The possible positions for the poles are strongly dependent on the filter structure. To illustrate this we shall consider the filter of Fig. 10.10.

The system function of this filter is:

$$H(z) = \frac{Y(z)}{X(z)} = \frac{b}{z^2 - 2az + a^2 + b^2} = \frac{b}{(z - a + jb)(z - a - jb)} \tag{10.21}$$

This system function can have two complex conjugate poles:

$$z_{1,2} = R\{\cos(\varphi) \pm j\sin(\varphi)\} \tag{10.22a}$$

Therefore:

$$H(z) = \frac{b}{[z - R\cos(\varphi) + jR\sin(\varphi)][z - R\cos(\varphi) - jR\sin(\varphi)]} \tag{10.22b}$$

From (10.21) and (10.22b) it follows that:

$$a = R\cos(\varphi) \quad \text{and} \quad b = R\sin(\varphi) \tag{10.23}$$

If we again quantize a and b to 4-bit coefficients (with one bit for the sign), we can realize poles inside the unit circle in the z-plane at the positions shown in Fig. 10.11. We see that the distribution of the poles in Fig. 10.11 is much more even than in Fig. 10.9. The effects of coefficient quantization can therefore be very different for the two filter structures.

As a final comment, we should like to state here that structures in direct form 1, direct form 2 or their transposed version are all exactly equivalent with regard to quantization of the coefficients. This is no longer true, however, for the limitation of the word length of intermediate results, which will be discussed in the next section.

10.6 LIMITATION OF THE WORD LENGTHS OF INTERMEDIATE RESULTS

The most complicated consequences of working with a finite word length are found in the limitation of the word lengths of intermediate results in digital systems.[48] By an intermediate result we mean the outcome of an addition or multiplication performed somewhere in the digital system. We shall assume here that we want to realize a digital system in which all quantities are represented as B-bit words in a fixed-point representation with only one bit (the sign bit) before the point. This means that we can only represent quantities with an absolute value less than 1. What happens if we now add two such numbers together or multiply them? We shall illustrate this by way of two simple examples with $B = 5$.

Example of addition:

Binary	Decimal
0.1101	+0.8125
0.1001	+0.5625
——————+	—————— +
01.0110	+1.3750

Example of multiplication:

Binary	Decimal
0.1101	+0.8125
0.1001	+0.5625
—————— ×	—————— ×
0.01110101	+0.45703125

We see that the addition of two B-bit words can give a total of $B + 1$ bits and that the product requires $2B - 1$ bits. The significant difference here is that addition gives an increase in the number of bits *before* the point whereas multiplication gives an increased number of bits *after* the point. If we want to go back to our original number representation, then for addition we may be concerned with *overflow*, while for multiplication we are only concerned with *quantization*.

We should like to illustrate this with the aid of a number of simple filters. Figure 10.12 shows a simple FIR filter, in which we impose no requirements on the word length. The filter has a B-bit input signal and B-bit coefficients. We see how the word length in the filter gradually increases and results in an output signal with a word length of $2B$ bits. If we do not want this increase, we must take steps to prevent it. We can choose the place where we limit the word length, and the way in which we do this. We now introduce

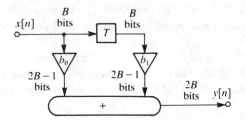

FIGURE 10.12 Increase in the word length in a simple FIR filter.

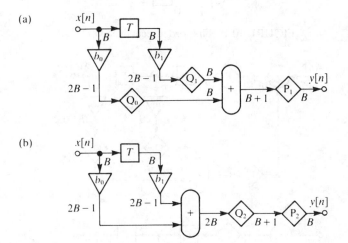

FIGURE 10.13 Simple FIR filter with limitation of the word length.

quantization (indicated by Q) or overflow (indicated by P) – or both – as discussed in section 10.3. Figure 10.13 shows two possible ways of obtaining an output signal of word length B. In one case we have two quantizers, in the other only one quantizer, but not so simple an addition. With nonrecursive filters, as in the preceding example, we can put up with the increase in word length in principle, since it always remains finite.

In recursive filters, however, the situation is completely different. Here we are *forced* to restrict the word length, since it would otherwise increase without limit. This can be seen from the first-order recursive filter in Fig. 10.14. This filter is described by the difference equation:

$$y[n] = v[n] + x[n] = a_1 y[n-1] + x[n] \qquad (10.24)$$

If we assume that a_1 and $y[n-1]$ are both represented as B-bit words, then in the absence of any special precautions $v[n]$ will be a $(2B-1)$-bit word and $y[n]$ a $2B$-bit word. In calculating the next output sample $y[n+1]$ we must multiply a_1 by a $2B$-bit word, so that we obtain a $3B$-bit output sample. And in the next calculation we obtain a $4B$-bit sample, and so on. In recursive filters limitation of the word length of the intermediate results is therefore unavoidable! Figure 10.15 shows a diagram of a purely recursive second-order filter in which word-length limiters (Q and P) have been included in two different ways.

The great problem in analyzing word-length limitation for intermediate results is really that, strictly speaking, the filter is now *nonlinear*. If we take great care in applying our special precautions, we shall indeed be able to describe our filter to a first approximation

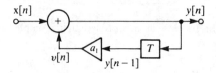

FIGURE 10.14 First-order recursive filter.

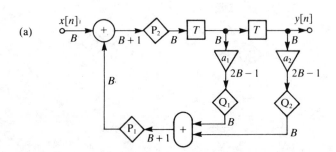

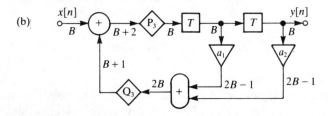

FIGURE 10.15 Recursive second-order filter with word-length limitation.

as the linear filter we set out to design. However, a number of specific effects can arise (such as oscillations in recursive filters), which betray the nonlinear nature of the filter. These effects occur for particular kinds of input signals, such as a constant input signal (especially $x[n] = 0$) or a periodic input signal (such as a sinusoidal signal). In addition, the nonlinearity due to overflow (P) causes different effects from the nonlinearity due to quantization (Q). Moreover, the resultant effects also depend on:

- The structure of the filter (recursive, nonrecursive, cascade structure, parallel structure, direct form, wave digital filter, etc.).
- The position of the word-length limiters.
- The characteristics of the word-length limiters (rounding, truncation, saturation, zeroing, etc.).

We shall try to introduce some order into this maze of different possibilities.

10.6.1 Overflow of intermediate results

Overflow in a digital filter can result in considerable errors in the output signal as compared with the output signal from the linear filter that we would really like to have. In a filter with a nonrecursive structure these errors are of finite duration (never longer than the duration of the impulse response). In a recursive filter the consequences of a single occurrence of overflow may be of unlimited duration and cause all kinds of undesirable effects. We do not want to give a detailed analysis of these effects here. We shall merely list some of them as an indication of the type of problems we may encounter in practice.

1. If the input signal $x[n]$ is equal to zero from a particular value of n *after* the occurrence of overflow, a permanent oscillation (*overflow oscillation*) can occur. This oscillation has a large amplitude that is related to the overflow level (see Exercise 10.6).
2. If the input signal $x[n]$ is periodic, completely different output signals can occur for the same input signal as a result of overflow and depending on the initial states of the filter registers.
3. If the input signal $x[n]$ is periodic, small changes in the input signal can result in large changes in the output signal as a result of overflow (*jump phenomena*).
4. If the input signal $x[n]$ is periodic, signals can occur at a lower frequency (subharmonics) as a result of overflow. Because of this effect an input signal in the stop band of a filter may produce a signal in the pass band!
5. If the input signal is sufficiently random, overflow will not readily produce clearly recognizable effects at the output. In spite of this, there will be additional errors as compared with the ideal linear filter. Not very much is known about this.

All in all, there are plenty of reasons for avoiding overflow. In principle this can be done by *scaling*. This means that the input signal of the filter is multiplied by a factor $S < 1$ such that overflow can no longer occur. It is preferable to use an integer power of 2 here (e.g. $2^{-2} = 0.25$) because multiplication then really only amounts to shifting the bits (this scaling is itself again combined with quantization to prevent an increase in the word length!). In this way we can eliminate the overflow nonlinearities P_1 and P_2 in Fig. 10.13. For Fig. 10.13(a) this results in the circuit of Fig. 10.16.

The scaling factor S can be combined with the coefficients b_0 and b_1 (Fig. 10.17). We can also eliminate the overflow nonlinearities in recursive filters in a similar way. This is illustrated in Fig. 10.18 for the filter we have already encountered in Fig. 10.15(a). Now, however, the scaling factor S *cannot* be combined with the filter coefficients, as this would change the entire frequency response of the recursive section! If general, the scaling

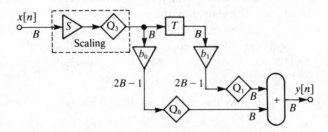

FIGURE 10.16 Scaling in a transversal filter, with the result that there is no overflow.

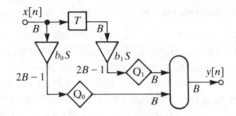

FIGURE 10.17 Scaling included in the coefficients.

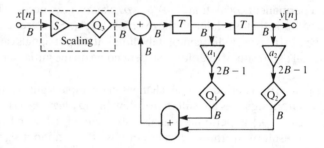

FIGURE 10.18 Scaling in a second-order recursive filter.

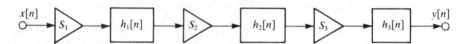

FIGURE 10.19 Scaling in a cascade structure.

factors can be combined with coefficients that determine the positions of the zeros and *not* with coefficients that determine the positions of the poles. If we are dealing with a filter that has a cascade structure, scaling is included between every two sections (Fig. 10.19). Scaling factors $S > 1$ can then occur if the output signal from a particular section is always very small.

The magnitude of each scaling factor must be selected very carefully. On the one hand its value must be such that overflow is prevented, while on the other hand too small a value results in the most-significant bits in the following filter section(s) not being used. In such a case the inaccuracies introduced by the quantizers Q play relatively too large a part. This gives a worse signal-to-noise ratio than possible at the output of the filter! To determine the correct value of S it is necessary to know both the signal to be scaled and the actual filter (or the filter section). Various ways of calculating S on this basis are given in the literature.[49,50] Several different criteria can be used:

1. *Worst-case criterion*: overflow must never occur as long as $|x[n]|$ remains below a certain value;
2. *Power criterion*: overflow must never occur as long as the power P_x remains below a certain value;

3. *Sine criterion*: overflow must never occur with sinusoidal signals of amplitude below a certain value.

If the scaling factor is determined by using the first criterion, overflow can never occur. If the second or the third criterion is used, this is not absolutely certain, but in general a shorter word length will suffice for a particular signal-to-noise ratio at the output. In a good design, overflow will be the exception. However, we have to make a conscious choice of the form of overflow that will occur in these exceptional cases. In general, saturation introduces fewer undesirable effects than the other types, such as zeroing or sawtooth overflow. A possible disadvantage here is that saturation requires a more complex operation (in two's complement sawtooth overflow is most easily realized). In the next section we shall assume that overflow has been effectively eliminated by the use of scaling, and we shall turn our attention to the residual nonlinearity due to the quantization of intermediate results.

10.6.2 Quantization of intermediate results

We shall now consider digital systems in which overflow cannot occur, because of correct scaling. Each quantization of an intermediate result corresponds to the introduction of an inaccuracy within the limits given in one of the equations (10.3), (10.4) or (10.5). If the successive values of the intermediate results are sufficiently random, we can think of these inaccuracies as an interfering signal (quantization noise) that is added to the useful signal *at the place where the quantizer is located*. This corresponds exactly to what we did for A/D conversion in section 10.4. We shall again restrict ourselves to quantizers that operate with rounding. (Magnitude truncation is dealt with in the literature.[56,57]) We can now make full use of the results in section 10.4. Considering the filter in Fig. 10.17 once again, we replace each of the quantizers by an adder with a noise source (Fig. 10.20). Each noise source produces quantization noise with a power:

$$P_e = q^2/12 \tag{10.25}$$

where q is the quantization step (equal to the value of the least-significant bit). If we assume that the noise signals are not correlated, we can combine the two noise sources

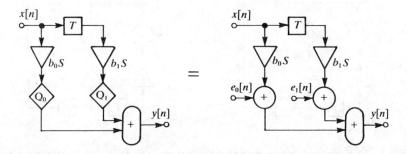

FIGURE 10.20 Quantizers modeled as quantization-noise sources.

$e_1[n]$ and $e_2[n]$ to form a single noise source $e_{tot}[n]$ (Fig. 10.21). The total quantization-noise power at the output of the filter then becomes:

$$P_{tot} = 2P_e = q^2/6 \qquad (10.26)$$

We can also apply this approximation to other kinds of filters. Figure 10.22 shows a second-order recursive filter with the direct-form-1 structure. It contains five quantizers Q. Next to it is a model containing five uncorrelated noise sources $e_0[n], \ldots, e_4[n]$. These five noise sources can be combined to form one total noise source $e_{tot}[n]$ with a noise power P_{tot}:

$$P_{tot} = 5P_e = 5q^2/12 \qquad (10.27)$$

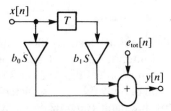

FIGURE 10.21 Transversal filter in which the quantization is represented by a single noise source $e_{tot}[n] = e_0[n] + e_1[n]$.

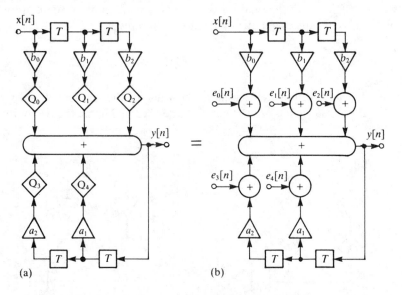

FIGURE 10.22 Recursive section with the direct-form-1 structure. (a) With quantizers and (b) with quantization-noise sources.

Is this the total noise power at the output of the filter? To find out we redraw Fig. 10.22 as in Fig. 10.23. We see that the quantization noise $e_{tot}[n]$ appears *filtered* at the output. However, the noise is not filtered by the total frequency response $H(e^{j\theta})$:

$$H(e^{j\theta}) = \frac{N(e^{j\theta})}{D(e^{j\theta})} = \frac{b_0 + b_1 e^{-j\theta} + b_2 e^{-2j\theta}}{1 - a_1 e^{-j\theta} - a_2 e^{-2j\theta}} \tag{10.28}$$

but only by:

$$\frac{1}{D(e^{j\theta})} = \frac{1}{1 - a_1 e^{-j\theta} - a_2 e^{-2j\theta}} \tag{10.29}$$

Using the results of (10.14) we find for the power P_u of the quantization noise at the output of the filter:

$$P_u = \frac{5q^2}{12} \cdot \frac{1}{2\pi} \int_{-\pi}^{\pi} \left| \frac{1}{D(e^{j\theta})} \right|^2 d\theta \tag{10.30}$$

Let us now use the same method for a second-order recursive filter with the direct-form-2 structure. This has been done in Fig. 10.24.

We see the noise signals $e_0[n]$, $e_1[n]$ and $e_2[n]$ directly at the output of the filter and they produce a noise signal $e_{tot1}[n]$ there with a noise power $P_{tot1} = 3q^2/12$. The noise signals $e_3[n]$ and $e_4[n]$ can also be replaced by a noise signal $e_{tot2}[n]$ with a noise power $P_{tot2} = 2q^2/12$. This noise signal is first filtered, however, before it appears at the output. A close examination shows that it is filtered by exactly the same frequency response $H(e^{j\theta}) = N(e^{j\theta})/D(e^{j\theta})$ as $x[n]$ (Fig. 10.25). On the basis of the results from section 10.4 we now find at the output of the filter a total quantization-noise power P_u given by:

$$P_u = \frac{3q^2}{12} + \frac{2q^2}{12} \cdot \frac{1}{2\pi} \int_{-\pi}^{\pi} \left| \frac{N(e^{j\theta})}{D(e^{j\theta})} \right|^2 d\theta \tag{10.31}$$

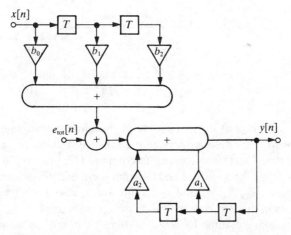

FIGURE 10.23 Recursive section as in Fig. 10.22, in which the quantizers are represented by a single noise source $e_{tot}[n] = e_0[n] + e_1[n] + \ldots + e_4[n]$.

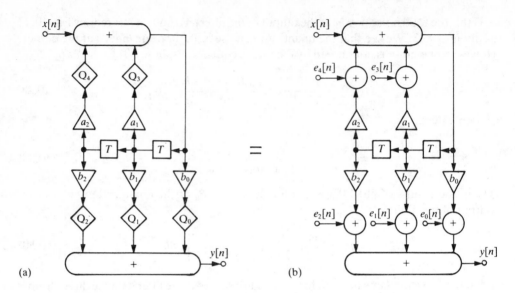

FIGURE 10.24 Recursive section with the direct-form-2 structure. (a) With quantizers and (b) with quantization-noise sources.

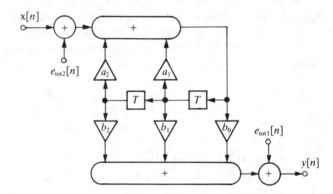

FIGURE 10.25 Recursive section as in Fig. 10.24, with the quantizers replaced by two noise sources $e_{tot1}[n] = e_0[n] + e_1[n] + e_2[n]$ and $e_{tot2}[n] = e_3[n] + e_4[n]$.

In applying this method to any given filter we therefore have to find out separately for each noise source what the frequency response is *between the location of the noise source and the output of the filter*. It is interesting to compare (10.30) and (10.31). We see that a second-order section in direct form 1 and a similar section in direct form 2 with exactly the same coefficients a_1, a_2, b_0, b_1 and b_2 – and hence also the same $H(e^{j\theta})$ – behave differently in relation to quantization of the intermediate results!

The above analysis relating to noise sources is based on assumptions that are sometimes not entirely correct (this applies especially to the assumption that the noise signals are not correlated with one another or with the signals before quantization).

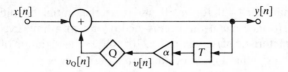

FIGURE 10.26 First-order recursive filter with $\alpha = -0.91$ in which limit cycles can occur.

TABLE 10.3 Limit cycle of Fig. 10.26

n	$v[n] = \alpha y[n-1]$	$y[n] = v_Q[n]$
−1	—	$+7q$
0	$-0.91 \times 7q = -6.37q$	$-6q$
1	$-0.91 \times -6q = +5.46q$	$+5q$
2	$-0.91 \times 5q = -4.55q$	$-5q$
3	$-0.91 \times -5q = +4.55q$	$+5q$
4	$-0.91 \times 5q = -4.55q$	$-5q$
etc.	etc.	etc.

However, there is the advantage that we are not in fact interested in an analysis with an accuracy of 100%. We can only vary the word length in steps of 1 bit, and this corresponds to steps of 6 dB in quantization noise. An analysis with an accuracy of 30 to 40% is therefore amply sufficient for determining the required word length. The above method is therefore perfectly adequate for input signals $x[n]$ of a sufficiently random nature. This situation changes, however, if the input signal has a constant value for a longer time or is of a periodic nature. The errors that are introduced then by the quantizers are certainly no longer uncorrelated and the output signal can have a constant or periodic error ('limit cycle'). A practical situation in which limit cycles are undesirable occurs if $x[n]$ is a digitized speech signal, for example. In the pauses ($x[n] = 0$) periodic interference signals, even at a very low level, soon become very annoying.

To obtain some idea of limit cycles we shall look at the first-order recursive filter of Fig. 10.26, where Q is a quantizer, with a quantization step q, and in which rounding takes place. Let us assume that $y[-1] = 7q$, $x[n] = 0$ for $n \geq 0$ and $\alpha = -0.91$. We can then simply find out how $v[n]$ and $y[n]$ vary for $n \geq 0$. This is shown in Table 10.3. We see that after some time $y[n]$ takes on a periodic behavior of period 2 and amplitude $5q$, even though the input signal $x[n]$ is equal to zero! The properties of the limit cycle depend on the initial state $y[-1]$, the value of α and the quantization characteristic of Q. In this first-order recursive filter limit cycles can always occur if the quantization characteristic is based on rounding and if $|\alpha| > 0.5$. Limit cycles can only be avoided for each $|\alpha| < 1$ if magnitude truncation is used in the quantization (Fig. 10.1(c)). This type of quantization gives more quantization noise than rounding, however; see (10.5).

Limit cycles can also occur in higher-order recursive filters. Figure 10.27 shows a second-order recursive filter with coefficients 1.9 and −0.98. It is described by the difference equations:

$$y[n] = x[n] + v_Q[n] \tag{10.32a}$$

$$v_Q[n] = \{1.9y[n-1] - 0.98y[n-2]\}_Q \tag{10.32b}$$

If we assume that $x[n] = 0$ for $n \geq 0$ and that the initial conditions are $y[-2] = y[-1] = 11q$, then once again the occurrence or not of a limit cycle depends on the quantization characteristic of Q. Figure 10.27(b) shows the variation of $y[n]$ obtained if Q is based on rounding; a limit cycle occurs with a period of $N = 20$ and an amplitude of $11q$. Figure 10.27(c) shows $y[n]$ for the case in which Q is based on magnitude truncation; here there is *no* limit cycle (verify this).

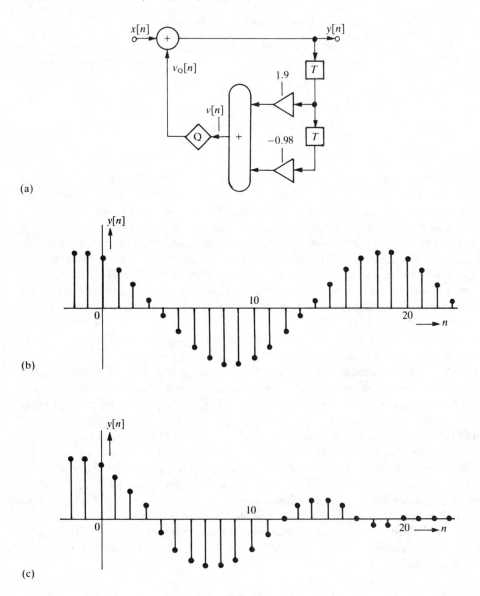

(a)

(b)

(c)

FIGURE 10.27 (a) Second-order recursive filter with coefficients 1.9 and −0.98. (b) In quantization based on rounding a limit cycle can occur here. (c) In quantization based on magnitude truncation no limit cycle occurs.

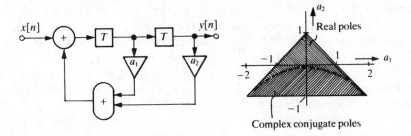

FIGURE 10.28 Purely recursive second-order section and the values of a_1 and a_2 for which this section (without quantization) is stable.

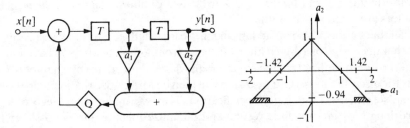

FIGURE 10.29 Purely recursive second-order section which is most suitable for preventing limit cycles (Q is a quantizer based on magnitude truncation; the hatched regions indicate conditions in which limit cycles can still occur).

The results published in the literature relate in the main to the second-order purely recursive filter section. This is interesting because of the fact that higher-order filters are often built up from second-order sections and the total filter can only be free of limit cycles if each of the second-order sections is itself free of limit cycles. Figure 10.28 shows such a section (without quantization) again. This has been done to introduce a new aid: the (a_1, a_2)-plane. In this plane we can indicate certain properties of the filter that depend on the values of a_1 and a_2. In Fig. 10.28, for example, the region inside the triangle corresponds to stable filters (poles inside the unit circle in the z-plane) and the region outside it corresponds to unstable filters. An indication of when the poles are real and when they are complex is also given.

We shall now return to the subject of limit cycles due to quantization. It turns out that there is one type of second-order purely recursive filter section which is most suitable for preventing limit cycles. This is shown in Fig. 10.29. It contains a single quantizer that operates with magnitude truncation. The hatched regions in the (a_1, a_2)-plane indicate when limit cycles are still possible.[48]

- *Comment*: This structure is not the optimum one for low quantization noise, in which the quantization would be based on rounding!

Unlike overflow oscillations, which we discussed earlier, limit cycles generally have a relatively small amplitude – perhaps of the order of magnitude of a few quantization steps

q. As the selectivity of the filter increases (poles moving closer to the unit circle in the z-plane or combinations (a_1, a_2) moving closer to the edges of the stability triangle) larger amplitudes can also occur. If we reduce the quantization step q of all the quantizers in a given filter in which limit cycles occur (by increasing the word length), the absolute amplitude (i.e. the amplitude expressed as an absolute level and not as a number of quantization steps) will also be reduced. If we then quantize the output signal $y[n]$ we can make a filter that is *apparently* free of limit cycles. There are also more complicated methods of reducing or eliminating limit cycles, with impressive names like 'randomized quantization', 'controlled quantization' and 'error feedback'.

10.7 CONCLUDING REMARKS ON WORD-LENGTH LIMITATION

Some important aspects of limiting the word length of intermediate results in recursive digital filters are listed in Table 10.4.

After all that has been said about limiting the word length of intermediate results it will be clear that the design of a digital filter will be far from complete when we have found the values of the non-quantized coefficients from the design procedures of Chapter 8 'Design methods for discrete filters'. Our next step will then be to quantize the coefficients in such a way that we still meet the design specification. Then we analyze the effects of limiting the word length of intermediate results and minimize them as far as possible. In the widely used cascade connection of second-order sections we still have four degrees of freedom:

1. *Pairing*: which poles and zeros are paired in a particular section?
2. *Ordering*: in what sequence are the sections cascaded?
3. *Structuring*: what does each second-order section look like? (e.g. direct form 1, direct form 2, number and position of quantizers, quantization characteristic, overflow characteristic).
4. *Scaling*: what constant factor should be used to multiply the signals between the different sections?

TABLE 10.4 List of the effects of limiting the word length of intermediate results in digital filters

	Quantization	Overflow
Input signal $x[n] = 0$	Limit cycles (relatively small amplitude)	Overflow oscillations (relatively large amplitude)
Periodic input signal (e.g. sinusoidal)	Limit cycles or quantization noise (in combination with desired output signal)	Jump phenomenon
Random input signal	Quantization noise (in combination with desired output signal)	Little known about this
Most suitable characteristic	• for quantization noise: rounding • for limit cycles: magnitude truncation	Saturation

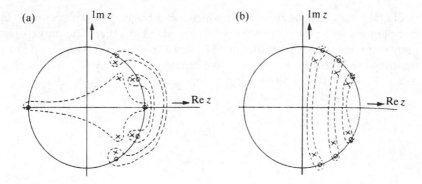

FIGURE 10.30 Pairing of poles and zeros of: (a) sixth-order band-pass filter; (b) sixth-order band-stop filter.

We have already said something about scaling and structuring. We can only consider ordering and pairing in very general terms here. A good rule of thumb for pairing is: realize poles and zeros that are located close together in the same section. Two examples of pairing by this rule are given in Fig. 10.30: the dashed lines show which poles and zeros are realized in the same section.

Sometimes it is better to cascade the different sections in order of ascending selectivity, sometimes in order of descending selectivity. Within the scope of this book we must make do with this brief comment; more information will be found in the literature.[8]

10.8 EXERCISES

Exercises to sections 10.2 and 10.3

10.1 We want to make a digital filter in which only 4-bit words occur in a fixed-point representation with two bits after the (binary) point.
   (a) Give a table of all the decimal numbers we can represent in this way and the corresponding binary words, for number representation by sign and magnitude.
   (b) As in (a), but now for two's-complement representation.
   (c) For one reason or another we want to change from 4-bit words to 2-bit words by means of quantization. We do this by neglecting the bits after the point. Draw the quantization characteristic for cases (a) and (b), and find out from this whether we are dealing with rounding, value truncation or magnitude truncation (use Fig. 10.1 here if necessary).

10.2 Consider the decimal values +2.375 and −2.375. We want to represent these values as binary numbers in a 6-bit fixed-point representation, with 3 bits after the point.
   (a) Do this in sign-and-magnitude representation.
   (b) We want to limit the number of bits after the point to one by quantization. Give the resulting decimal and binary values for rounding, value truncation and magnitude truncation.
   (c) Repeat (a) and (b) for two's-complement representation.

10.3 A digital filter contains an adder in which two 3-bit numbers (represented in two's complement with no bits after the point) are added together. The sum $x[n]$ is a 4-bit number in the same representation. The word length of $x[n]$ is reduced to 3 bits by omitting the bit furthest to the left. We can describe this by an overflow characteristic x_P (see Fig. 10.31). Draw x_P.

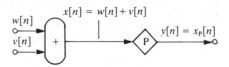

$$x[n] = w[n] + v[n]$$

$$y[n] = x_P[n]$$

FIGURE 10.31 Exercise 10.3.

Exercises to section 10.5

10.4 A second-order recursive filter has a direct-form-1 structure and the system function $H(z)$:

$$H(z) = \frac{1 + b_1 z^{-1}}{1 - a_1 z^{-1} - a_2 z^{-2}}$$

This filter must have a zero on the unit circle in the z-plane.
(a) What values can b_1 assume?
(b) Draw a block diagram of this filter.
(c) Can the zero of $H(z)$ be affected by quantization of the coefficients? Why?

10.5 A second-order recursive filter has a direct-form-2 structure and the system function $H(z)$:

$$H(z) = \frac{1 + b_1 z^{-1} + b_2 z^{-2}}{1 - a_1 z^{-1} - a_2 z^{-2}}$$

This filter must have two zeros on the unit circle in the z-plane.
(a) What values can b_1 and b_2 assume?
(b) Can the zeros of $H(z)$ be affected by quantization of the coefficients? Why?

Exercises to section 10.6

10.6 A second-order digital filter is shown in Fig. 10.32. The input signal $x[n] = 0$ for $n \geq 0$; Q has a quantization characteristic x_Q based on rounding with a quantization step q; P has an overflow characteristic x_P based on zeroing: $x_P = x$ for $|x| \leq 128q$ and $x_P = 0$ for $|x| > 128q$. The filter coefficients are $a_1 = 1.9$ and $a_2 = -0.9995$. The initial state of this filter is given by $y[-2] = -111q$ and $y[-1] = -29q$.
(a) Determine $y[n]$ for $-2 \leq n \leq 10$.
(b) What is the period of the overflow oscillation?

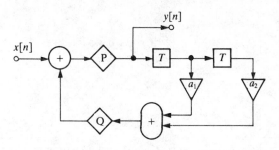

FIGURE 10.32 Exercises 10.6 and 10.8.

10.7 Consider the first-order digital filter shown in Fig. 10.26. The input signal $x[n] = 0$ for $n \geq 0$; Q has a quantization characteristic with a quantization step q. Find out what happens if:
(a) $y[-1] = 10q$; $\alpha = -0.91$; Q based on rounding;
(b) $y[-1] = 5q$; $\alpha = -0.91$; Q based on rounding;
(c) $y[-1] = 10q$; $\alpha = +0.91$; Q based on rounding;
(d) $y[-1] = 10q$; $\alpha = -0.91$; Q based on magnitude truncation;
(e) $y[-1] = 5q$; $\alpha = -0.91$; Q based on value truncation.

10.8 A second-order digital filter is shown in Fig. 10.32. The input signal $x[n] = 0$ for $n \geq 0$; $a_1 = 1.6$; $a_2 = -0.99$. Determine the output signal $y[n]$ for $n \geq -2$ and determine the period of the oscillation if $y[-1] = y[-2] = 11q$ and:
(a) Q is based on rounding;
(b) Q is based on magnitude truncation.

PART II

Chapters 11–15 are reprinted applications articles from *Philips Technical Review*, published by Philips Research Laboratories.

11

An integrated switched-capacitor
filter for viewdata

A. H. M. van Roermund and P. M. C. Coppelmans[†]

Any ordinary colour television set can be adapted for the display of 'videography' by adding a special decoder circuit. We use the general term 'videography' as a name for a large number of different systems in which text and simple graphics are converted into a digital signal in accordance with fixed rules. This digital signal can be transmitted as part of a television signal (teletext) or sent upon request to one particular subscriber via an ordinary telephone line (videotex or, more specifically, viewdata). A subscriber to a viewdata system needs a 'viewdata modem' for the connection between his television set and the telephone line. An important component in the modem is the receive filter. The authors of this chapter have realized this filter as a single integrated circuit by making use of the rather new technique of switched-capacitor filters.

11.1 INTRODUCTION

The theory and design of electrical filters have long had an important part to play in electronic engineering. Filters enable the different frequency components of a signal to be treated in different ways. The great majority of filters are used to stop the transmission of certain frequencies (or frequency bands) while others are passed. Filters are therefore very widely used in all kinds of telecommunication equipment (in radio, television, telephony and so on). At first filters were built up from the three basic elements encountered in the theory of electrical networks: inductors (L), resistors (R) and capacitors (C). As time went by, whole classes of filters evolved, each with their advantages and disadvantages. There are for example doubly terminated LC filters, built entirely from inductors and capacitors except for the two terminating resistors. The great

[†]Philips Research Laboratories, Eindhoven. This chapter is reprinted from *Philips Tech. Rev.* **41** No. 4: 105–23, 1983/84.

advantage of these filters is the relatively small change in filter characteristics when the element values depart from the theoretically ideal values. These filters are therefore said to have low parameter sensitivity. They do have one great disadvantage, however: they include inductors. These are relatively large, heavy and expensive, especially for applications at low frequencies. Also, it is almost impossible to make inductive elements as part of an integrated circuit. New filter types in which there are no inductors have therefore been sought for many years.[61-63] This objective can be achieved by using active elements, such as amplifiers. Well-known examples are active RC filters, composed of resistors, capacitors and operational amplifiers, and gyrator filters. Unfortunately, active RC filters have poorer parameter sensitivity. On the other hand, the low parameter sensitivity of LC filters can be retained, if the rules for design are properly 'translated' for other types of filter. The LC filters therefore still remain of considerable interest, as we shall see later in this chapter.

All the filters we have mentioned so far are *analog* filters. The signals in these filters can change in value at any moment and can take any value between two extremes. These analog signals are thus continuous in time and in amplitude (Fig. 11.1(a)). Another kind of filter that has been known for some time is the *digital* filter. These filters rather resemble small computers: the signals consist of series of numbers and the filter

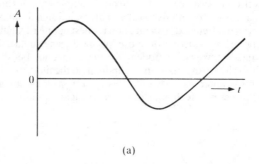

(a)

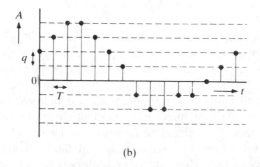

(b)

FIGURE 11.1 (a) An *analog signal* can have any amplitude A at any time t; it is continuous in time and continuous in amplitude. (b) A *digital signal* is discrete in time and discrete in amplitude; it can be completely represented by a series of numbers that can each take only a finite number of different values. A digital signal is usually represented graphically by a sequence of points on a fixed 'grid' as shown here. q quantization step, T sampling interval.

operations are implemented as a succession of simple arithmetical operations such as multiplication, storage and addition. Since these digital signals are built up from clearly separated successive numbers ('samples') that can each take only a finite number of different values, they are discrete in time and in amplitude (Fig. 11.1(b)). A digital filter is in general highly suitable for design in integrated-circuit form and represents an interesting method of making filters.

Recently a very promising third category of filters has attracted increasing attention:

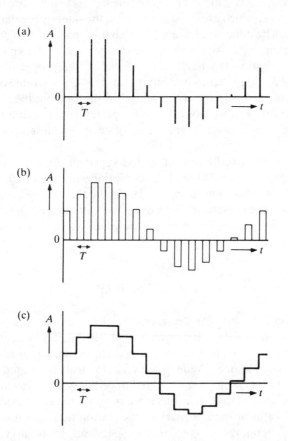

FIGURE 11.2 A *sampled-data signal* can take any value A, but only at certain times, separated by T, the sampling interval. It can be obtained by sampling the value of a given analog signal at regular intervals (a). The level of the sampled-data signal is often held constant for a time after sampling (b) and (c). In all these cases the signal can be considered as a series of signal elements of fixed shape and fixed duration T, but of arbitrary amplitude. These signals are therefore said to be discrete in time, but continuous in amplitude. The exact shape of the signal elements is *not* of importance for the information (e.g. speech or music) carried by the signal. The signals in (a), (b) and (c) are fully equivalent in this respect.

However, we should note that a sampled-data signal can also be considered as an analog signal that is subject to certain constraints: its value is indeed defined for every time t, but cannot vary freely. If we consider a sampled-data signal in this way and process it accordingly, e.g. in an analog filter, then the exact form of the signal elements *is* of importance for the final result. The same analog processing of the signals (a), (b) and (c) can then lead to slightly different final results.

sampled-data filters. In these filters the signals can change value at certain instants only, but then they can take any value (between two extremes). Signals of this type can be obtained by periodically taking samples of an analog signal with a sampling circuit (Fig. 11.2(a)). The signal level after sampling is often held constant for a time by a holding circuit. We then obtain signals like those shown in Fig. 11.2(b) or (c). All these signals can be considered to consist of a series of signal elements of fixed shape and duration but of arbitrary value. To put it another way, they are discrete in time and continuous in amplitude. In most sampled-data filters now in use the signal elements are represented by packets of electric charge, which are passed on within the filter at regular intervals and thus processed. This can be done for example by means of capacitors that are periodically switched to give a continuing redistribution of the charge packets. This gives a switched-capacitor filter, often abbreviated to SC filter. Some of the properties of SC filters correspond to those of analog filte·s, while others more closely resemble those of digital filters. We shall have more to say about this later. For the moment the most important feature is that SC filters are readily fabricated in integrated-circuit technology, although for the present the frequency range in which they offer the greatest advantage goes no higher than 100 kHz.

In this chapter we shall describe an integrated viewdata filter fabricated in MOS technology[†] as an example of an SC filter. The description will include a discussion of the fundamental principles of these filters and of the essentials of the design process that resulted in a filter on a single chip, which is ready for use without any adjustment. First of all, however, we must say something about the viewdata system, for which the filter is intended.

11.2 VIDEOGRAPHY

A conventional analog TV signal can represent practically any picture, complete with movement. In videography, on the other hand, a binary digital signal – i.e. a succession of bits ('ones' and 'zeros') – is used to give an encoded representation of letters, numbers and simple graphic symbols only. Videography is also usually limited to stationary pictures and its resolution is rather less than that of ordinary TV. However, much less information has to be transmitted. It is therefore possible to accommodate broadcast videography (*teletext*[64]) in an unused part of the transmitted television signal or to provide interactive videography (*videotex*, or more specifically *viewdata*) via the ordinary telephone system.

Although a teletext picture may look remarkably like a videotex picture, the two systems are quite different. In teletext the pictures to be added to the ordinary TV signal at a fixed rate are selected centrally at the studios. The viewer only decides which of the 'passing' teletext pictures will be displayed. To keep down the average waiting time, the number of different pictures (or 'pages') that can be consulted is restricted to a few hundred at the most. These pages therefore mainly contain subjects of general interest,

---

[†]It is interesting to note that the same viewdata filter has been described as an example in ref. [62]. However, there a completely different approach was used to obtain an integrated-circuit version. The filter was an *analog* filter, based on *gyrators*, and fabricated in a *bipolar* process.

such as the latest news, the weather forecast, sports results, traffic information, summaries of radio and TV programmes or subtitles for TV programmes transmitted simultaneously.

In videotex the viewer orders each picture separately from a central computer via the telephone network; hence the designation *interactive* videography. There is no direct relation between the number of different pages available and the average waiting time. In principle this number is therefore unlimited.

The videotex system can make all kinds of data bases accessible to anyone, with contents from stock-exchange prices and tourist information to catalogues of mail-order companies and encyclopaedias. Videotex offers a number of facilities that teletext does not have. It is possible for example to charge the subscriber for a transmitted page, since the information is only provided on demand. It is also possible to make certain pages available only to certain users (e.g. regular subscribers). Also, in certain circumstances a user can respond by booking a journey, say, or placing an order ('teleshopping'). It is also possible for videotex subscribers to send certain messages (e.g. greetings telegrams) to one another. So in this way videotex offers a simple form of electronic mail.

The introduction of videography has been accompanied by a flood of new terms, which sometimes are very confusing to both outsiders and insiders. The definitions for videography, teletext and videotex, as given above, have recently been proposed by some international bodies, including the IEC (International Electrotechnical Commission) and the ITU (International Telecommunication Union). *Teletext* has different names in different countries: Teletekst (the Netherlands), Videotext (Germany), Ceefax and Oracle (United Kingdom), Didon-Antiope (France). For *videotex* the situation is even more complicated. The original idea for videotex came from Britain, where it was first called viewdata. Then systems with different technical aspects and different names began to appear in other countries. At the same time some of the operating organizations introduced new names for technically identical systems. So now we have Viditel (the Netherlands), Bildschirmtext (Germany), Prestel and, shortly, Panda (United Kingdom), Teletel (France), Videotel (Italy), Telidon and Vista (Canada), Datavision (Sweden), Teledata (Denmark and Norway), Telset (Finland) and Captain (Japan).

To increase confusion even further a service called teletex has recently become available in some countries. However, this has nothing at all to do with videography, but is an improved version of the long-established telex.

In this chapter we shall pay particular attention to a filter from a modem for the original viewdata system. Such a filter is suitable for a large number of the other videotex systems, but is not automatically suitable for all of them. To emphasize this point, we shall continue to use the term viewdata in this chapter, even though this name seems to be falling out of favour in some quarters.

11.3 VIEWDATA MODEM

An essential condition for the use of videotex is the existence of a two-way connection between the subscriber and the videotex computer. This connection is set up via the telephone network (Figs 11.3 and 11.4). To connect the subscriber's television set to the telephone network there must be a modem (from modulator/demodulator). In the

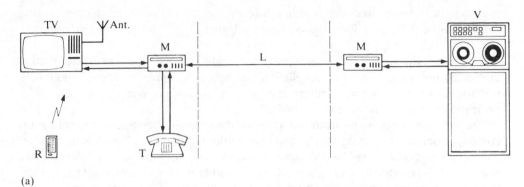

(a)

FIGURE 11.3 (a) Basic diagram of a videotex connection. The TV receiver is connected to the videotex computer V, which contains all the information, via two modems M and the telephone line L. The telephone T is used for making the connection. Once this has been made, the user communicates with the computer via the TV remote control R. Ordinary television programmes, including teletext signals, can be received with the antenna Ant. (b) A videotex system in use. The telephone handset has been replaced on its cradle, since the connection has been made and the user provides all further commands via the remote control.

viewdata system, the series of ones and zeros that represent the subscriber's commands are first of all encoded in the modem into series of tones at 390 or 450 Hz. These series of tones can be sent to the computer via the telephone connection. Another series of tones, at 1300 or 2100 Hz, is received in the modem and translated back into the ones and zeros that represent the viewdata pictures. Information can be sent from the subscriber to the computer at the rate of 75 bit/s and simultaneously in the reverse direction at 1200 bit/s.

FIGURE 11.4 The business user often prefers to use special versions of videotex equipment. The colour TV receiver is then replaced by a simpler colour monitor and a complete keyboard is substituted for the remote control. The photograph also shows special equipment for making a print of the information on the screen. The modem can be seen below the telephone.

An important component of the viewdata modem is the receive filter. This filter must pass the signal arriving for the subscriber with the minimum of distortion and at the same time it must attenuate all interfering signals (particularly 50-Hz mains hum, the signal sent by the subscriber himself and noise) as much as possible. In more specific terms this requires:

- low distortion due to attenuation and group delay in the passband (1100 to 2300 Hz);
- more than 50 dB of attenuation at 50 Hz;
- more than 50 dB of attenuation between 390 and 450 Hz, and
- more than 28 dB of attenuation above 3400 Hz.

A computer program that can simulate a viewdata connection has been used to find a filter transfer function that meets this specification. The result was a 9th-degree transfer function with the attenuation characteristic and group-delay characteristic shown in Fig. 11.5. Other computer programs that have been available for a long time could then be used to calculate a fully specified filter from these curves. In Fig. 11.6 we show as an example a doubly terminated *LC* filter that has been derived in this way. In principle there are a large number of *LC* filters with the desired characteristics. One of the special features of the filter shown here is the small ratio of the largest to the smallest values of the different elements. The circuit of Fig. 11.6, which meets all the stated filter requirements and also has the desirable low parameter sensitivity of the doubly terminated *LC*

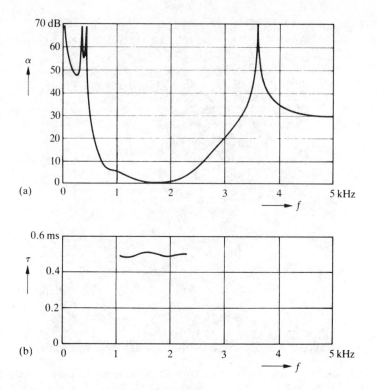

FIGURE 11.5 (a) Attenuation characteristic of the receive filter of a viewdata modem. The figure clearly shows the passband (centre 1700 Hz) for the received signal and the stopband (centre 420 Hz) for the subscriber's own signal. Attenuation is also necessary at lower and higher frequencies because of 50-Hz mains hum and noise. (b) Group-delay characteristic of the same filter in the passband. The distortion of the received signal is determined by the shape of both curves in this band. α attenuation, τ group delay, f frequency.

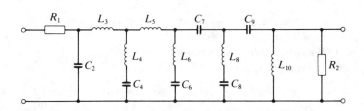

FIGURE 11.6 A possible implementation of the characteristics of Fig. 11.5 in the form of a doubly terminated LC filter. L inductance, C capacitance, R resistance. For this filter $L_3 L_4 + L_3 L_5 + L_4 L_5 = 0$, which implies that at least one of the three inductances L_3, L_4 and L_5 must be negative. This can be achieved by combining L_3, L_4 and L_5 in a transformer.[†]

[†] All the computer calculations for the simulation of the viewdata connection and for the specification of the filter of Fig. 11.6 were performed by our colleagues B. Huber, J. Kunze and R. Lücker of the TEKADE Company. Nürnberg, West Germany. They also decided on the signal flow diagram for the viewdata filter and carried out the final optimization.

filter, could be directly applied in the viewdata modem. But because of the inductors (and especially the coupled inductors L_3, L_4 and L_5) it cannot be made as an integrated circuit. However, we shall show later that this circuit is very useful as the starting point for the design of an integrated SC filter.

Recently our colleagues at Mullard Application Laboratory, Mitcham, Surrey, England, have made a 'two-chip' version of a videotex modem. One of these chips contains an SC filter with roughly the same characteristics as the ones given in Fig. 11.5. The main differences from the integrated filter of this chapter are: their filter has an 8th-degree transfer function, it is formed from four second-order sections ('biquads') and it has been fabricated in NMOS technology.[65]

11.4 SC FILTERS

11.4.1 Fundamentals

SC filters are readily integrated because they are built up from three kinds of components that are highly suitable for fabrication in a single process. These are switches, capacitors and amplifiers, which can all be integrated successfully in the well-known MOS (Metal-Oxide Semiconductor) technology.[66,67] In these components use is made of the following unique combination of features, which MOS circuits have but bipolar circuits do not.

- An MOS transistor used as a switch has a very high impedance between the source S and the drain D in the switched-off state (Fig. 11.7). Electric charges on small capacitors (e.g. 1 pF) connected to the source or drain will therefore continue to be stored for a relatively long time (up to several milliseconds). In a bipolar transistor the minority carriers cause recombination currents and hence a change in the stored charge.
- The coupling between the gate G (Fig. 11.7) of an MOS transistor and the source and drain is purely capacitive. This has two consequences. If the transistor is used as a *switch*, there will be no loss of charge from G to S or D. If it is used as an *amplifier*, the quantity of charge stored in a capacitor can be read out continuously and non-destructively via G.
- An MOS transistor can be made completely symmetrical, so that the source and drain cannot be distinguished from one another and have completely interchangeable roles. It therefore makes no difference to the transistor whether the voltage difference between S and D is positive, negative or zero. (To emphasize this point we have used a symmetrical symbol for the MOS transistor; there is also an asymmetrical symbol in which G is shown opposite to S.)
- Both the dimensions and the parasitic capacitances of an MOS transistor can be very small.

As the name indicates, SC filters contain switched capacitors. These can be compared with resistors in some respects. Figure 11.8(a) shows the combination of a changeover switch with a capacitor of capacitance C. If we assume that in a period T the switch first takes the left-hand position and then the right-hand position, the capacitor will be charged first to a value

$$q_1 = Cv_i$$

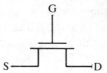

FIGURE 11.7 Symbolic representation of an MOS transistor. S source, G gate and D drain. If the transistor is used as a *switch*, a gating signal is applied to G; this gating signal is used to make or break the connection between S and D as required. If the transistor is used as an *amplifier*, the signal to be amplified is generally applied to G.

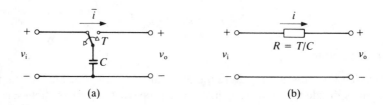

(a)   (b)

FIGURE 11.8 (a) A switched capacitor C is charged alternately to a voltage v_i and a voltage v_o at a frequency $1/T$. The charge transfer thus produced corresponds to a *mean* current $i = C(v_i + v_0)/T$. (b) At low frequencies this circuit is equivalent to a resistor $R = T/C$.

and then to

$$q_2 = Cv_o$$

In time T a total charge Δq will travel from left to right, where

$$\Delta q = q_1 - q_2 = C(v_i - v_o)$$

This corresponds to a *mean* current $\bar{i}$ given by

$$\bar{i} = \frac{\Delta q}{T} = \frac{C(v_i - v_o)}{T}$$

If the frequencies in which we are interested are sufficiently low compared with the switching frequency $1/T$ (we usually speak of the *sampling rate* or *clock rate*), then we can consider the switched capacitor in Fig. 11.8(a) as a resistance of value $R = T/C$ (Fig. 11.8(b)). We can make the changeover switch in Fig. 11.8(a) with the aid of two MOS transistors that are switched alternately on and off by clock signals ϕ_1 and ϕ_2 of period T (Fig. 11.9).

An RC network of the 1st order with a time constant $\tau_A = R_1 C_2$ can be approximated by combining a switched capacitance with a fixed one (Fig. 11.10). The associated time constant τ_{SC} is then given by

$$\tau_{SC} = \frac{C_2}{C_1} T$$

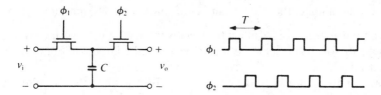

FIGURE 11.9 In MOS technology the switched capacitor of Fig. 11.8(a) can be made with a fixed capacitor C and two switching transistors, operated by two different clock signals, ϕ_1 and ϕ_2. Each transistor only makes its connection when its clock signal has a high value. The two clock signals are never allowed to have a high value both at the same time.

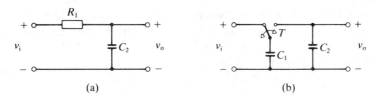

FIGURE 11.10 (a) Passive 1st-order RC filter and (b) the SC filter derived from it with $C_1 = T/R_1$. The SC filter is a good approximation to the RC filter for frequencies that are low with respect to the switching frequency or sampling rate $1/T$.

(if the sampling rate $1/T$ is sufficiently high). This demonstrates two general properties of SC filters: time constants are proportional to *capacitance ratios* and inversely proportional to the *clock rate*. In modern integration processes capacitance ratios can be reproduced to an accuracy of 0.1% without difficulty and clock rates can be even more accurately set. In addition the capacitance ratios of integrated capacitors vary only slightly with temperature or capacitor voltage, especially if the capacitors are close to one another on the chip. If moreover we decide on a type of filter that intrinsically has low parameter sensitivity, then we can obtain the following impressive combination of desirable features:

- Filter can be fully integrated.
- Very high accuracy, without trimming (e.g. deviations of only 0.05 dB in the passband).
- Very low temperature coefficients.
- Processing conditions for manufacture are not critical, so that SC filters can easily be combined on a single chip with other circuits, such as logic circuits, oscillators, rectifiers and comparators.
- Small dimensions (e.g. $0.2\,\text{mm}^2$ per filter pole) so that high-order filters can be produced on a single chip.
- Low power consumption (e.g. 0.1 to 1 mW per filter pole).
- No analog-to-digital or digital-to-analog converters are necessary for applications with analog signals.
- Frequency response can be scaled by changing the clock rate.
- No quantization as in digital filters and hence none of the accompanying negative effects it introduces (stability problems, preference for certain filter structures).

Beside these advantages there are a number of limitations. The most important ones are:

- The desirable features of SC filters (especially those relating to surface area, power and accuracy) are for the present only available in a limited frequency range (for signals up to about 100 kHz).
- An extra analog filter may be necessary before or after the SC filter to suppress frequency components higher than half the sampling rate.
- If an SC filter is the only component in a system that requires a clock signal, this implies a certain complication.
- SC filters introduce noise into the signal (typical signal-to-noise ratios are 70 to 80 dB).
- Like all active filters, SC filters take some electrical power from the supply; usually this is very low, however.
- SC filters can introduce more distortion than *LC* filters. (Distortion, power consumption and signal-to-noise ratio can be traded against one another at the design stage.)
- SC filters can off-set the d.c. level of the signal (by 10 to 100 mV).

11.4.2 SC filters with amplifiers

With the aid of Fig. 11.8 we showed how a switched capacitor could be modelled as a resistor. In principle with such a model we could replace any resistor from any passive *RC* filter by a switched capacitor and hence derive an SC version of that filter. Indeed, that was what we did for an *RC* filter of the first order (Fig. 11.10). However, this approach will not take us very far. In the first place our model is only an approximation. Moreover, we can only apply the technique to passive *RC*-filter prototypes with their inherently low selectivity. Finally, this approach yields SC filters that turn out to be very sensitive to the parasitic capacitances encountered in a practical integrated circuit. To avoid these difficulties and to make use of as many of the advantages listed in the previous subsection as we can, we have to include special amplifiers in our SC filters. In the literature these amplifiers are usually called 'operational amplifiers', or just 'op-amps'. Since we think the term 'operational' is something of a misnomer[†] here, we shall not apply it for these amplifiers in the rest of the chapter.

   The use of amplifiers in SC filters can be illustrated with the aid of Fig. 11.11. This figure shows first of all an analog integrator, as widely used in active *RC* filters. It consists of a resistor, a capacitor and an amplifier. Next to it an SC approximation to it is shown. Making use of the complex variable p and the Laplace transforms $V_i(p)$ and $V_o(p)$ of the

---

[†]An operational amplifier is a universally applicable amplifier with two inputs. The difference voltage between these two inputs acts as the input signal, and there is usually a single output. The input impedance, the gain and the bandwidth are very high, while the output impedance is extremely low. Between the output and the (inverting) input negative feedback is almost always used to give the desired amplifier response.

   The differential amplifiers that we use in SC filters have a high input impedance and a reasonably high gain (e.g. a voltage gain of 10 000), but a fairly high output impedance and a bandwidth that must be separately 'tailored' for each amplifier, and should not be larger than strictly necessary, because of noise (e.g. 400 kHz). These amplifiers clearly have a number of features that are specific to their application, and are not really the same as the universal operational amplifiers. We shall return to these special amplifiers later.

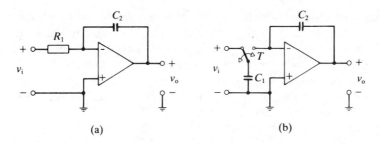

(a)   (b)

FIGURE 11.11 (a) Active 1st-order RC filter or analog integrator and (b) the SC filter derived from it with $C_1 = T/R_1$. Here again the properties of the two filters only correspond well for frequencies that are low with respect to $1/T$.

input and output voltages v_i and v_o, we can describe the analog integrator[63] in terms of the *transfer function* $H(p)$:

$$H(p) = \frac{V_o(p)}{V_i(p)} = \frac{-1}{pR_1C_2} \tag{11.1}$$

The transfer characteristics of the analog integrator for a sinusoidal signal of angular frequency ω rad/s are given by the *frequency response* $H_A(\omega)$. We can derive $H_A(\omega)$ from $H(p)$ by putting $p = j\omega$, where j is the imaginary unit $(= \sqrt{-1})$. We find

$$H_A(\omega) = \frac{-1}{j\omega R_1 C_2} \tag{11.2}$$

The subscript A indicates 'analog'.

To determine the frequency behaviour of the SC filter from Fig. 11.11 we first consider its operation in the time domain. During clock phase 1 (i.e. when the clock signal ϕ_1 is 'high', see Fig. 11.9) the switch is in the left-hand position. At the end of this phase, e.g. at time $t = nT$, the charge on the capacitor C_1 is $C_1 v_i(nT)$. At that the moment there is a charge $C_2 v_o(nT)$ on the capacitor C_2 as the result of all the previous events in the filter. During clock phase 2 the switch is in the right-hand position. The amplifier always tries to bring the voltage difference between its two inputs back to zero. This is done by compensating the charge on C_1 with charge from C_2. The charge on C_2 therefore diminishes by $C_1 v_i(nT)$ and then remains constant during the rest of the period T, i.e. until $t = nT + T$. The relation between the charges on C_2 at the times nT and $nT + T$ is given by:

$$C_2 v_o(nT + T) = C_2 v_o(nT) - C_1 v_i(nT) \tag{11.3}$$

By varying the number n, which is an integer, we can use this equation to determine the behaviour of the SC filter at all times separated by a multiple of T; we shall not consider what exactly happens between these times and we shall therefore consider the SC filter purely as a sampled-data filter. In fact (11.3) is an example of the description of a discrete-time system by a *difference equation*. Just as we can directly convert the description of a continuous-time system in terms of a *differential equation* into a description in terms of the complex variable p (the 'Laplace transform'), we can convert a difference equation into a description in terms of the complex variable z (the 'z-transform'). To do this we replace $v(nT)$ by $V(z)$, $v(nT+T)$ by $V(z) \cdot z$ and in general $v(nT+mT)$ by $V(z) \cdot z^m$. The equation above then becomes

$$C_2 V_o(z) \cdot z = C_2 V_o(z) - C_1 V_i(z) \tag{11.4}$$

The ratio of $V_o(z)$ to $V_i(z)$ is called the *transfer function* $G(z)$ of the discrete-time system, and in this case is given by

$$G(z) = \frac{V_o(z)}{V_i(z)} = -\frac{C_1}{C_2} \frac{1}{z-1} \tag{11.5}$$

From this $G(z)$ we can derive the transfer characteristics for a sinusoidal signal of angular frequency ω rad/s in the form of the *frequency response* $G_D(\omega)$. We find $G_D(\omega)$ from $G(z)$ by putting $z = e^{j\omega T}$, where j is again the imaginary unit, e the base of the natural logarithms and $1/T$ the sampling rate of the discrete-time system. Therefore:

$$G_D(\omega) = -\frac{C_1}{C_2} \frac{1}{e^{j\omega T} - 1} \tag{11.6}$$

The subscript D indicates 'discrete'. For frequencies that are very much smaller than the sampling rate $1/T$ (i.e. $\omega T \ll 1$), $e^{j\omega T}$ is approximately equal to $1 + j\omega T$.

(The validity of this approximation can easily be seen by starting from $e^{j\omega T} = \cos\omega T + j\sin\omega T$ and applying the approximations $\cos\omega T \approx 1$ and $\sin\omega T \approx \omega T$ for small values of ωT.)

In this way we find the frequency response $G_{SC}(\omega)$ of the SC integrator of Fig. 11.11 at low frequencies:

$$G_{SC}(\omega) = -\frac{C_1}{C_2} \frac{1}{j\omega T} \tag{11.7}$$

A comparison with $H_A(\omega)$, found earlier in equation (11.2), shows us that at low frequencies the two circuits in Fig. 11.11 are equivalent, provided that $R_1 = T/C_1$. This corresponds to our earlier derivation with Fig. 11.8. Also, we see again that the frequency behaviour is determined by a capacitance ratio and the clock rate.

In equations (11.3), (11.4) and (11.5) we have given descriptions in the time domain and the complex-frequency domain ('z-domain') of the SC integrator of Fig. 11.11 as a purely discrete-time system. These equations can also be represented by block diagrams (discrete-time equivalent circuits). This has been done in Fig. 11.12(a) and (b). A delay of T in the time domain corresponds to a multiplication by z^{-1} in the z-domain, as we have indicated above. For the block diagram of Fig. 11.12(a) we have:

$$v_x(nT) = -\frac{C_1}{C_2} v_i(nT) + v_o(nT)$$

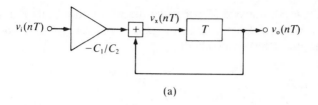

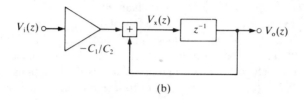

FIGURE 11.12 Discrete-time equivalent circuit for the SC integrator of Fig. 11.11 in the time domain (a) and in the complex-frequency domain or z-domain (b). We clearly recognize three basic elements of discrete-time systems: adder, delay element (indicated by T and z^{-1}) and constant-factor multiplier (here $-C_1/C_2$).

and

$$v_o(nT + T) = v_x(nT)$$

Eliminating $v_x(nT)$ from these two equations gives equation (11.3).

For the block diagram of Fig. 11.12(b) we have:

$$V_x(z) = -\frac{C_1}{C_2}V_i(z) + V_o(z)$$

and

$$V_o(z) = z^{-1} \cdot V_x(z)$$

Eliminating $V_x(z)$ here gives equation (11.4) directly. Equivalent diagrams like those of Fig. 11.12 are a very useful aid in the analysis of SC filters as discrete-time systems. They are particularly useful for filters of complicated structure.

In going from the analog integrating circuit of Fig. 11.11(a) to the SC integrator of Fig. 11.11(b) we have gone from a description in terms of $H(p)$ to a description in terms of $G(z)$. We have also seen that for low frequencies both integrators are equivalent if $R_1 C_1 = T$. With this condition we can characterize this particular transition as the substitution of $(z-1)/T$ for p in $H(p)$; we then obtain $G(z)$ directly. This transition is called a *transformation* or *mapping*. In Fig. 11.11 we are concerned with 'forward-difference mapping'. If in any active RC filter of transfer function $H_{tot}(p)$ every analog integrator is replaced by the SC integrator of Fig. 11.11(b), we obtain the resulting $G_{tot}(z)$ by substituting $(z-1)/T$ for every p in $H_{tot}(p)$. The transformation rule thus completely establishes the nature of the relationship between the frequency behaviour of the original filter and that of the new filter. In practice different transformations from the one in Fig.

11.11 are usually used, since that particular transformation entails relatively large differences between the frequency responses corresponding to $G_{\text{tot}}(z)$ and $H_{\text{tot}}(p)$. We shall return to this in the following section.

11.4.3 SC integrators

The SC integrator is a good starting point for building complicated SC filters. We have already encountered one possible version of such an integrator in Fig. 11.11. There are a countless number of variations, however, and a few of them are shown in Fig. 11.13, with a discrete-time equivalent circuit and the associated transfer function $G(z)$. The switches in this figure are shown in the position that they take during clock phase 1, i.e. when the clock signal ϕ_1 is 'high'. The other position is associated with clock phase 2 (ϕ_2 'high').

We see that in some cases input signal and output signal occur during the same clock phase but not in other cases. We must therefore take proper care when connecting different integrators together. Signals that occur during clock phase 2 are indicated by an asterisk. If the input signal occurs during clock phase 1 and the output signal during clock phase 2, the associated transfer function has also been indicated by an asterisk: $G^*(z)$. A difference in clock phase between input and output is manifest by the occurrence of fractional powers of z in $G^*(z)$. In practice there are no output switches, but the input switches of the following integrators will automatically fulfil this function by operating them in the required clock phase. We shall keep to the same convention in all the succeeding figures.

The integrators in Fig. 11.13 differ from one another in several respects; the following are the most important:

- The frequency behaviour of an analog integrator is approximated in various ways. This can be seen from the fact that each of the SC integrators has a different transfer function ($G(z)$ or $G^*(z)$).
- The sensitivity to parasitic capacitances, which are always present in practical circuits, differs considerably. If the sensitivity is high, a practical version of an SC-filter design can behave completely differently from the calculated version.

Of the SC integrators that we have considered so far, the ones from Fig. 11.13(a) to (d) are the most interesting, since they are the least affected by parasitic capacitances.[63,68] We can explain this better with the aid of Fig. 11.14. Here the parasitic capacitances that have to be taken into account in practical circuits have been included for two different integrator structures. In Fig. 11.14(a) there is one such parasitic C_{par}, which is completely described by a departure from the desired value of the capacitance C_1. In Fig. 11.14(b) there are two parasitics, but neither of them causes any trouble. The charge that is

---

FIGURE 11.13 Various types of SC integrator with corresponding discrete-time equivalent circuit and transfer function in z. Comparing (a) with (b) and (c) with (d) shows that the clock phase with which the output signal is read out clearly affects the behaviour of the circuit. All the switches are shown in the position they adopt during clock phase 1. Signals that occur during clock phase 2 are indicated by an asterisk. Circuits (b) and (d) are of the LDI type (LDI stands for lossless discrete integrator). Circuit (e) represents the bilinear transformation of an analog integrator.

(a)

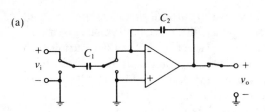

$$G(z) = \frac{-C_1}{C_2} \frac{1}{1 - z^{-1}}$$

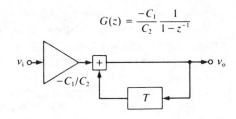

(b)

$$G^*(z) = \frac{-C_1}{C_2} \frac{z^{-1/2}}{1 - z^{-1}}$$

(c)

$$G(z) = \frac{C_1}{C_2} \frac{z^{-1}}{1 - z^{-1}}$$

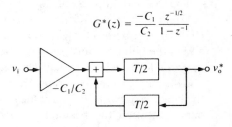

(d)

$$G^*(z) = \frac{C_1}{C_2} \frac{z^{-1/2}}{1 - z^{-1}}$$

$$G(z) = \frac{-C_1}{C_2} \frac{1 + z^{-1}}{1 - z^{-1}}$$

(e)

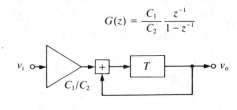

collected on C_{par1} during clock phase 1 is discharged to earth during clock phase 2 and does not reach C_2. Since the amplifier keeps the voltage difference between its two inputs as close to zero as possible, C_{par2} is connected to the same constant voltage during both clock phases; this parasitic will therefore also have no effect on the charge transfer from C_1 to C_2. It can be shown in a similar way that the integrator of Fig. 11.13(e) is very sensitive to parasitic capacitance and therefore less suitable.

In the integrators of Fig. 11.13(b) and (d) the input and output switches are in antiphase. In a cascade of such integrators the clocks ϕ_1 and ϕ_2 of adjacent stages alternate in role.

The SC integrators of Fig. 11.13(b) and (d) are 'lossless discrete integrators' (LDIs),[36,38] and the SC integrator of Fig. 11.13(e) is a 'bilinear integrator' (BI).[68] In the LDI transformation $G^*(z)$ is obtained from $H(p)$ by substituting $K_1(z^{1/2} - z^{-1/2})/T$ for p and in the bilinear transformation $G(z)$ is obtained from $H(p)$ by substituting $K_2(z - 1)/(z + 1)T$ for p, where K_1 and K_2 are constants. These transformations are the only two known simple frequency transformations in which a direct relation can be found between the imaginary axis ($p = j\omega$) of the p-plane and the unit circle ($z = e^{j\omega T}$) of the z-plane and hence between the frequency responses before and after transformation. They are therefore highly suitable for 'translation' of analog filters into discrete-time filters. There are a number of clear differences between the two transformations, however.[69] One difference is that in the bilinear transformation the *entire* imaginary axis is mapped on to the unit circle, whereas in the LDI transformation there is a direct relation only if $|p| < |2K_1/T|$. In the first case the entire frequency response of the analog filter translates

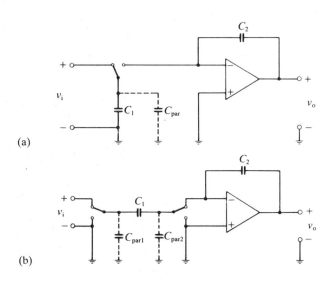

(a)

(b)

FIGURE 11.14 When an SC integrator is fabricated as an integrated circuit, unintended capacitances C_{par} have to be taken into account. In some configurations these parasitics have much more effect on charge transfer from C_1 to C_2 than in others. The configuration of Fig. 11.11(b) is not very good in this respect (a); the configuration of Fig. 11.13(a), (b), (c) and (d) is much less affected by parasitics (b).

to the frequency band from 0 to $1/2T$ Hz in the discrete-time filter. In the second case the frequency behaviour of the analog filter below $K_1/\pi T$ Hz translates to the frequency band from 0 to $1/2T$ Hz in the discrete-time filter. The bilinear transformation is also widely used in the design of digital filters.

We can readily extend any SC integrator to make it into a circuit module that offers even more scope. This can be done in various ways; an integrator can for example be provided with more than one input (Fig. 11.15), so that a linear combination of a number of signals can be integrated. For Fig. 11.15:

$$V_o(z) = -\frac{C_1}{C_2}\frac{V_1(z)}{1-z^{-1}} - \frac{C_3}{C_2}\frac{V_2(z)}{1-z^{-1}}$$

where we have made use of the z-transforms of the input and output signals. The combination integrator/adder is also widely used. This is shown in Fig. 11.16 for the circuit of Fig. 11.13(b). Such a combination circuit can not only be used for integration but can also be used to 'weight' and add signals. For Fig. 11.16 we have

$$V_o^*(z) = -\frac{C_1}{C_2}\frac{z^{-1/2}}{1-z^{-1}}V_1(z) - \frac{C_3}{C_2}V_2^*(z)$$

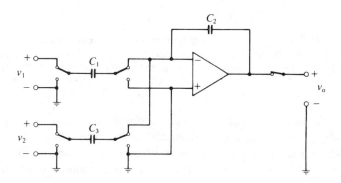

(a)

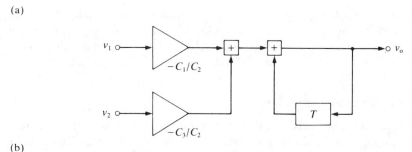

(b)

FIGURE 11.15 An SC integrator can be extended in various ways to form a more complex module; here the integrator of Fig. 11.13(a) has been extended by adding a second switched capacitor at the input. The input signal now consists of a linear combination of the signals v_1 and v_2 (a). This can clearly be seen from the discrete-time equivalent circuit (b).

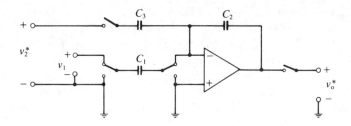

(a)

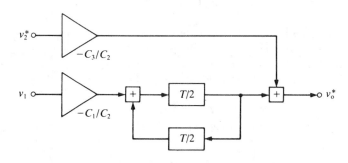

(b)

FIGURE 11.16 The combination of a summation circuit or 'adder' and an LDI integrator with discrete-time equivalent circuit. Such circuits form a versatile module for making complicated SC filters (a). The discrete-time equivalent circuit gives a clear representation of the operation (b).

We have to remember here that we are considering the output signal during clock phase 2. This signal contains a contribution from the preceding signal values of v_1 during clock phase 1 and from the current signal value of v_2^* during clock phase 2. To emphasize this we have provided C_3 with an input switch, but in practice there would not be a separate switch.

Another extension frequently found in practice is obtained if the output signal of an amplifier is used for further processing during both clock phases. A simple example is shown in Fig. 11.17. In fact this represents a combination of two simpler circuits, for the circuit of Fig. 11.17 can be considered as the combination of Fig. 11.13(a) and (b).

11.4.4 Continuous-time and discrete-time

So far we have placed the emphasis on the discrete-time nature of SC filters: we have shown how the filter operation can be described by considering only times separated by an integral number of sampling periods (or sometimes half-periods). This discrete-time approach is the natural method of describing SC filters built up from smaller constituent parts as a complete entity, e.g. so that the overall transfer function can be determined. One characteristic discrete-time property of SC filters is that the overall transfer function can be varied linearly with the clock rate (this is also true for digital filters). Another

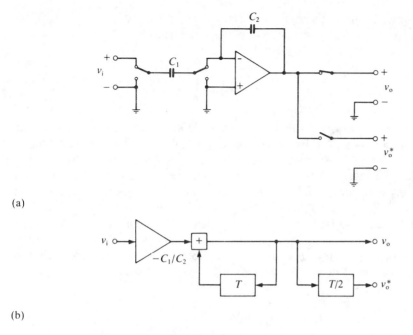

(a)

(b)

FIGURE 11.17 In some cases the output signal of an SC integrator is used during both clock phases (a). This gives a combination of two simpler circuits, in this case from Fig. 11.13(a) and 13(b), as can clearly be seen from the discrete-time equivalent circuit (b).

discrete-time aspect is the requirement that the bandwidth of the signals we wish to filter must satisfy the sampling theorem. They must contain no frequencies higher than half the sampling rate; otherwise there could be trouble from 'aliasing' in the SC filter. This usually requires a simple 'pre-filter'. In addition, the frequency spectrum of the sampled-data signal at the output of an SC filter is in principle unlimited in bandwidth. Therefore at that point we often have need of an analog filter that suppresses frequency components higher than half the sampling rate (a 'post-filter'), as in digital filters.

On the other hand, some of the things that can happen in an SC filter are impossible in true discrete-time filters. For example, an SC filter can contain closed loops in which there are no delay elements, yet some signal processing can take place. Also, in addition to various switched paths, an unswitched (and hence continuous-time) path can exist between input and output during one or both clock phases. In these cases a discrete-time approach gives an incomplete and sometimes unsatisfactory description of the filter.

Something related occurs if we wish to process the output signal of an SC filter as an analog signal (see the caption to Fig. 11.2): here again the discrete-time description as a sampled-data signal is inadequate. In this case the exact waveform of the signal elements must be taken into account. For example, the amplitude characteristic $|G_D(\omega)|$ calculated by discrete-time methods must still be multiplied by a frequency function corresponding to this exact waveform. For the signals from Fig. 11.2(b) and (c) this is a $\sin(x)/x$-function, with $x = \omega T/4$ and $x = \omega T/2$ respectively.

(It is easily shown by Fourier analysis that a signal $b(t)$ consisting of a single rectangular pulse of width τ and height 1 has an amplitude spectrum given by

$$|B(\omega)| = \left| \frac{2\sin(\omega\tau/2)}{\omega} \right| = \tau \left| \frac{\sin x}{x} \right|$$

where $x = \omega\tau/2$.)

It could be said that the discrete-time treatment of SC filters represents a 'macroscopic' approach, which confines itself to a description *at* (or really just after) the switching or sampling times. On the other hand, there is also a 'microscopic' treatment, which is concerned with the electrical behaviour *between* the switching times. This kind of treatment is of use in the design of the components (switches, amplifiers) of SC filters and in analyzing an SC filter with maximum accuracy. In these cases the electrical behaviour of an integrated circuit that is to be fabricated is calculated on a computer, with as many practical constraints as possible (such as parasitic elements and noise) taken into account. The resulting computer programs are fairly complicated,[70,71] since the switched capacitors give rise to continuous-time *periodically varying* systems.

11.5 SC-FILTER DESIGN

11.5.1 Signal flow graphs

If we take an analog filter as the starting point, we have various ways of designing an SC filter built up from integrators. One of the most widely used procedures[38] is based on the derivation of a *signal flow graph* of the analog filter. A filter description of this type comes somewhere between the abstract description by means of a transfer function and the practical description in terms of a detailed circuit diagram. A signal flow graph is more detailed than the transfer-function description but more abstract than the circuit diagram. In addition it is a non-unique description: for any one filter there is an unlimited choice. For our purposes we require a signal flow graph that only includes integrating operations, additions and multiplications by a constant – the operations that can be performed by the SC modules of the preceding section. We shall illustrate the principle of working with signal flow graphs with the aid of Fig. 11.18. This gives a diagram of a simple analog filter with the transfer function

$$H(p) = \frac{V_o(p)}{V_i(p)} = \frac{1}{p^2 LC + pCR + 1}$$

We can also describe this filter by a number of equations that each represent at the most a single integrating operation (corresponding to a multiplication by $1/p$):

$$V_1 = V_i - RI_1, \qquad I_1 = \frac{1}{pL}(V_1 - V_2),$$

$$V_2 = \frac{I_1}{pC}, \qquad V_o = V_2$$

(a)

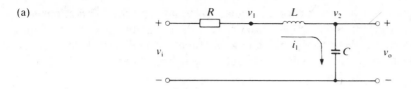

(b)

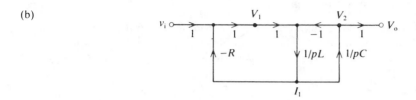

(c)

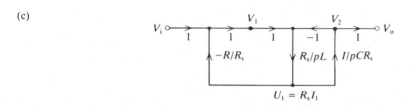

FIGURE 11.18 Any filter can be represented in many ways by a signal flow graph. A simple analog filter of the 2nd order (a) is represented here by a signal flow graph in which the Laplace transforms of the actual voltages and currents can be identified (b) and as a signal flow graph with voltages alone (c). R_s is a scaling factor with the dimension of resistance. The signal flow graphs are chosen in such a way that all the frequency-dependent operations correspond to integration (multiplication by $1/p$) and can therefore readily be carried out by SC integrators.

These equations can be represented graphically in the form of the signal flow graph of Fig. 11.18(b). By introducing an arbitrary scaling resistance R_s we can put $I_1 = U_1/R_s$, where U_1 is a fictitious voltage proportional to I_1. The filter of Fig. 11.18(a) is then described by:

$$V_1 = V_i - \frac{R}{R_s}U_1, \qquad U_1 = \frac{R_s}{pL}(V_1 - V_2),$$

$$V_2 = \frac{U_1}{pR_s C}, \qquad V_o = V_2$$

The nodes of the associated signal flow graph represent voltages only (Fig. 11.18(c)).

(The dimensions of the quantities indicated by the nodes of a signal flow graph are not of essential interest. The treatment above was based upon voltages and thus we found a scaling factor R_s with the dimensions of *resistance*. If it had been based upon currents, we would have found a scaling factor with the dimensions of *conductance*.)

In principle the design of the SC filter is now complete: each of the integrating operations is performed by an appropriate SC integrator, the multiplications by a constant are obtained by a correct choice of capacitance ratio and the summations are produced by the correct interconnections. If we replace the integrating operations in the signal flow diagram by the corresponding discrete-time block diagrams of the SC modules, we have immediately a complete discrete-time model of the practical SC filter at our disposal.

11.5.2 Signal flow graph of viewdata filter

We now return to the doubly terminated *LC* filter from Fig. 11.6 that satisfies all the specifications of the viewdata filter. This filter is shown again in Fig. 11.19(a), with a number of currents and voltages now introduced. The filter can be completely described by the set of equations:[72]

$$V_2 = \frac{1}{pC_2}\left[\frac{1}{R_1}(V_i - V_2) - I_3\right] \tag{a}$$

$$I_3 = \frac{1}{\rho}\left[I_5 + \frac{1}{pL}(V_2 - V_4)\right] \tag{b}$$

$$V_4 = \frac{1}{\rho}\left[V_2 + \frac{1}{pC}(I_3 - I_5)\right] \tag{c}$$

$$I_5 = I_7 + \frac{V_4 - V_6}{pL_6} \tag{d}$$

$$V_6 = \frac{1}{pC_6}(I_5 - I_7) \tag{e}$$

$$I_7 = I_9 + \frac{V_8}{pL_8} \tag{f}$$

$$V_8 = V_4 + \frac{1}{p}\left(-\frac{1}{C_{78}}I_7 + \frac{1}{C_8}I_9\right) \tag{g}$$

$$I_9 = V_{10}\left(\frac{1}{R_2} + \frac{1}{pL_{10}}\right) \tag{h}$$

$$V_{10} = V_8 + \frac{1}{p}\left(\frac{1}{C_8}I_7 - \frac{1}{C_{89}}I_9\right) \tag{i}$$

where $\rho = 1 + L_3/L_4$, $C = C_4/(\rho - 1)$, $L = -L_5 = L_3 L_4/(L_3 + L_4)$, $C_{78} = C_7 C_8/(C_7 + C_8)$ and $C_{89} = C_8 C_9/(C_8 + C_9)$. The corresponding signal flow graph is shown in

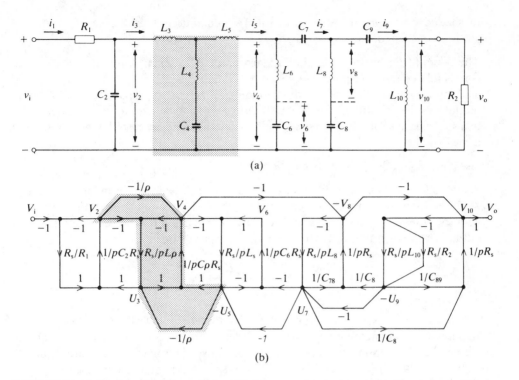

FIGURE 11.19 The viewdata filter of Fig. 11.6, but now with currents and voltages shown (a), so that a signal flow graph can be set up (b). In the signal flow graph every current I_j is shown as a fictitious voltage $U_j = I_j R_s$. The value of R_s is fixed, but can in principle take any value. As an aid to the explanation the fourpole $L_3 L_4 L_5 C_4$ in the filter circuit and the corresponding part of the signal flow graph are shown shaded.

Fig. 11.19(b), in which we again represent each current I_j as a fictitious voltage U_j by multiplying by a freely chosen resistance R_s. To explain our design procedure more clearly we have selected a part of the filter, the fourpole consisting of L_3, L_4, L_5 and C_4 (Fig. 11.19(a)). The electrical behaviour of this fourpole is completely described by the equations (b) and (c) above. The corresponding part of the signal flow graph is shown shaded in Fig. 11.19(b). This contains two integrating operations, both included in a single loop; this pattern can be found repeated many times in the signal flow graph.

11.5.3 Choice of SC integrator

The determination of the signal flow graph brings the design of the SC filter to an advanced stage. A crucial decision still has to be made, however: the choice of the type or types of SC integrator to be used. Earlier in this chapter we encountered a number of basic types, and we considered the most important differences (approximation of the analog integrator and sensitivity to parasitic capacitances). This led to a preference for the integrators of Fig. 11.13(a) to (d). How can we narrow the choice further? The necessary criteria can be derived from the signal flow graph: let us examine Fig. 11.19(b) again. We find that the pattern of a closed loop containing two integrators is repeated many times (the shaded part contains such a loop). Two conditions must be satisfied for the realization of such loops:

- The overall transfer function of all the branches of an (open) loop must be negative (this is related to the required stability of the circuit).
- The input switches of either of the two integrators must be capable of acting as the output switch of the other one (if not, the effect is as if the integrators are connected by two switches in series that are never both 'on' at the same time, so that no closed loop can exist).

These conditions allow of only two options:

- A closed loop contains one integrator as in Fig. 11.13(a) and one as in Fig. 11.13(c).
- A closed loop contains one integrator as in Fig. 11.13(b) and one as in Fig. 11.13(d), with the role of the clocks ϕ_1 and ϕ_2 exactly interchanged for the two integrators. In both cases the overall transfer function for the (open) loop is

$$G_L(z) = \frac{-Kz}{(z-1)^2}$$

where K is a positive constant whose value is determined by capacitance ratios.

In converting an analog filter of transfer function $H_{tot}(p)$ into an SC filter of transfer function $G_{tot}(z)$ a complication arises: since at least two kinds of integrators are required, the relation between $H_{tot}(p)$ and $G_{tot}(z)$ is not as simple as it might have appeared so far. For every closed loop containing *two* integrators the transfer function is $G_L(z)$ as given above. Now for both of the cases quoted we can interpret this as the LDI transformation of a combination of two continuous-time integrators. At the beginning and end of the signal flow graph, however, a special situation arises: there is always one loop with only *one* integrator (this is associated with the real terminating impedances R_1 and R_2). If we were to realize this integrator in the way shown in Fig. 11.13(a) or (c), the LDI-transformation rule would not apply to this part of the signal flow graph. On the other hand, we cannot use the LDI circuit of Fig. 11.13(b) or (d), since there would never then be a closed loop: the input and output switches of an LDI are always out of phase. This provides an interesting problem, for which various solutions exist.[73,74] Sometimes, however, they yield a slight modification of the transfer function. If this is unacceptable, a filter designer always has the option of abandoning the signal-flow-graph approach and using another design method in which this problem does not arise.[63,68,75]

In our viewdata filter we mainly use the integrators of Fig. 11.13(a) and 11.13(c). As an illustration we show in Fig. 11.20(a) what the SC circuit derived for the shaded part of Fig. 11.19 looks like. The capacitances in Fig. 11.20(a) and the quantities in Fig. 11.19(b) are related as follows:

$$\frac{C_{15}}{C_{12}} = \frac{C_{16}}{C_{14}} = \frac{1}{\rho}, \quad \frac{C_{11}}{C_{12}} = \frac{R_s T}{L\rho}, \quad \frac{C_{13}}{C_{14}} = \frac{T}{CR_s\rho}$$

where $1/T$ is again the switching frequency of the capacitors. We can give a fully discrete-time model for Fig. 11.20(a); this is shown in Fig. 11.20(b).

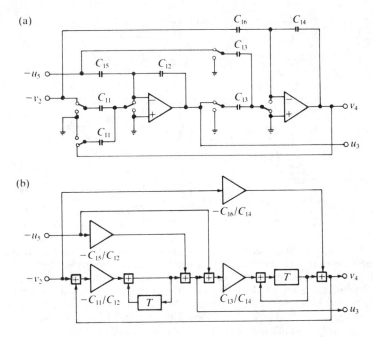

FIGURE 11.20 An SC circuit derived for the shaded part of Fig. 11.19 with the aid of SC integrators (a) and a corresponding discrete-time equivalent circuit (b).

11.6 SC-FILTER COMPONENTS

The most important components used in the construction of an SC filter are switches, capacitors and amplifiers. We shall now give some attention to the realization of each of these components.

11.6.1 The switches

As we have seen, an SC filter contains many changeover switches, each formed from two on/off switches (Fig. 11.8). These on/off switches are operated by two clock signals ϕ_1 and ϕ_2 (Fig. 11.9). Each on/off switch can consist of a single NMOS transistor, a single PMOS transistor or a combination of both. In comparison with a PMOS transistor an NMOS transistor has a lower resistance in the conducting state, but (in the fabrication process that we used) it also has a higher parasitic capacitance. The combination of both types gives a switch that has fewer limitations to signal voltage swing, but it requires two clocks of opposite polarity and has an even higher parasitic capacitance. In our viewdata filter we use PMOS switches.

11.6.2 The capacitors

The frequency-dependent behaviour of an SC filter is primarily determined by *ratios* of capacitances. It is very important to keep this in mind at the design stage. We base our

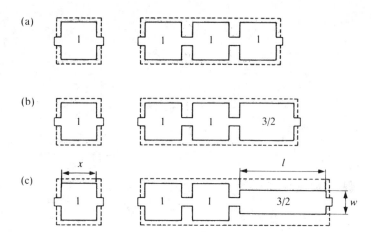

FIGURE 11.21 (a) To obtain accurate capacitance *ratios*, in spite of small dimensional errors in the fabrication process, larger capacitances are built up by stringing together a row of unit capacitors. (b) For non-integral ratios, such as 1:3.5, one of the sub-capacitors is given a different shape. (c) Even better results follow if the ratio of the periphery to area is always kept constant. This can be done by choosing length l and breadth w suitably. If the side of a unit capacitor is equal to x, then in our example $w = 0.634x$ and $l = 2.366x$. The top plate of each capacitor is represented by a continuous line, the lower plate by a dashed line.

design on a 'unit capacitor', which represents the smallest capacitance used. Larger capacitances are obtained by stringing unit capacitors together (Fig. 11.21(a)). An error that is essentially proportional to the *periphery* of a capacitor will now also be proportional to the *area* and hence its capacitance. In this way certain variations in the etching process in the production will give a change in the *absolute values*, but not in the *ratio* of two capacitances. Since not all the capacitance ratios will generally correspond to an integer, we cannot pursue this system to the utmost. A ratio of 1:3.5, for example, can be obtained as shown in Fig. 11.21(b). A better solution, however, is the one shown in Fig. 11.21(c), where we make a capacitor with a relative capacitance of 3.5 by combining two unit capacitors with a capacitor of relative capacitance 1.5 that has the same ratio of periphery to area as the unit capacitor. This approach gives a capacitor that is almost unaffected by the production variations mentioned earlier.

Another cause of error is the occurrence of parasitic capacitances: in an integrated capacitor these appear both between the upper plate and the substrate and between the lower plate and the substrate. The first can be largely eliminated by making the upper plate slightly smaller than the lower plate; this can be seen clearly in Fig. 11.21. The effect of both kinds of parasitic capacitance can be greatly reduced at the design stage by ensuring that both capacitor plates are switched only between earth and a low-impedance voltage source or between two points of equal voltage (Fig. 11.14(b)).

To make the chip on which an SC filter is fabricated as small as possible, we should keep both the *ratio* and the *absolute values* of the capacitances small. We can influence the ratios through the shape of the signal flow graph, the value of the scaling resistance R_s and the magnitude of the sampling rate $1/T$. However, the signal flow graph and the scaling resistance also determine the dynamic range of the signals in the filter; similarly there are some constraints on the choice of the sampling rate. There is a lower limit to the absolute value of the capacitances, because the unit capacitor must have at least some minimum dimensions to ensure a certain accuracy. A high absolute value for the capacitances has the advantages of insensitivity to parasitics, noise, interference from clock signals and off-set voltages, but high absolute values are also associated with high power dissipation and slower circuits. We see that there are various conflicting factors, which have to be set against one another in the design process.

The capacitance per unit area depends upon the fabrication process. In our integrated viewdata filter we used capacitors with a capacitance of $340\,\text{pF/mm}^2$.

11.6.3 The amplifiers

The most critical component is the amplifier. This determines to a great extent the specifications that a particular filter can or cannot meet. The amplifiers that are used in SC filters have a number of special features, so from now on we shall call them SC amplifiers. An SC amplifier always has a capacitive load, for example, and never a purely resistive load. Since a capacitance has a high impedance at low frequencies, it is not necessary to use an output buffer, which would have an undesirable effect on the noise performance and stability of the amplifier. Another special feature is that the output voltage of an SC amplifier only has to reach the correct value at the instant when the input switch of the following stage is switched to 'on'. After any change in input voltage, the SC amplifier therefore has a certain time available in which the output voltage can settle to the corresponding value (Fig. 11.22).

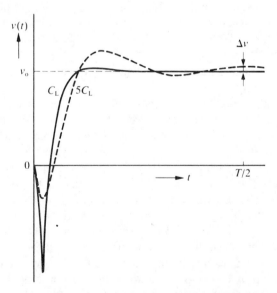

FIGURE 11.22 Typical variation of the output voltage $v(t)$ of an SC amplifier as a function of time for two different capacitive loads C_L and $5C_L$. It can clearly be seen that the load of $5C_L$ is too large for a stabilized final value v_o to be reached within the desired half clock period $T/2$: at $t = T/2$ there is still a difference Δv.

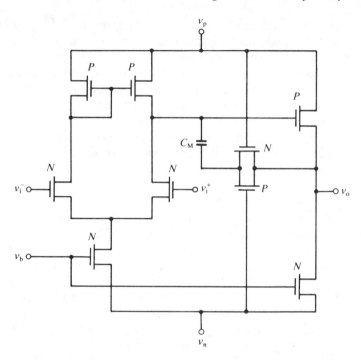

FIGURE 11.23 SC amplifier for the integrated SC filter for viewdata. A CMOS process (CMOS = complementary MOS) is used in the fabrication, so that both PMOS transistors (P) and NMOS transistors (N) can be included. The amplifier consists of a differential stage with inputs v_i^- and v_i^+ and a final stage with output v_o. Between the two stages compensation is provided by a Miller capacitance C_M and a resistance formed by the combination of an NMOS and a PMOS transistor. v_p positive supply voltage, v_n negative supply voltage, v_b bias voltage.

Since we have fabricated our integrated viewdata filter by a LOCMOS process,[67] we can build up an SC amplifier from NMOS and PMOS transistors. The result is shown in Fig. 11.23. The complete amplifier consists of an input stage that operates as a differential amplifier and a final stage formed by a PMOS transistor and an NMOS transistor, with the NMOS transistor acting as a constant-current source. The compensation that produces the desired gain characteristic takes place between the two stages. It is realized on the chip by a Miller capacitance C_M in series with a resistance formed by the combination of a PMOS transistor and an NMOS transistor.[76]

(In the ordinary way the Miller capacitance would provide direct negative feedback from the output to the input of the second stage. Since this output has a high impedance, the Miller capacitance then also provides feedforward. This has a significant and undesirable effect on the performance of the amplifier (more phase shift). We have corrected this by including a resistance in series with the Miller capacitance.)

The area taken up by this amplifier in the integration is about $0.1 \, \text{mm}^2$ and for most applications its power dissipation is between 0.1 and $5 \, \text{mW}$. The exact dimensioning of the SC amplifier is closely dependent on the surrounding circuits. A large capacitive load requires a relatively large output transistor, for example, which also dissipates a relatively high power.

11.7 INTEGRATED VIEWDATA FILTER

In the previous section we saw that too large a capacitive load can upset the proper operation of an SC amplifier. The requirements that the amplifiers have to meet can become less exacting as this capacitive load is reduced. Now it is found that a direct realization of the signal flow graph of Fig. 11.19(b) as an SC filter is not particularly satisfactory in this respect; the trouble seems to be mainly caused by the paths connecting points that are not directly adjacent and having a transfer function of -1. This is because these paths are formed with the aid of a summation circuit (see Fig. 11.16) with $C_3 = C_2$. Consequently there is a large capacitive load, especially for the *preceding* SC amplifier. It would therefore be preferable to reduce the ratio C_3/C_2. This is done in the integrated filter by first making a number of changes in the signal flow graph (Fig. 11.24). Although the result looks more complicated than the original signal flow graph, the number of integrating operations remains the same and the maximum summation ratio has been reduced from -1 to -0.178. Furthermore, this is accompanied by a reduction in the total capacitance contained in the filter of no less than 30%. A detailed circuit diagram of the integrated viewdata filter is shown in Fig. 11.25. Finally, Fig. 11.26 is a photograph of the integrated SC version of the viewdata filter. The separate components are clearly recognizable in the photograph; they include the strings of unit capacitors, the long 'residual' capacitors and ten SC amplifiers. The measured attenuation and group-delay characteristics are very close to the calculated characteristics for the SC filter: the

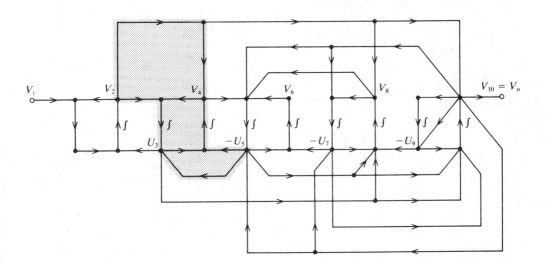

FIGURE 11.24 Modified version of the signal flow graph of Fig. 11.19(b), in which the transfer between non-adjacent nodes has been minimized. This enabled the maximum capacitive load on the SC amplifiers to be reduced by a factor of about six, greatly easing the conditions that these amplifiers have to satisfy. For simplicity the exact values of the transfer between all the nodes have been omitted. Although this signal flow graph looks more complicated than the original one, the number of integrating operations ($\int$), and hence the number of SC amplifiers, remains the same. Moreover, the total capacitance required is 30% smaller.

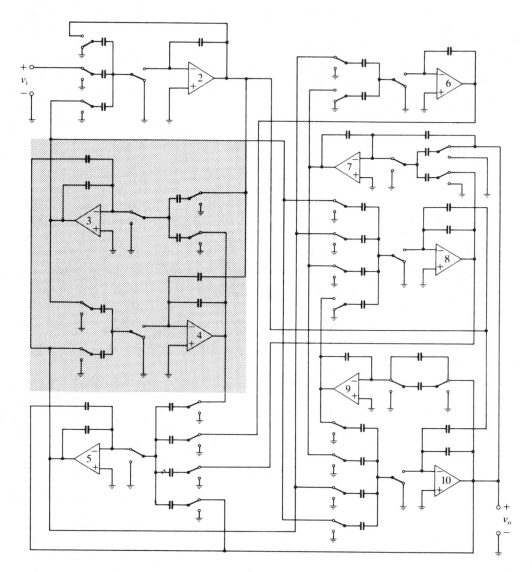

FIGURE 11.25 Detailed circuit diagram of the viewdata filter corresponding to the signal flow graph of Fig. 11.24. Each integrating operation is performed with the aid of an SC module, which is characterized by a low sensitivity to parasitic capacitance. The number of each amplifier corresponds to the subscript of the output voltage in the previous figure. We should also point out that some of the amplifiers (Nos 2, 6 and 8) form part of more than one type of integrator, since the input capacitors are switched in more than one way. The shaded part again corresponds to the shaded part in previous figures.

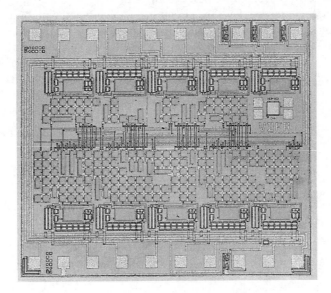

FIGURE 11.26 The integrated viewdata filter. The strings of unit capacitors can clearly be seen on either side of the interconnection pattern, which is almost central. At the top and bottom of the picture there are five SC amplifiers, readily visible because of the large Miller capacitors. Nine of these amplifiers correspond to the amplifiers of Fig. 11.25. The tenth (top right, close to the letters VDFC) is used for testing only. The area of the chip is about 3.3 mm^2.

deviation from the attenuation characteristic in the passband is smaller than 0.05 dB and never more than 3 dB in the stopband (at 430 Hz); the maximum deviation for the group delay is 12 μs. The area of the integrated circuit (not including bonding pads) is about 3.3 mm^2. The power dissipation is about 10 mW at a supply voltage of 10 V. Recent developments in SC amplifier design and IC technology indicate that the required power can be reduced even further, to about 1 mW.

11.8 SUMMARY

In the last few years switched-capacitor filters (or 'SC filters') have been found very suitable for making fully integrated electrical filters. Since the signals in these filters are discrete in time but continuous in amplitude ('sampled-data signals'), the filters behave in some ways like analog filters and in others like digital filters. SC filters consist of switches, capacitors and amplifiers. All these components are readily integrated in MOS technology. The starting point for the design can be a doubly terminated *LC* filter. By setting up a signal flow graph in which the only frequency-dependent operations are integrations, a filter built up from SC integrator/adder modules can be derived fairly directly. The method gives a filter on a single chip, with very small dimensions, low parameter sensitivity and low power consumption. No trimming is necessary in production, temperature effects are very low, the circuit can be combined with others on the same chip and analog signals can be processed without analog-to-digital conversion. The chief limitation of SC filters is the frequency range in which they can be used to advantage (up to about 100 kHz at present). A number of the above aspects are illustrated with the aid of an integrated viewdata filter, fabricated in a LOCMOS process. Its area is about 3.3 mm^2. The power dissipation is about 10 mW at a supply voltage of 10 V.

12

Digital signal processing in television receivers

M. J. J. C. Annegarn, A. H. H. J. Nillesen and J. G. Raven[†]

12.1 INTRODUCTION

To convert a scene into electrical signals all conventional television systems employ periodic image scanning in a line pattern. The scan is made in much the same way as our eyes follow the lines on the pages of a book when reading. Little has changed here in the hundred years or more since the very first ideas about television were put forward. The German patent filed by P. Nipkow in 1884, in which he proposed a form of mechanical scanning using a rotating disc perforated with a spiral pattern of holes (the 'Nipkow disc'), is of course well known.

Since that time there has really been only one fundamental change in the scanning pattern used. This is 'interlacing', the system in which all the even lines are scanned first and then all the odd lines, producing two alternate 'fields' of even and odd lines, which combine to make up a complete picture, or frame. This is done to prevent flicker in the display.

Television first began to make real headway when a completely electronic system became possible. The first broadcasts (in monochrome) took place at the end of the thirties, e.g. in the United Kingdom and the United States. In 1936 in Germany there was even an experimental television broadcast with a portable transmitter from the Olympic Games in Berlin.[77] Since that time there has been a continuous development, yielding steady improvements of quality and additional features, such as large-screen displays, colour reproduction, improved image and sound quality, stereo sound, teletext and remote control.

Figure 12.1 shows the block diagram of a modern television receiver. With the exception of parts of the remote-control unit, the remote-control receiver and the teletext decoder,

[†]Philips Research Laboratories, Eindhoven; Elcoma Division, Philips NPB, Eindhoven, and Consumer Electronics Division, Philips NPB, Eindhoven, respectively. This chapter is reprinted from Philips Tech. Rev. **42** No. 6/7: 183–200, Apr. 1986.

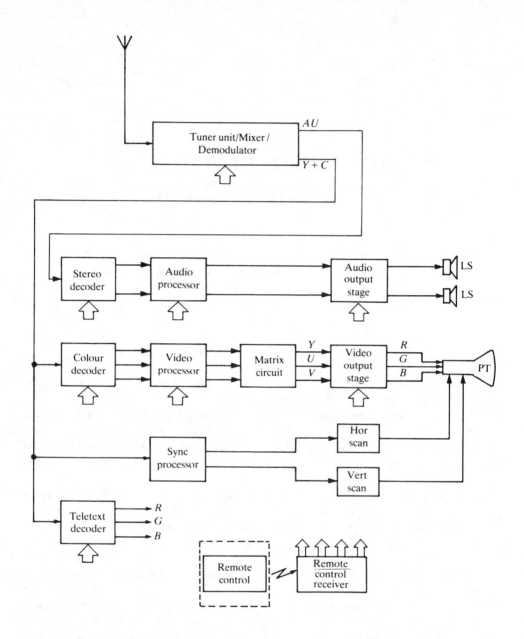

FIGURE 12.1 General block diagram of a modern colour television receiver. The incoming signal, after tuning, mixing, amplification and demodulation, is split into a sound signal AU and a composite picture signal $Y+C$. The sound signal is separately decoded, processed and reproduced via the loudspeakers LS. The signal $Y+C$ is processed in several stages to provide the colour signals R, G and B (for red, green and blue) for producing a picture on the picture tube PT. Horizontal and vertical deflection signals are required for the formation of the picture. These are generated by means of a sync processor and the horizontal and vertical scanning circuits Hor scan and Vert scan. In most present-day receivers digital circuits are only used for the remote control, the remote-control receiver and the teletext decoder. In the near future, however, other blocks will increasingly be digitized. At present this does not apply to all units because analog circuits still have certain advantages, connected with the high frequencies, high voltages or high powers in some circuits.

most receivers are still based entirely on analog circuits. There is however a distinct trend towards the application of digital circuits in other parts of the sets.[78] This applies in the first place to the decoders for image, sound and teletext and to the processors for image, sound and synchronization. (Existing analog receivers either do not contain a video processor, or contain only a very rudimentary form of video processor. In digital receivers, on the other hand, this processor plays an extremely important role, as will appear in this chapter. (For technical reasons digitization will not for the time being extend to the circuits for tuning, mixing, i.f. amplification, demodulation, output amplification and deflection, owing to the high frequencies, high powers or high voltages in these circuits.

The advantages to be expected from this advancing digitization of television receivers are to be found in three main fields:

- New features
- Improved quality
- Greater economy

The new features that digitization will generate are closely connected with the use of digital field or picture memories. Some examples of these are still pictures ('freeze-frame'), combined pictures ('picture-in-picture') and teletext background memory. We shall deal with these later in more detail.

The second category of advantages arising from the digitization of television receivers comes about because present-day colour TV systems can really provide a better picture quality than analog methods now give.[79] To some extent this is due to undesirable interaction between chrominance and luminance signals and to the relatively crude methods used for correction. In addition there may be noise, and buildings and mountains can give signal reflections (echoes). There can also be negative effects due to the scanning pattern used, such as flickering pictures (large-area flicker) or jittering lines (line flicker). With larger and brighter television screens, effects of this kind have become more of a nuisance. Digital signal processing, however, can do much to improve this situation.

The economic advantages can be summarized under 'lower price' or 'lower costs'. To some extent these stem from the reduced number of components and the replacement of relatively expensive components (for example analog glass delay lines) by inexpensive ones (digital memories). Production costs have been brought down by better trimming procedures and better stock control, and also by the increased reliability of the products.

In this chapter we shall confine our attention to the digital processing of the composite signal $Y + C$ (see Fig. 12.1), which contains the actual picture information and consists of the luminance signal Y and the modulated colour subcarrier signal C, referred to in this chapter as a 'chroma' signal. We shall start by describing in some detail the characteristics of the frequency spectrum of this composite signal as defined by the PAL and NTSC system standards and we shall also describe the operations performed on it in the receiver. From this information we shall show how important the choice of sampling rate is here. We shall also show that certain filters that are particularly well suited to digital implementation, such as comb filters, are eminently useful for television applications. We shall then explain how digital signal processing can be used to obtain pictures of better quality and to add new features to the television receiver.

12.2 BACKGROUND

Colour television is based on the fact that a colour picture can be formed from three basic colours: red, green and blue. In the television camera a scene is therefore converted into three electrical signals R, G and B, each representing one basic colour. A replica of the original picture is recovered from these signals in the picture tube of the receiver. The colour signals R, G and B are not, however, sent directly from transmitter to receiver. This would require three TV channels, and in any case none of these three signals would be suitable for reproduction as a monochrome picture on a monochrome screen. In the studio, therefore, a matrix circuit is used for deriving from R, G and B a brightness or luminance signal Y and two colour-difference or chrominance signals U and V. (Y, U and V are sometimes referred to as component signals.) The signal Y now contains all the monochrome information and the signals U and V can be combined by modulation to produce a single chroma signal C, which can be added to Y relatively easily. This results in the composite signal $Y + C$, which can be sent via a single television channel to the receivers.[80] The composite signal also contains the signals required for synchronization in the receiver, but these do not really come within the scope of this chapter.

Major differences between the three colour television systems PAL (Phase Alternation Line), NTSC (National Television System Committee) and SECAM (SEquentiel Couleur A Mémoire), which are world standards,[81] are to be found in the modulation process used in generating the chroma signal C. PAL and NTSC use a combined form of phase and amplitude modulation (quadrature modulation), and SECAM uses frequency modulation. The PAL system is mainly used in Western Europe (except in France), the NTSC system is the standard in the United States of America and Japan, and the SECAM system is used in Eastern Europe and France.

12.2.1 Frequency spectra in NTSC and PAL

In dealing with the composite signals we shall confine our attention in this chapter to the two standardized colour television systems known as NTSC and PAL. However, much of the digital processing performed after colour decoding is in principle also suitable for signals in the SECAM system.

Figure 12.2 shows the spectra[82] of composite signals in NTSC and PAL. From the general diagram of Fig. 12.2(a) it is clear that the chroma signal C and the luminance signal Y share the upper part of the frequency band, around the subcarrier frequency f_{sc}. The more detailed view on an expanded frequency scale (Fig. 12.2(b)) shows that Y consists of separate frequency components situated at multiples of the line frequency f_l. With appropriate modulation of the U and V signals, however, the frequency components of C can be made to appear at different locations, which makes it possible to separate Y and C in the receiver. The PAL system even uses different frequencies for the U and the V information.

The spectral structures of Fig. 12.2(b) are strictly speaking only applicable to stationary pictures in which all the successive lines are identical. This would be very unusual, of course. When differences do exist between the lines (but there is still no movement) then a second expansion of the frequency scale gives the spectra shown in Fig. 12.2(c). Around each frequency component from Fig. 12.2(b) there are other Y, U,

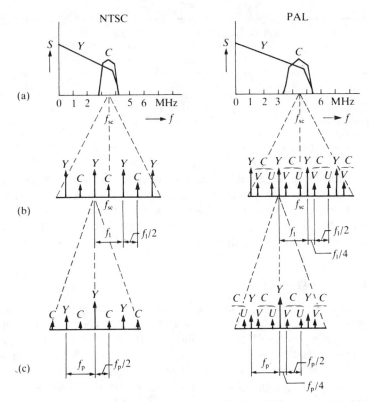

FIGURE 12.2 Frequency spectra of the composite signals $Y + C$ in the standardized NTSC and PAL colour-television systems. S signal power per Hz bandwidth. Some distinct similarities and differences can be seen. (a) In general terms it can be seen that in both cases the colour information, in the form of the chroma signal C, lies in the same frequency band as the highest frequencies of the luminance signal Y. This is produced by modulating a subcarrier of frequency f_{sc} with the colour information. (b) If the frequency scale is expanded it can be seen that Y and C consist of separate frequency components spaced by an amount corresponding to the line frequency f_1. In the PAL system it is in fact possible to distinguish between the U and V information in the spectrum. Strictly speaking, these spectra are applicable only to stationary pictures in which all the television lines are identical. (c) If there are differences between the lines, a further expansion of the frequency scale shows an even finer spectral structure, which is associated with the picture frequency f_p. (In this figure the important aspect is the relative locations of the different frequency components, rather than the exact representation of their relative amplitudes. Since the expansions in (c) are made around a Y component from (b), this component is clearly stronger than the other ones. At other places in the spectrum other components will dominate. The frequency positions are classified in the same way throughout the complete spectrum, however.)

V or C components at frequency spacings that are related to the picture frequency f_p. If we also wish to take randomly moving images into consideration, we then find that the elementary line structure of the spectra remains, but that the lines, depending on the movement, become broader or 'spread out'. The values of the principal frequencies that occur in standard signals of the NTSC or PAL type and the relations between them[81] are summarized in Table 12.1.

TABLE 12.1 The principal frequencies and frequency relationships in two standardized NTSC and PAL colour television systems

		NTSC	PAL
Picture frequency	f_p	29.97 ($\approx$ 30) Hz	25 Hz
Field frequency	f_f	59.94 ($\approx$ 60) Hz $= 2 f_p$	50 Hz $= 2 f_p$
Line frequency	f_l	15 734 Hz $= 525 f_p$	15 625 Hz $= 625 f_p$
Colour subcarrier frequency	f_{sc}	3 579 545 Hz $= 227\frac{1}{2} f_l$	4 433 618 Hz $= 283\frac{3}{4} f_l + f_p$

In the receiver the composite signal $Y + C$ is recovered from the antenna signal after a number of operations such as tuning, mixing, amplification and demodulation (Fig. 12.1). This composite signal must be subjected to a second demodulation or decoding (at least in a colour television receiver) to recover the individual Y, U and V signals, which are then converted, by a matrix circuit, into the R, G and B signals ultimately required.

12.2.2 Digital TV signal processing

The analog processing of TV signals is mainly based on the characteristics of the overall spectrum (Fig. 12.2(a)). The separation of Y and C is then achieved by using, say, a bandpass filter and a bandstop filter ('notch filter'), both centred on f_{sc}. It is assumed that the first filter suppresses the Y signal sufficiently and the second filter suppresses the C signal sufficiently. However, this is only partly true, and in fact high frequencies of the Y signal interfere with the C signal, causing the coloured streaks referred to as 'cross-colour'. Conversely, upon closer examination some of the C signal may be observed as a variation in luminance in the form of fine dots ('cross-luminance'), especially at the edges of coloured transitions. In this form of analog filtering only successive signal values are combined, corresponding to closely adjacent picture elements (pixels) in the same line, so it is therefore called *horizontal filtering*. If line memories are used it is possible to combine signals separated by exactly one or more lines, and therefore corresponding to pixels above and below one another in the picture. This is referred to as *vertical filtering*; it is seldom used in analog processing. The filters employed have frequency character- istics with a fine structure of the order of magnitude of f_l. In digital technology it is possible to make filter characteristics with a fine structure of the order of f_p. This requires delays of a field period or a picture period, which can be achieved by means of a field memory or a picture memory. This is referred to as *temporal filtering* since corresponding pixels occurring in successive fields or pictures are combined.[83,84]

In the rest of this chapter we shall base our treatment of the processing of $Y + C$ to R, G and B on the diagram in Fig. 12.3. First, the signal $Y + C$ is converted into the signal $y_n + c_n$ (n is the discrete-time variable; see footnote on p. 40) in an analog-to-digital converter with sampling rate $f_s = 1/T_s$. In the digital colour decoder the 'crude' component signals y_n^*, u_n^* and v_n^* are obtained as an intermediate result; after further digital processing they give y_n, u_n and v_n. These are applied to digital-to-analog converters to give the required analog component signals Y, U and V, from which the R, G and B signals are derived with a matrix circuit.

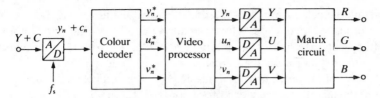

FIGURE 12.3 Diagram illustrating the digital conversion of the analog composite signal $Y + C$ into the analog colour signals R, G and B. A/D analog-to-digital converter, f_s sampling rate. The digital circuit Colour decoder produces the 'raw' component signals y_n^*, u_n^* and v_n^* from the digital composite signal $y_n + c_n$ as an intermediate result. These raw component signals are processed in the digital circuit Video processor to provide the 'pure' digital component signals y_n, u_n and v_n. The digital-to-analog converters D/A produce the analog component signals Y, U and V.

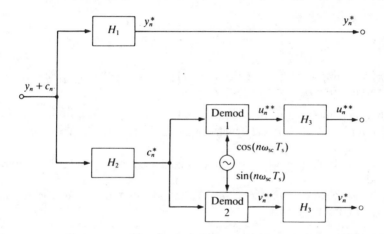

FIGURE 12.4 Block diagram of a digital colour decoder. H_1 band-stop filter, centred on the subcarrier frequency $f_{sc} = \omega_{sc}/2\pi$. H_2 band-pass filter, also centred on f_{sc}. Demod1, Demod2 demodulators. H_3 low-pass filter. $y_n + c_n$ digital composite signal. y_n^*, u_n^*, v_n^* digital component signals. c_n^*, u_n^{**}, v_n^{**} digital intermediate results. T_s sampling interval.

The R, G and B signals are simple linear combinations of the Y, U and V signals and vice versa. The R signal, for example, is given by:

$$R = aY + bU + cV$$

where a, b and c are constants, which are different in every colour television system. In principle it would also be easy to design the matrix circuit as a digital circuit. The D/A converters would then be located *after* this circuit instead of *in front of it*. This has deliberately not been done in the diagram shown in Fig. 12.3, because the signals U and V have a much smaller bandwidth than Y. Use is made of this in digital processing by applying a lower sampling rate for u_n^* and v_n^* than for y_n^*. The same applies for the corresponding D/A converters. The signals R, G and B, on the other hand, all have the same wide bandwidth as Y. The solution of Fig. 12.3 therefore offers a number of advantages.

A more detailed block diagram of a possible digital colour decoder is given in Fig. 12.4. The digital composite signal $y_n + c_n$ is first split into y_n^* and c_n^* by a bandstop filter H_1 and a bandpass filter H_2. Next, the signals u_n^* and v_n^* are obtained from c_n^* by demodulating it with two carriers in quadrature at a frequency $f_{sc} = \omega_{sc}/2\pi$ and using two lowpass filters H_3. In Fig. 12.4 there is only horizontal filtering. All the vertical and temporal filtering takes place in the next block, the video processor (Fig. 12.3).

Although Fig. 12.4 gives the essential functions of the colour decoder, some important simplifications have been made. The generation of the colour subcarrier, for example, is not shown explicitly; it must in some way be brought into synchronism with the received signal. Also, a practical circuit includes operations such as amplitude control of the chroma signal, 'colour killing' for monochrome transmissions, and periodic phase-switching for one of the subcarrier signals in PAL (PAL switch).

12.3 DIGITAL FILTERS FOR TELEVISION APPLICATIONS

Filters play a very important part in digital signal processing.[52] In television applications we can list three categories:

- Filters in which the length of the individual delays amounts to only one or a few sampling intervals $T_s = 1/f_s$ and which are characterized by 'simple' coefficients (e.g. 1/4, 1/2, 1, 2, 4).
- Transversal comb filters in which the individual delays correspond to a line period or a field period.
- First-order recursive comb filters in which the individual delays also correspond to a line period or a field period.

With non-recursive filters it is possible to achieve an exactly linear phase characteristic. This is often important in television applications and is one of the advantages of digital solutions compared with analog ones. We shall now take a closer look at all three categories and discuss some examples.

12.3.1 Filters with simple coefficients

Because of the relatively high sampling rate (10 to 20 MHz) digital filters for television should perform as few multiplications as possible, since these require relatively complex circuits and considerable electrical power. One way of accomplishing this is to use filter coefficients that as far as possible are integer (positive and negative) powers of 2.

FIGURE 12.5 (a) Some examples of digital filters H_1, H_2 and H_3 with simple coefficients, which can be used in the block diagram of Fig. 12.4. These filters have the following system functions:
$H_1(z) = (1 + \frac{15}{16}z^{-1} + z^{-2})(1 + z^{-1})^2(1 - 4z^{-1} + z^{-2})^2$,
$H_2(z) = (1 - z^{-2})^2(1 - z^{-1})^4(1 - 4z^{-3} + z^{-6})(1 + z^{-3})^3$ and
$H_3(z) = (1 + z^{-1})^3(1 + z^{-2})^2(1 + z^{-1} + z^{-2})$.
(b) Amplitude characteristics A_1, A_2 and A_3 of the filters H_1, H_2 and H_3. The sampling rate here is $f_s = 1/T_s = 13.5$ MHz. These filters have a linear phase characteristic. The bandstop and band-pass characteristics of H_1 and H_2 are clearly seen around $f_{sc} = 4.43$ MHz.

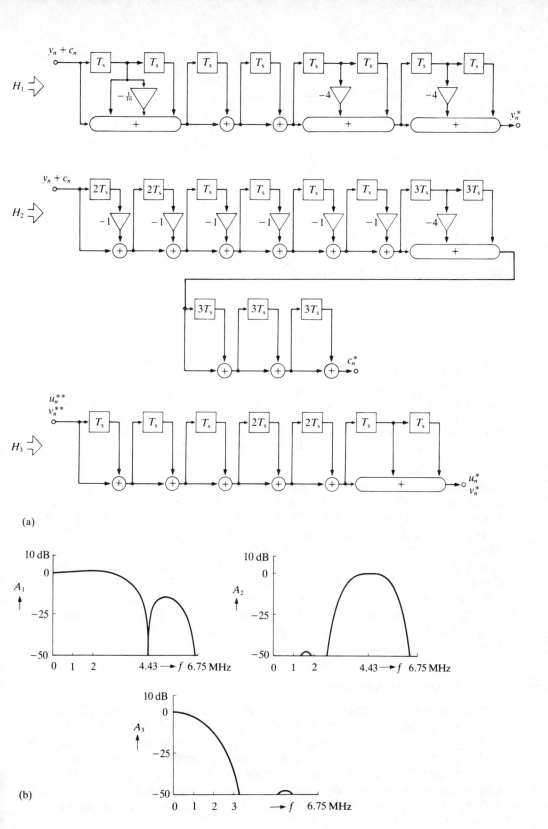

(a)

(b)

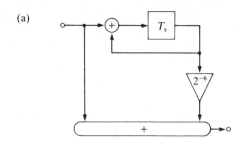

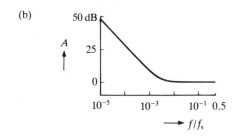

FIGURE 12.6 (a) Simple filter, suitable for application in a digital phase-locked loop in synchronization circuits. Its system function is $H(z) = \{1 - (1 - 2^{-6})z^{-1}\}/(1 - z^{-1})$. (b) Amplitude characteristic A of this filter. For clarity the frequency is shown on a logarithmic scale. The filters in Figs 12.5 and 12.6 form part of the digital multi-standard decoder developed at VALVO, Philips GmbH, Hamburg, West Germany.

Multiplication then amounts to no more than shifting the bits of a binary number by an integer number of bit places, with a change of sign if necessary (for negative coefficients). Figure 12.5(a) shows how the filters H_1, H_2 and H_3 of the colour decoder in Fig. 12.4 can be realized. They consist of a cascade arrangement of simple transversal filters. A linear phase characteristic is obtained by ensuring that the filter coefficients always satisfy the appropriate rules of symmetry. Figure 12.5(b) shows the corresponding amplitude characteristics A_1, A_2 and A_3, based on $f_s = 1/T_s = 13.5\,\text{MHz}$.

Figure 12.6(a) gives an example of a recursive filter with simple coefficients. Filters of this type can be used in digital synchronization circuits (phase-locked loops or PLLs). Finally, Fig. 12.7 shows an integrated circuit of a simple digital transversal filter made a few years ago for an application in colour television.[85]

12.3.2 Transversal comb filters

The characteristic line structure of the frequency spectrum of PAL and NTSC can be used for separating the different signal components. Comb filters are particularly useful for this. Some widely used types of comb filter are shown in Fig. 12.8. The length of the delay KT_s of the digital delay elements is deliberately not specified here ($K = \text{constant}$, $T_s = 1/f_s$) because it depends on the exact application; we shall give some examples shortly. We can make a rough distinction between delays of a line period T_l (vertical comb filters or line comb filters), delays of a field period T_f (temporal comb

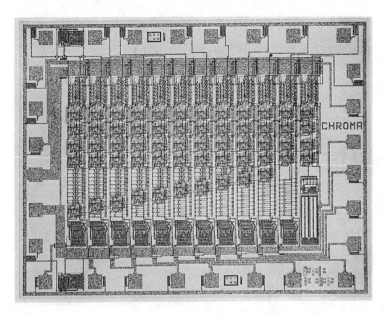

FIGURE 12.7 Integrated circuit that can be used to make a digital transversal filter with the system function $H(z) = -1 + 2z^{-2} + 6z^{-4} + 2z^{-6} - z^{-8}$ or $H(z) = 1 + 2z^{-2} + z^{-4}$. The area of this chip is about 10 mm^2. Made in two-phase NMOS technology, it operates at a maximum sampling rate of 40 MHz. The input signal consists of 9-bit samples and the output signal of 11-bit samples.[85]

filters or field comb filters) and possible combinations of both. It is also quite common for the comb characteristic to apply to only a part of the spectrum when a comb filter is combined with 'ordinary' filters.[24]

Comb filters are particularly useful for improving the separation of the signals y_n^*, u_n^* and v_n^* to reduce cross-colour and cross-luminance.

12.3.3 First-order recursive comb filters

Interesting filters for video applications can also be realized with simple first-order recursive structures. A few examples are shown in Fig. 12.9. The filters in Fig. 12.9(a) and (b) with $KT_s = T_1$ have been proposed earlier[80] in analog form for splitting the composite signal into its luminance and chroma components. The amplitude characteristics of these filters resemble those of the comb filters discussed in the previous subsection; by varying the one filter coefficient g the shape of the characteristic can be modified slightly. These filters, however, do *not* have a linear phase characteristic and are stable only for $-1 < g < 1$.

Another recursive filter, which we shall deal with at some length later, is shown in Fig. 12.9(c). Completely different frequency characteristics can be obtained by taking the value of g between 0 and 2; for other values of g the filter is unstable. It is also possible to switch g between two fixed values, e.g. $g = 1$ and $g = 1/8$. In one case ($g = 1$) there is apparently only a direct connection between input and output. In the other case ($g = 1/8$) the circuit acts as a comb filter (Fig. 12.10).

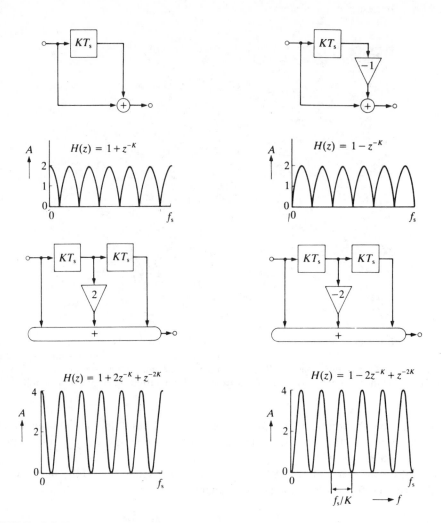

FIGURE 12.8 Transversal comb filters that are particularly suitable for separating the frequency components of colour-television signals (see Fig. 12.2). In addition to a block diagram for each filter, the system function $H(z)$ and the amplitude characteristic A are also shown. For the amplitude characteristics shown here $K = 6$, so the characteristics have only six 'teeth'. In reality KT_s corresponds to a line period T_1 or a field period T_f. At a sampling rate of about 13.5 MHz the characteristics then contain nearly 900 or 250 000 teeth in the fundamental frequency interval from 0 to $f_s = 1/T_s$.

FIGURE 12.9 Some examples of first-order recursive comb filters that are very suitable for the digital processing of colour-television signals. The system function $H(z)$ and the corresponding amplitude characteristic A and phase characteristic ϕ can be changed by varying the values of the coefficients g. A few examples are given for each filter. The values chosen for the upper two filters are $g = -\frac{2}{3}$, 0 and $\frac{2}{3}$. Those for the lower filter are $g = \frac{3}{2}$, 1, $\frac{3}{4}$ and $\frac{1}{4}$. Owing to their recursive structure, these filters do not in general have a linear phase characteristic, although the deviations are not necessarily large. A linear phase characteristic is found only for $g = 0$ in (a) and (b) and for $g = 1$ in (c). (In these three special cases, however, the filter is not really recursive.) For clarity the value $K = 3$ has been taken in this figure, so that the comb filter characteristics have only three 'teeth'.

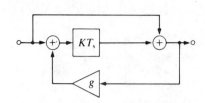

$$H(z) = \frac{1 + z^{-K}}{1 - gz^{-K}}$$

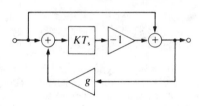

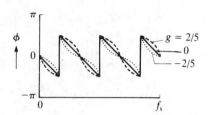

(a)

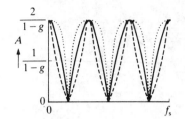

$$H(z) = \frac{1 - z^{-K}}{1 + gz^{-K}}$$

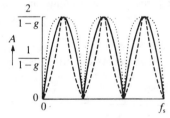

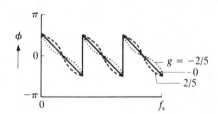

(b)

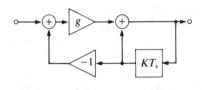

$$H(z) = \frac{g}{1 - (1 - g)z^{-K}}$$

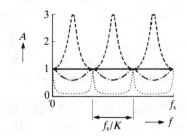

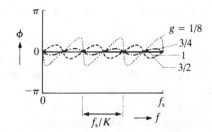

(c)

FIGURE 12.10 Integrated digital circuit that takes two 10-bit input signals a_n and b_n to form an 11-bit signal $c_n = ga_n + (1 - g)b_n$, where $g = 0, \frac{1}{8}, \frac{2}{8}, \ldots, \frac{7}{8}, 1$. The addition of a delay KT_s results in the filter shown in Fig. 12.9(c), but this integrated circuit has many other applications. Made in two-phase NMOS technology, the chip has an area of $15\,\text{mm}^2$. The maximum sampling rate is 40 MHz.[86]

These recursive filters are especially useful for reducing cross-colour, cross-luminance and noise.

12.4 THE SAMPLING RATE

In the digital processing of television signals the choice of the sampling rate f_s is more significant than in the digital processing of other types of signal, such as speech or music. A relatively small change in the value of f_s can have considerable consequences. This is closely related to the special features of TV signals such as the breakdown of a scene into lines and fields and the way in which the colour information is coded.

Table 12.1 shows the main frequencies that occur in a composite signal in the NTSC or PAL systems. If the interrelationships shown are satisfied, the signals are referred to as 'standard signals'. In television broadcasts these are virtually the only types of signal used.

After recording and playback of such a signal, e.g. with a video cassette recorder, there may be considerable departures from these frequencies. The same applies to certain video signal sources such as video games or home computers. We then refer to 'non-standard signals'.

The ratio f_l/f_p, even if it deviates from the standard value of 625 or 525, will often be a fixed integer number. In general, however, there will be no connection at all between f_{sc}

and f_l or f_p. Since the digital processing of video signals must also include non-standard signals, it has to be accepted that no fixed relation exists between f_l and f_{sc}. That is really a great pity, as we shall show.

The great difference between standard signals and non-standard signals appears most clearly from a comparison of the tolerances applicable to certain important parameters. In standard PAL, for example, where the number of lines per frame is $N = f_l/f_p = 625$, the line frequency f_l has a tolerance of 0.0001%, and $f_{sc} = 283\frac{3}{4}f_l + f_p$. For non-standard PAL, however, we find $607 \leq N \leq 643$, f_l has a tolerance of 4% and f_{sc} is given by $f_{sc} = 4\,433\,618 \pm 200\,\text{Hz}$.

12.4.1 Locking f_s to f_{sc}

A very simple decoder can be designed if the sampling rate f_s is in a fixed ratio to (is 'locked' to) f_{sc}. If $f_s = 4f_{sc}$, demodulating with say $d_n = \cos(n\omega_{sc}T_s)$ and $e_n = \sin(n\omega_{sc}T_s)$ amounts to multiplying by $-1, 0$ and $+1$. These operations can be performed *without* a true multiplier circuit, which usually requires a fair number of basic digital elements. Attractive solutions for the multiplying operation can also be found for some other rational ratios of f_s to f_{sc}. A further advantage of such a ratio, which we mention here in passing, is that the linearity of the A/D converter preceding the colour decoder (Fig. 12.3) is not so critical.

12.4.2 Locking f_s to f_l

It is necessary to lock the sampling rate f_s to the line frequency f_l of a television signal (so that $f_s = Mf_l$, where M is an integer) if digital memories are to be used in the signal processing to obtain delays of exactly one line period, one field period or multiples of these. We have seen some examples of this with the filters of Figs 12.8 and 12.9. Other examples will be encountered in section 12.10 'New features'.

12.4.3 Choice of f_s

Since, as noted, we must assume that no fixed relation exists between f_l and f_{sc}, we have to opt for locking f_s to either f_l or f_{sc}. But which is best?

Locking f_s to f_{sc} with a fixed ratio is the optimum choice for the colour decoder. However, the signals y_n^{**}, u_n^{**} and v_n^{**} obtained in this way, especially if they are non-standard signals, cannot easily be processed with digital memories. This first requires a change ('conversion') of the sampling rate to lock to the line frequency (Fig. 12.11(a)). For non-standard signals, however, there is not necessarily any fixed common multiple of f_l and f_{sc}, and this makes the conversion complicated and unattractive. Such a solution does not therefore seem a good choice if digital signal operations have to be performed on the output signals of the colour decoder.

On the other hand, a fixed ratio of f_s to f_l in the decoding of colour information requires 'real' multiplications because the carrier samples d_n and e_n are now no longer limited to the values 0 and ± 1, but may assume any value. A further and more serious complication arises from the very high accuracy required of the colour subcarrier samples to ensure high-quality colour decoding. (In an analog version the carrier must therefore possess

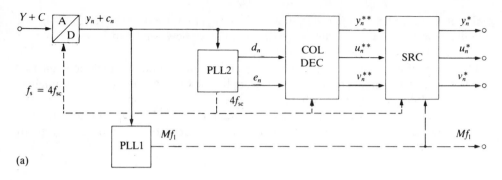

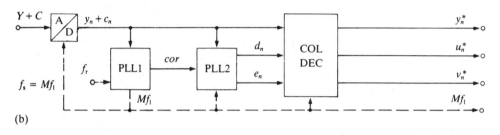

FIGURE 12.11 (a) Locking the sampling rate f_s to the subcarrier frequency f_{sc} (here $f_s = 4f_{sc}$) considerably simplifies the colour decoding, since demodulation only requires simple multiplication factors. (The carrier samples d_n and e_n only need to have the values -1, 0 and $+1$ here.) For a large number of other digital operations, however, it is necessary to couple f_s to f_1. In this arrangement a fairly complicated sampling-rate converter SRC is therefore necessary for changing the sampling rate from $4f_{sc}$ to Mf_1. (b) Locking f_s directly to the line frequency f_1 of the incident signal (so that $f_s = Mf_1$) requires special precautions to ensure that the carrier samples $d_n = \cos(n\omega_{sc} T_s)$ and $e_n = \sin(n\omega_{sc} T_s)$ are generated with sufficient accuracy. This is done by synthesizing f_s from a stable reference frequency f_r and generating a correction signal *cor*. A/D analog-to-digital converter. PLL1, PLL2 digital phase-locked loops. COL DEC colour decoder. M constant factor (≈ 900). y_n^{**}, u_n^{**}, v_n^{**} digital signals that are only significant as intermediate results; the other signals have the same significance as in the other figures.

'crystal stability'.) Locking f_s to f_1, however, gives the sampling rate f_s (and hence $T_s = 1/f_s$) the same relative inaccuracy as the line frequency f_1 of the incident composite signal. This inaccuracy may be unacceptably large, especially for non-standard signals. The required accuracy of d_n and e_n can then only be achieved by taking special measures (Fig. 12.11(b)). Starting with a stable reference frequency f_r, a sampling rate f_s locked to the incident f_1 (i.e. $f_s = Mf_1$) is derived in a phase-locked loop PLL1. This sampling rate should be accurately known and because of its relationship to f_r the magnitude of any error can be determined and expressed as a correction signal. This signal is applied to a second digital phase-locked loop PLL2 in which the carrier for the colour decoder is generated. Here the errors in the sampling rate are exactly compensated. This is done to ensure that in spite of the variations in the sampling times, the carrier samples d_n and e_n always have the exact value corresponding to each actual sampling time.

In principle there is considerable freedom in the choice of the constant factor $M = f_s/f_1$, but in practice several factors have to be considered. In the first place it will usually be desirable to satisfy the sampling theorem, which gives a lower limit for the value of M. (Sometimes the theorem is *not* satisfied, by deliberate choice. Reference is then made to 'sub-Nyquist sampling'.[24]) While an increase in M relaxes the specifications for the analog prefilter in analog-to-digital conversion and for the analog post-filter in digital-to-analog conversion,[52] it also increases the number of digital operations required per second as well as the memory capacity necessary for storing the signals. The sampling theorem indicates that a minimum value between 10 and 12 MHz is required for the frequency f_s, depending on the television system being considered (so that $M > 650$ to 750). A good choice for both NTSC and PAL seems to be $f_s = Mf_1 = 13.5$ MHz. Looking at the systems of Table 12.1, this means that $M = 858$ for NTSC and $M = 864$ for PAL. An additional argument favouring this choice is that the standard sampling rate for studio applications is also 13.5 MHz. Keeping to this value can have certain advantages, especially if transmission from studio to receiver is to be entirely digital in the future.

In the SECAM system the methods of coding and decoding the colour information do *not* indicate any preference for locking f_s to f_{sc}. This leaves only the preference for locking to f_1, particularly if field or picture memories are used for the signal processing. In spite of the larger bandwidth of the luminance signal (6 MHz) than in NTSC and PAL, a good choice for SECAM will probably also be a line-locked f_s of 13.5 MHz.

At the present time the digitization of video equipment is still in its infancy. The choice we now make for the sampling rate f_s will have a long-term impact, however: the first sets will contain a number of different digital ICs that operate at f_s. In addition, we may expect a gradual but sustained progress towards new and improved chips. But since we shall require compatibility with the existing chips, these will have to be based on the same f_s. It therefore seems likely that the 'life span' of f_s will be much longer than that of a single generation of chips. This emphasizes yet again the importance of a well-considered choice of f_s in the present initial phase.

12.5 IMPROVING THE PICTURE QUALITY

There are many ways in which the picture quality can be improved with the digital filters in Fig. 12.5, 12.8 and 12.9, primarily by reducing cross-colour, cross-luminance and noise. The wide variety of possible methods is connected with the existence of the following alternatives:

- The colour coding system used (NTSC or PAL).
- The type of digital filter (recursive or not).
- The amount of delay (one or more line periods, one or more field periods).
- The location of the filters (before or after colour decoding).

Complete coverage in this chapter is not really possible, so we shall confine ourselves to giving a few outstanding examples of vertical filtering and temporal filtering.

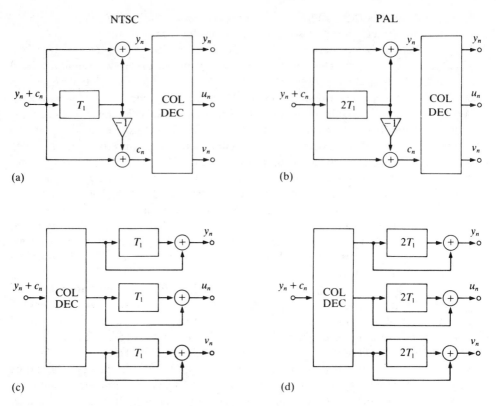

FIGURE 12.12 Some examples of possible applications of transversal line comb filters for reducing cross-effects in NTSC and PAL. In principle the filters can be placed either before the colour decoder COL DEC (a) and (b) or after it (c) and (d). In the first case different filters are used for reducing cross-colour and cross-luminance; in the other case the filters are the same, but three are necessary instead of two. With the actual spectral relationships (Fig. 12.2) delays of T_1 are required for NTSC and $2T_1$ for PAL, where T_1 is the period of one TV line.

12.5.1 Vertical filtering

Figure 12.12 shows a number of comb filters based on vertical filtering, with which cross-colour and cross-luminance can be reduced. With NTSC a delay of one line duration T_1 in the composite signal is sufficient to produce an improvement (Fig. 12.12(a)). This is because the frequency of the colour subcarrier is exactly an odd multiple of half the line frequency; see Fig. 12.2(b). With PAL this is not the case and for the same improvement a minimum delay of two line periods would be necessary (Fig. 12.12(b)).

In principle, cross-effects can also be reduced after colour decoding, both for NTSC and for PAL. This requires separate operations on the three component signals, however. In each of these signals the required frequency components are always concentrated around integer multiples of the line frequency. The unwanted (cross-colour or cross-luminance) frequency components lie in between. The same type of comb filter

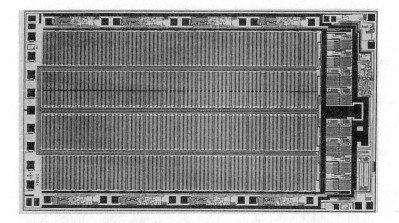

FIGURE 12.13 Integrated digital line memory. This memory can serve as an elementary module in the construction of circuits for vertical filters in which delays of a line period or multiples of this are required. It has a capacity of 8×1024 bits, is a product of dynamic NMOS technology and has an area of almost $13\,\text{mm}^2$. This chip is suitable for sampling rates up to $40\,\text{MHz}$.

can therefore be used for all three signals, for NTSC again with a delay of T_1 and for PAL with a delay of $2T_1$ (Fig. 12.12(c) and (d) and Fig. 12.13).

The operation of the two comb filters for PAL signals given here is not ideal. The required delays of two line periods cause a perceptible reduction of the definition of the picture in the vertical direction because the signals corresponding to relatively widely spaced lines are combined. In addition the cross-effects are only strongly reduced if the combined picture lines are identical, that is to say if the picture has a distinct vertical structure. Both for NTSC and for PAL a loss of picture definition occurs in the diagonal direction because diagonal picture structures give rise to frequency components that are strongly attenuated by line comb filters. This can be improved somewhat by using a rather more complicated comb filter structure, where the actual 'combing' of the luminance signal does not take place at low frequencies but is limited to the frequency range in which the chroma signal is situated.[24]

The comb filter in Fig. 12.12(a) has long been used in *analog* systems, mainly for reducing cross-colour. All that is required is a delay line whose performance is not critical except in the frequency band of the chroma signal (on account of the bandpass characteristic). Conventional analog glass delay lines give adequate results here.

12.5.2 Temporal filtering

Completely new ways of improving the picture quality can be obtained by basing filters on delays of about one field period T_f or multiples of T_f. Until now this has hardly been possible with analog methods, but recently the use of integrated digital memories has made such delays an economic proposition (Fig. 12.14).[87–89] Temporal filtering is best

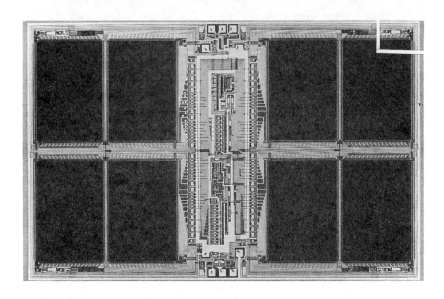

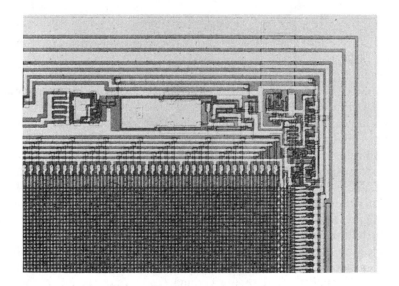

FIGURE 12.14 Advancing technological developments have recently opened up the prospect of relatively low-cost fully integrated digital field and picture memories. The use of such memories is essential if the many potential advantages of the digital processing of television signals are to be fully realized. The upper photograph shows an integrated memory of the CCD type (charge-coupled device) with a capacity of 308 kbit. The chip (about 35 mm^2) is a product of 2-μm NMOS technology. Seven such chips can be combined to make a complete field memory for the component signals of all current television systems. To make the fine details visible, the part of the upper photograph in the white rectangle is also shown on a greatly enlarged scale.[87]

implemented after colour decoding. First of all, there is no longer any need to take account then of the exact value of f_{sc}. In the second place the signal operations are then largely independent of the colour television standard in use. We shall therefore confine ourselves to signal operations performed after colour decoding.

As we have seen, vertical filtering combines the information from pixels that are closely adjacent to each other in successive lines of the same field. Temporal filtering, on the other hand, combines the information from pixels that are close together in successive fields or even exactly coincident in successive pictures. In the main, vertical filtering depends on properties of the signal spectrum on the scale of Fig. 12.2(b) and temporal filtering depends on those on the scale of Fig. 12.2(c).

The greatest advantage of temporal filtering is that the degree of cross-effect reduction is not so strongly dependent on the structure of the picture, and is therefore not so very different for sharp vertical, horizontal or diagonal colour transitions. We shall now present some actual examples of temporal filtering.

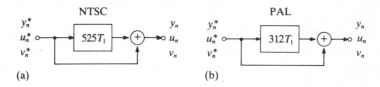

(a)   (b)

FIGURE 12.15 Temporal transversal filter for reducing cross-effects in (a) NTSC signals and (b) PAL signals. These filters are designed for insertion after the colour decoder. The delay in both cases is such that cross-colour or cross-luminance can be reduced by the adding operation.

Figure 12.15(a) shows a filter that removes cross-effects from the component signals in the NTSC system (with 525 lines) and Figure 12.15(b) shows a similar filter for the PAL system with 625 lines. An effect of the $312T_1$ delay for the PAL system is that two pixels directly one above the other in two successive fields are combined in the adder. If f_{sc} were equal to $283\frac{3}{4}f_1$, then (because 312 is a multiple of 4) the cross-effects in the adder would be in phase and would reinforce each other – exactly what we do *not* want. But since, at least with standard signals, f_{sc} also has a frequency-offset of 25 Hz (see Table 12.1), a delay of $312T_1$ represents an effective additional phase shift of 180° for f_{sc}, and the cross-effects in the adder will in fact be reduced.

The exact phase shift is $(312/312.5) \times 180°$. In principle the delay of $312T_1$ for the PAL system can be replaced by a delay of $1250T_1$ (corresponding to the duration of four fields, or two complete pictures). Since 1250 is two plus a multiple of four, the subcarrier phase is shifted by an integer number of periods *plus* a half period, while the 25-Hz offset now gives no extra phase shift. Once again the result is therefore the extinction of cross-effects. The advantage is that in this case we combine coincident pixels in the adder (whereas with the delay of $312T_1$ there is a slight difference in the vertical position). A disadvantage, however, is the time difference of two complete pictures, which can cause blurring effects with moving scenes.

In the circuit for the NTSC system (Fig. 12.15(a)) the delay of $525T_1$ corresponds to one picture (two fields). The odd number of line periods means that there is a delay for f_{sc} of an integer number of subcarrier periods *plus* a half period. Here again cross-effects are reduced in the addition.

Temporal filters can also be obtained with the recursive structures in Fig. 12.9 by giving the delay an appropriate value. If we make KT_s equal to one picture period (i.e. $625T_1$ for PAL and $525T_1$ for NTSC) we can for example use the filter in Fig. 12.9(c) for reducing cross-colour or cross-luminance in the component signals.

12.6 MOVEMENT DETECTION

In all temporal filters the values of certain pixels are combined that were recorded in the camera with rather long intervals between them ($312T_1$, $625T_1$, etc.). For stationary pictures this has no consequences, of course, but if the picture moves, the movements cause changes in the signal value that have nothing to do with cross-colour or cross-luminance, although they are treated as if they had. This can lead to movement blur that completely negates the positive result of the reduction of cross-effects and can even make matters worse.

In this respect recursive filter structures are inferior to non-recursive types; the effects of picture delays will be worse than those of field delays, and for purely perceptive reasons the luminance signal will be more vulnerable than the colour-difference signals. Experiments have shown that one of the few cases in which there is virtually no perceptible movement blur with temporal filtering is when the filter in Fig. 12.15(b) is used for cross-colour reduction in the PAL system.

When temporal filters are used, it is therefore usual to add a movement detector, which decides for each pixel whether there has been a movement with respect to the preceding field or picture. Temporal filtering is only applied for those parts of the picture in which there is no movement; the other parts are not processed in this way. In the structure of

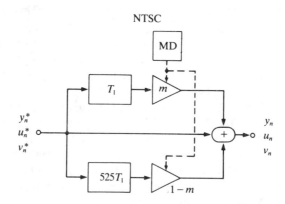

FIGURE 12.16 Reduction of cross-effects in NTSC signals with a movement detector MD. When there is movement in the picture, $m = 1$ and there is vertical filtering; with no movement, $m = 0$ and there is temporal filtering.

Fig. 12.9(c) a movement detector can ensure that $g = 1$ for moving parts of the picture (it is not a recursive filter then but a direct connection). All other temporal filters can be similarly extended with a movement detector. A second example is given in Fig. 12.16, which shows a circuit for cross-colour reduction in the NTSC system. Here we can recognize a combination of reduction by vertical filtering (Fig. 12.12(c)) and reduction using temporal filtering (Fig. 12.15(a)). Under the control of a movement detector a decision is made between $m = 0$ and $m = 1$ for each pixel and thus between the two methods of filtering.

We shall not go into the details of movement-detector design in this chapter; in principle they all make use of the fact that the frequency spectrum of a truly stationary picture consists of lines at known positions (see Fig. 12.2). If the spectrum contains energy at other positions, this must be the result of movement. Various types of comb filters are again used for detecting this.

12.7 NOISE REDUCTION

Any form of cross-colour or cross-luminance reduction based on filtering out unwanted frequency components will usually be associated with a reduction in the noise originating from external sources of interference, and therefore with an improvement in the signal-to-noise ratio (although this does not necessarily imply a subjective improvement in picture quality). The extent of this improvement, however, will differ from one type of filter to the other. Let us for convenience assume that the noise has a flat power spectrum (i.e. white noise). Comb filters of the upper two types in Fig. 12.8 then give an improvement of 3 dB. Comb filters of the lower two types in Fig. 12.8 give as much as 4.3 dB. An even greater improvement in signal-to-noise ratio can be achieved with some recursive filter structures. With the filter of Fig. 12.9(c) a gain of 2.2 dB is obtained for $g = \frac{3}{4}$ and 11.8 dB for $g = \frac{1}{8}$.

For stationary pictures it is easy to see how the noise can be reduced with this type of recursive filter, if KT_s is a delay of an entire *picture period* (i.e. $625T_1$ or $525T_1$). The signal values corresponding to one particular pixel are constant and are simply added together, whereas the noise contributions vary and are therefore averaged out.

A noise reduction can also be achieved, however, by using a memory that gives a delay of only about one *field period*. Here use is made of the fact that the signal values for every two pixels lying immediately one above the other in successive fields do not as a rule change very much. Owing to the recursive structure, the averaging takes place over a large number of fields. If in PAL a fixed field memory of $312T_1$ or $313T_1$ were to be taken, pixels on lines increasingly far apart would then gradually be combined (Fig. 12.17). To avoid this, the delay must be made to alternate between $312T_1$ and $313T_1$ in this case. This can be done with a field memory of $312T_1$ to which an extra line delay T_1 is periodically added; during the successive fields the switch S is alternately in the upper and lower positions.

It is possible to use the same picture or field memory simultaneously for reducing cross-effects and for additional improvement of the signal-to-noise ratio. We shall give two examples of this. Figure 12.18 shows a possible solution for NTSC and is in fact a

PAL

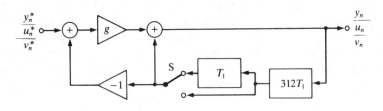

(a)

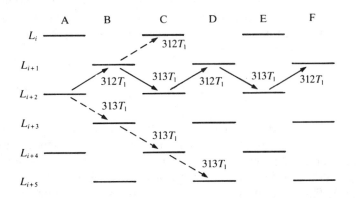

(b)

FIGURE 12.17 (a) Noise reduction for each of the PAL component signals y_n^*, u_n^* and v_n^*, by using a first-order recuesive comb filter with only one *field* memory ($312T_1$ or $313T_1$). (b) Half of the successive lines L_i, L_{i+1}, L_{i+2}, . . . of a television picture occur during the odd fields A, C, E, . . . and the other half occur during the even fields B, D, F, ... With a fixed delay of either $312T_1$ or $313T_1$ the result with this circuit would be that lines more and more distant from each other would be combined (dashed arrows), causing an unacceptable deterioration of definition in the vertical direction. This is corrected by making the delay during the successive fields alternately equal to $312T_1$ and $313T_1$ (solid arrows). This is done by means of the switch S.

NTSC

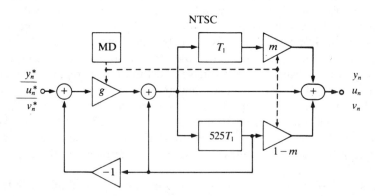

FIGURE 12.18 Reduction of cross-effects and noise for each of the NTSC component signals y_n^*, u_n^* and v_n^*, with a *picture* memory and a movement detector MD. If there is movement in the picture, $g = m = 1$. If there is no movement, $0 < g < 1$ and $m = 0$.

combination of Fig. 12.9(c) and Fig. 12.16. A complete *picture* memory is used in this case. Figure 12.19 shows a solution for PAL that needs only one *field* memory. This is in fact a combination of the circuits in Fig. 12.15(b) and Fig. 12.17.

As we have seen, most circuits in which temporal filtering is used include a movement detector. One result of this is that there is no noise reduction at the pixels where movement is detected. This is not so serious as it may seem, however, since the noise at these pixels is perceptually less of a nuisance.

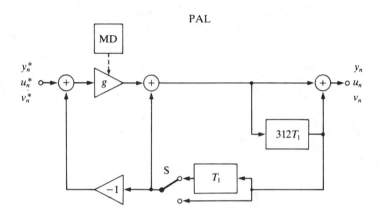

FIGURE 12.19 Reduction of cross-effects and noise for each of the PAL component signals y_n^*, u_n^* and v_n^*, with a *field* memory and a movement detector MD. If there is movement, $g = 1$; otherwise $0 < g < 1$.

12.8 REDUCTION OF LARGE-AREA FLICKER

One of the imperfections of present-day television receivers is most clearly observed on uniform picture areas in the TV systems operating at 25 pictures per second. These areas show a visible periodic variation in luminance, referred to here as 'large-area flicker', even though its frequency is doubled to 50 Hz by using interlaced fields. The higher luminous output of modern picture tubes, which enables us to watch television with more background lighting and in daylight, does make this effect much more troublesome. The reason is that faster variations can be observed when the luminance is higher. A solution to this problem is to display each field twice, giving a field repetition rate of 100 Hz, which is much more acceptable to the human eye. This field doubling can be obtained by using a memory that stores only one field (half a picture). The control of such a memory is rather complicated, however. It becomes much simpler if two field memories are used in which, alternately, the fields are written in slowly and read out quickly (Fig. 12.20). Denoting the successive original fields by A, B, C, D, . . . we obtain with this circuit a signal with successive fields, A, A, B, B, C, C, D, D, ... To display the signal properly, the receiver unit has to be modified so that:

• the signal bandwidth is doubled;

- each picture line lasts for only 32 μs instead of 64 μs and the duration of each field is only 10 ms instead of 20 ms; the field scan is thus twice as fast in both the horizontal and the vertical directions;
- the successive fields are displayed *in pairs* (instead of alternately) at the same position in the frame.

The effect of large-area flicker can be completely suppressed in this way.

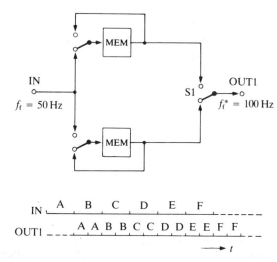

FIGURE 12.20 *Large-area* flicker can be reduced by doubling the field frequency (from f_f = 50 Hz to f_f^* = 100 Hz here). This can be done by using two field memories MEM, in which the signals are alternately written in slowly and then read out twice at double speed. The original field sequence A, B, C, D, E, ... is thus changed to A, A, B, B, C, C, D, D, ... This is known as 'field doubling'.

12.9 REDUCTION OF LINE FLICKER

In addition to large-area flicker there is also *line flicker*. This also results from the kind of scanning pattern, and especially from the interlacing. A sharp image transition that is exactly parallel to the lines of the picture occurs at different heights in consecutive fields and therefore seems to dance up and down at a frequency of 25 Hz. With the circuit in Fig. 12.20 nothing has been done about this; so it still has a line flicker at 25 Hz. By adding a third field memory (Fig. 12.21) we can have a 100-Hz field frequency in which the sequence of the fields is A, B, A, B, C, D, C, D, E, F, E, F, ... We then have, just as in Fig. 12.20, a field frequency of 100 Hz, but any line flicker now occurs at 50 Hz and is therefore barely perceptible, if at all. However, this circuit gives undesirable effects with moving pictures, since the original sequence of the fields has been altered so that the older field A, say, appears again after the display of the newer field B. In essence we would like to combine the advantages of one 100-Hz field sequence (OUT1) with that of the other 100-Hz field sequence (OUT2). However, the signals OUT1 and OUT2 have become relatively displaced in time, so that direct switching between them is not possible. This problem has been solved in the circuit of Fig. 12.22. The signal OUT3 is a delayed version of OUT1. The switch SM, controlled by a movement detector, can now be switched at any instant between OUT2 and OUT3, depending on whether there is any movement or not in the picture.

In switching between OUT2 and OUT3 even fields are repeatedly replaced by odd fields and vice versa. Since the even and odd fields correspond to slightly different positions on the television screen, some means of correction is required. The problem can be solved by the simple expedient of inserting a line memory and an adder. We shall not pursue the matter further here, however.

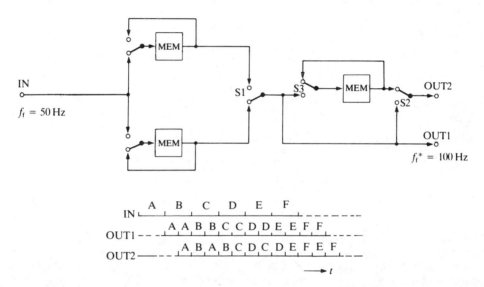

FIGURE 12.21 The addition of a third field memory MEM to the circuit in the previous figure gives a signal OUT2 with twice the field frequency and a field sequence A, B, A, B, C, D, C, D, ... This is known as 'frame doubling'. In this way both *line* flicker and *large-area* flicker are reduced. This circuit has some adverse effects, however, if the picture contains movement. (The OUT1 signal is not used here.)

Also shown in Fig. 12.22 is a switch S4 that can provide the signal OUT4, which is delayed by exactly one *picture* period with respect to the original input signal IN. This signal OUT4 is used to good effect in the video processor block for temporal filtering, as dealt with in the previous sections. The circuit in Fig. 12.22 can therefore improve the quality of the picture in five respects: large-area flicker, line flicker, cross-colour, cross-luminance and noise.

The methods described here for reducing large-area flicker and line flicker leave the interlaced structure of the picture essentially unchanged: each field received is displayed without modifications. With the availability of field memories, however, the lines of each two successive fields can be interwoven ('field addition'), so that all the lines of one picture (two fields) are displayed consecutively from top to bottom. Then interlacing is eliminated, and we have obtained another method of reducing the two types of flicker.

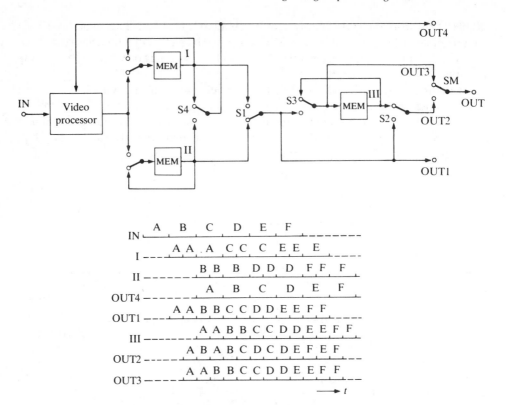

FIGURE 12.22 By adding a switch SM, operated by a movement detector (not shown), the advantages of the circuits in Fig. 12.20 and Fig. 12.21 can be combined. The signal OUT has the field sequence A, A, B, B, C, C, D, D, . . . or A, B, A, B, C, D, C, D, . . ., depending whether there is movement or not. The sequence can be chosen separately for each pixel. The addition of a switch S4 makes the signal OUT4 available. This is a version of the input signal delayed by a complete picture period. It can be used as the starting point for any desired form of temporal filtering in the circuit Video processor. In all, the circuit in this figure can be used for the combined reduction of large-area flicker, line flicker, cross-colour, cross-luminance and noise.

12.10 OTHER IMPROVEMENTS

The improvements discussed above are just a few of those offered by the introduction of digital signal processing in television receivers. We shall now briefly mention two other cases in which digital filters play an important part. In the first place, picture definition in the horizontal and vertical directions can be increased by horizontal and vertical filtering with the appropriate filters. This is referred to as 'aperture correction', because it corrects some of the blurring due to certain optical and electrical limitations in the receiver.[90,91] In this kind of aperture correction it is often of great importance that digital filters can have a perfectly linear phase characteristic.

A second example of more advanced digital signal processing is echo cancellation. Because of multiple reception of reflected signals, e.g. from high buildings or mountains, a television picture may contain annoying echoes or 'ghost images'. In principle these echoes can be removed by echo cancellation with adaptive filters. Adaptive echo cancellation is not only applied to video signals. An *analog* integrated circuit for adaptive echo reduction in teletext is described in ref. [64]. A *digital* integrated circuit for adaptive echo reduction in data transmission is described in ref. [19].

12.11 NEW FEATURES

Now that it is becoming an economic proposition to digitize television pictures and to store them in a memory in a television receiver, a number of new features can be introduced. Some of these are fairly obvious but are still a welcome addition to the existing features. In the first place there is of course the facility of stopping ('freezing') a particular picture, so that a stationary picture can be observed. A picture can also be stored for shorter or longer periods of time for later display or for printing out on paper (hard copy). A part of a stored picture can be displayed as a complete picture (i.e. magnified), by a process referred to as 'zooming'. The memory can also be used for simultaneously displaying a second picture in the corner of the screen ('picture-in-picture') or even a combination of (say) nine pictures ('multi-picture-in-picture'). This latter facility is particularly convenient for making a choice from a large number of programmes (Fig. 12.23).

FIGURE 12.23 The use of field memories also makes it possible to combine parts of pictures. The example given here is 'multi-picture-in-picture', where the signals originating from several different transmitters are displayed simultaneously. This can be a very useful facility when making a selection from a large number of programmes.

If digital signal processing is used, it becomes simpler and therefore cheaper to make a receiver designed to receive television signals that have been coded in different standards (PAL, NTSC, SECAM).

With a field or picture memory 'slow-scan TV' operation can be obtained, so that signals from a telephone line, for example, can be displayed on a conventional television screen as a series of slowly changing pictures. One application of this facility is found in remote visual monitoring. In combination with a video cassette recorder, video disc player or video camera, countless other applications come to mind. If memories are used, signals originating from non-standard sources can be synchronized with each other, or with standard signals, permitting simple switching between them ('editing').

A very useful application is the background memory for teletext. In teletext all the available pictures ('pages') are transmitted in a continuously repeated sequence as a small part of an ordinary broadcast television signal. At the present time only the page actually being displayed on the screen is stored in a memory of limited capacity in the receiver. This means that, after selecting a new page, the viewer has to wait until that particular page is received again before he can see it displayed. In a cycle of 300 pages this can take as long as a minute. If the user wants to change pages frequently, i.e. to browse through them as in an information search, this can be a long time to wait. Since one complete teletext page in coded form only occupies a fraction (about 0.3%) of a television field, many hundreds of teletext pages can be stored in only one field memory. All these pages can be displayed with a delay of half a second at most – almost instantaneously.[92]

Economic digital field and picture memories ultimately open up the prospect of fundamental changes in the signal standards for colour television. By virtue of these memories, chrominance and luminance signals, which must be *simultaneously* available for display, can nevertheless be transmitted and received *consecutively*. This can provide a fundamental solution to the existing problems of cross-colour and cross-luminance. In addition, each component signal can have exactly the right amount of the total transmission capacity allocated to it to give the optimum overall result. Systems of this kind, known as MAC ('Multiplexed Analog Components'),[91,93] are the subject of steadily growing interest.

12.12 SUMMARY

The modern trend of replacing analog electronic circuits by digital ones is now starting to take effect in colour television receivers. Besides bringing economic advantages, digitization can substantially improve the quality of the present television picture. The improvements include a better separation of chrominance and luminance information (reduction of cross-colour and cross-luminance), noise reduction and the reduction of large-area flicker and line flicker. New features such as 'freeze-frame', combined pictures ('picture-in-picture'), the magnification of parts of a picture (zoom) and teletext memory also appear. This chapter describes the background and present capabilities of digital television signal processing. Some characteristic filter operations are discussed, including vertical filtering and temporal filtering, as well as characteristic types of filter, such as comb filters. It is found that in very many cases there is a need for economic field or picture memories. The importance of making the right choice of sampling rate is underlined. This chapter is concerned mainly with the NTSC system (America, Japan) and the PAL system (Europe).

13

Digital audio: examples of the application of the ASP integrated signal processor

E. H. J. Persoon and C. J. B. Vandenbulcke[†]

13.1 INTRODUCTION

Electronic processing of audio signals is not new. Tone control by means of one or more analog filter stages is a well-known example from hi-fi technology – the technology of the faithful reproduction of speech and music. Another example is the improvement of audio quality by altering spatial perception with the aid of stereo effects and reverberation. The basic knowledge of acoustics and perception required here has long been available, but until now many practical applications have been restricted to professional use in the studio, e.g. for records, radio, TV and film.

However, great changes are on the way, for we are now in the digital age of audio technology, and this includes equipment for the 'ordinary' consumer. For the best results, sound is now recorded in digital form on tape or disc (the Compact Disc[14]). This can be done virtually without distortion and with signal-to-noise ratios previously thought impossible. From digital recording to digital processing is but a single step. All the advantages of digital signal processing are then assured.

With digital methods it is easy to make an ideal memory. This is of primary importance for audio technology, for signals can then be delayed as desired. Another important feature is that processing, once selected, remains fixed and immutable – for example, feedforward controls always remain stable, even in the long term. On the other hand, various settings of the equipment are easy to change automatically by digital methods ('adaptive control'), so that an optimum result is always obtained by an appropriate self-adjustment.

†Philips Research Laboratories, Eindhoven, and Philips Consumer Electronics Division, Eindhoven, respectively. This chapter is reprinted from *Philips Tech. Rev.* **42** No. 6/7: 201–16, Apr 1986.

For the listener, the introduction of digital signal processing in audio[94] provides a number of new facilities that were previously too expensive or technically too difficult, such as:

- dynamic-range compression or expansion – reducing or increasing the difference between the levels of loud and soft passages as required;
- reverberation – adding delayed versions of the audio signal;
- equalization – detailed adaptation of the frequency characteristics of the reproducing equipment to the acoustics of the room.

The above operations can be performed digitally and with high quality by circuits that can be incorporated in one or more ICs. The filters in these circuits can be completely free of undesirable phase distortion, something that is not really possible with conventional filter circuits.

Digital circuits can be designed for external programming, so that the same circuit can perform different operations. This has led us to develop a programmable integrated circuit for processing audio signals. We call it ASP (Audio Signal Processor). A system consisting of a few of these ASPs will perform operations such as those described above.

In this chapter we describe the architecture of the ASP. In the context in which this chapter appears, however, we should first like to give more attention to the background, the characteristics and some practical implementations of the examples of advanced audio signal processing mentioned above.

13.2 COMPRESSION AND EXPANSION OF THE DYNAMIC RANGE

13.2.1 Necessity for adapting the dynamic range

The human ear has a formidable dynamic range. It is so sensitive that it can almost detect the Brownian motion of the air molecules against the eardrum (the threshold of hearing, 0 dB), but it is not overloaded until the sound intensity level is 10^{12} times higher (the threshold of pain, 120 dB). These extreme limits are reproduced in Fig. 13.1.

In everyday life there is always a background of sound. Music or speech that we want to listen to must stand out above it. In a quiet concert hall the sound background is very low (30 dB above the threshold of hearing). The quietest musical passages are also on that level. On the other hand, the *fortissimo* of a symphony orchestra reaches about 110 dB, so that the dynamic range is 80 dB.

In a domestic setting the background is usually louder. In a living room a level of 40 dB is typical. Reproduction of the original dynamic range of the symphony orchestra would give levels of 120 dB in the room, which is not acceptable. In general levels above 100 dB seem unpleasantly loud, particularly in a small room. The difference in volume between the loudest and quietest passages must therefore be reduced: this is called compression of the dynamic range.

Even greater compression is desirable if the music is only to be used as background. In this case people must be able to speak to one another, at a conversational level of about 65 dB; the loudest passages of music must not exceed this level. Very large compression is also desirable in a car, where the background level can be 80 dB.

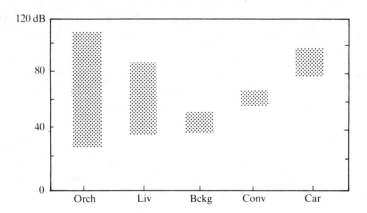

FIGURE 13.1 Schematic presentation of the dynamic range of sound under various listening conditions. *Orch*, the dynamic range of a symphony orchestra in a concert hall. *Liv*, the desirable dynamic range for quality reproduction in a living room. *Bckg*, the desirable dynamic range for background music in a living room; it should not swamp the conversation, *Conv*. *Car*, the desirable dynamic range in a moving car.

The new Compact Disc has an unprecedentedly large dynamic range; in principle it can record music with the large dynamic range of a concert hall extremely faithfully. However, each listening environment requires its own dynamic range for the optimum final reproduction (Fig. 13.1). Now that the Compact Disc is here, the desirability of adapting the dynamic range to the listening conditions has become even greater. In most cases dynamic range *compression* is required.

Under certain conditions it may be preferable to have a greater dynamic range than that offered by, say, a particular conventional record. In such cases the opposite of compression – *expansion* – of the dynamic range is useful.

13.2.2 How can the dynamic range be compressed (or expanded)?

There are analog circuits for dynamic range compression (or expansion). These include a peak detector that provides a control signal $r(t)$, that reduces or increases the gain A in the signal path (Fig. 13.2(a)). We shall be mainly concerned with compression here.

The typical reaction of a compression system to a varying signal $x(t)$ is represented in Fig. 13.2(b). After an increase in the amplitude of $x(t)$ the gain falls very rapidly and after a reduction in the amplitude the gain gradually returns to the initial value (the release effect).

Some 'sluggishness' is desirable so that the gain does not follow the fine structure of the audio signal too closely and thus eliminate all of the dynamic variation. For the best result, the start of the control action should be somewhat delayed, so that the reaction to the signal $x(t)$ becomes like that shown in Fig. 13.2(c).

All this can also be carried out by digital methods. The rectification, which forms a part of the peak detection, is then replaced by for example one or more simple operations on the binary representation of the signal. A digital filter delays the start of the control and also limits the control bandwidth to a few hertz. In dynamic-range compression and

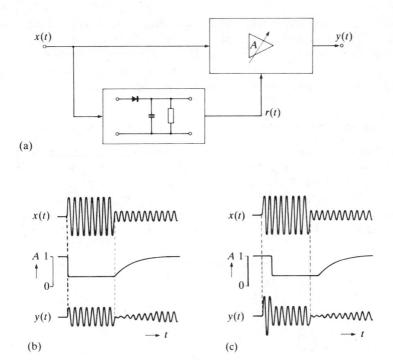

(a)

(b)

(c)

FIGURE 13.2 Dynamic-range control. (a) A control signal $r(t)$ is derived from the applied signal $x(t)$ by peak detection and used to vary the gain A. If the gain increases with increasing $r(t)$ we obtain expansion, in the opposite case compression. $y(t)$ is the resulting signal. (b) Example of dynamic-range compression with instant response. t time. (c) Example of dynamic-range compression with delayed response.

expansion with the ASP the delayed start, the release effect and the compression/ expansion ratio can in principle be set independently.

13.3 ARTIFICIAL REVERBERATION

When recordings are made in the studio, artificial reverberation is usually added. This is obtained with the aid of a reverberation chamber, a reverberation plate (a fairly large metal plate in which an electrically excited oscillation is allowed to die away) or – nowadays – an electronic reverberation circuit. Home-recording enthusiasts would also like to have reverberation for their recordings, which are usually made in rooms with not enough reverberation, if they could afford it. And in reproducing music recorded elsewhere, it is sometimes desirable to liven up dead acoustics with artificial reverberation. (In this way the special character of a live performance can be enhanced.)

Reverberation occurs because of the combination of long delays and sound reflections in some enclosed spaces (concert halls, churches, bathrooms). The sound reaches the listener with different delays from different directions. These delayed presentations of the sound can now be achieved with electronic equipment of acceptable size, because of the miniaturization of semiconductor memories. Electronic reverberation is thus becoming available to the consumer and is also one of the applications of the ASP.

We are especially interested in artificial reverberation that sounds natural. Therefore we should examine the phenomenon of reverberation more closely.

13.3.1 The characteristics of reverberation

If someone in a hall is listening to a source of sound on the stage, the sound coming directly from the source reaches him first. A little while later this is followed by the early reflections, reflected once by walls, ceiling and floor around the proscenium. In a large hall these early reflections take longer to arrive than in a small hall and a listener, influenced by experience, unconsciously connects their delay with the dimensions of the hall. This initial delay is generally smaller, and usually much smaller, than 150 ms.

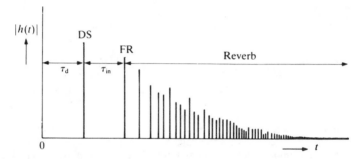

FIGURE 13.3 Schematic rendering of the impulse response $h(t)$ of a concert hall (absolute value), as observed at an arbitrary location in the hall. The original impulse occurs at time $t = 0$. The direct sound DS arrives after a time τ_d, the first reflection FR arrives after a second period τ_{in} followed by a series of reflections of exponentially decreasing amplitude and spaced more and more closely together (Reverb).

The sound spreads out into the hall and is reflected many more times there. After the first, early reflections the listener hears multiple reflections from all sides. These reach him at irregular times. Their density increases continuously (Fig. 13.3), on average as the square of the time. In view of the attenuation at each reflection their amplitude decays exponentially with time. The rate of decay increases with the attenuation at each reflection, i.e. as the reflecting surface absorbs more energy from the sound.

The time in which the level of the reverberation falls by 60 dB is called the reverberation time T_{60}. The reverberation time is not the same for all audio frequencies, because the absorption characteristics of the different materials are not the same at all frequencies. In general the absorption is higher at high frequencies – so that the reverberation time is shorter – than at low frequencies.

13.3.2 Electronic excitation of reverberation

A suitable aid for the excitation of artificial reverberation is the recursive digital comb filter of the Nth order (Fig. 13.4(a)). This consists of a series of N unit-delay elements

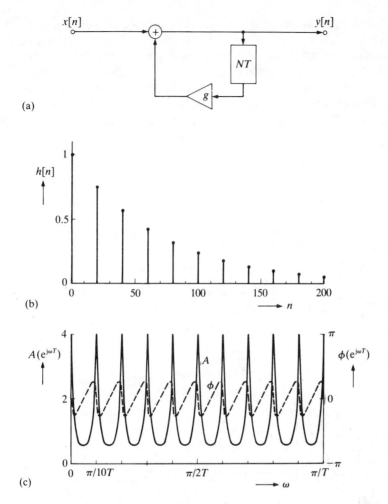

(a)

(b)

(c)

FIGURE 13.4 Recursive digital comb filter of the Nth order with one filter coefficient g. (a) Block diagram. The output of a delay line (delay NT) is attenuated ($|g| < 1$) and fed back to the input. (b) Impulse response $h[n]$. A pulse presented to the input is repeated at the output after every N sampling intervals (here $N = 20$); each repetition is weaker than the previous one by a factor g (here $g = 0.75$). (Output samples of value 0 are not explicitly reproduced in this and following figures because there are very many. Also $h[n]$ in principle lasts for an infinite time.) (c) Associated amplitude characteristic $A(e^{j\omega T})$ and phase characteristic $\phi(e^{j\omega T})$. These characteristics are periodic with a period $2\pi/NT$. There are therefore exactly N periods in the fundamental interval $0 \leq \omega < 2\pi/T$. Here we have shown only half of the fundamental interval, i.e. $0 \leq \omega \leq \pi/T$.

whose output is fed back to the input (via an attenuation $|g| < 1$). A pulse presented to the input will therefore appear regularly at the output, with intervals equal to N sampling intervals (Fig. 13.4(b)). The attenuation in the feedback loop ensures that the pulse grows weaker at each reappearance. The amplitude characteristic of the filter has N peaks spaced regularly throughout the fundamental interval $0 \leq \omega < 2\pi/T$ (Fig. 13.4(c)). This explains the name of the filter.

The behaviour of this comb filter is similar to that of reverberation in a hall. In both cases there is continuous repetition of a single pulse with a gradually decreasing amplitude. The rate of decrease of this amplitude can be altered by means of the attenuation factor in the feedback path. If it is desirable for the higher-frequency components to die off more rapidly, as is usual in natural reverberation, the frequency-independent attenuation in the feedback path can be replaced by a lowpass filter, e.g. that of Fig. 13.5. We shall call the result a 'modified comb filter' (Fig. 13.6). Examples of

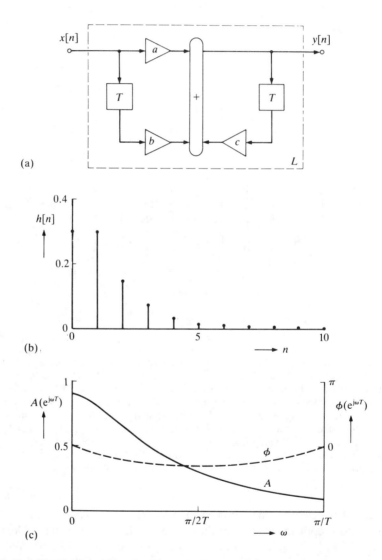

FIGURE 13.5 Simple digital filter L of the 1st order. (a) Block diagram. (b) Impulse response $h[n]$ for $a = 0.3$, $b = 0.15$ and $c = 0.5$. (c) Associated amplitude characteristic $A(e^{j\omega T})$ and phase characteristic $\phi(e^{j\omega T})$.

the associated impulse response and phase and amplitude characteristics are also given in Fig. 13.6.

One difference from 'natural' reverberation is the regularity with which the pulses recur. This gives the periodicity in the frequency characteristic. The frequency interval between successive peaks in this characteristic is constant, just like the frequency interval between the components of a musical tone. This means that our hearing interprets a series of pulses with such a frequency characteristic as a 'tone'. Somewhat incorrectly it is

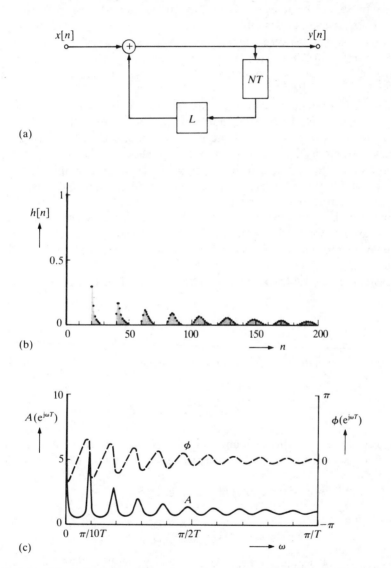

FIGURE 13.6 Modified recursive digital comb filter. (a) Block diagram; the constant attenuation g in the feedback loop of Fig. 13.4 has now been replaced by the filter L from Fig. 13.5. (b) Impulse response $h[n]$ for $N = 20$. (c) Associated amplitude characteristic $A(e^{j\omega T})$ and phase characteristic $\phi(e^{j\omega T})$.

said that the reverberation produced with a comb filter suffers from 'colouring'. In fact this is not a matter of sound colour but of pitch. If a complex of different tones is presented, the tone that is 'in tune with' the comb filter is strongly enhanced.

Such regular impulse responses are often met when metal objects, e.g. metal springs, are acoustically excited. Our hearing makes associations with what is familiar from experience, so that the reverberation produced by a comb filter is often said to be 'metallic'. This is undesirable, of course. If the delay in the filter is very large the individual pulses can be detected. The rapid rattling sounds then heard are called 'flutter echoes'. These are also undesirable.

One way of correcting these difficulties is to connect several comb filters in parallel, each with a different delay. This increases the density of the 'reflections' and masks the regularity of each individual comb filter. Another way is to add phase-shifters ('all-pass filters').[95,96,97]

A phase-shifter is a filter that introduces a frequency-dependent phase shift and has a flat amplitude characteristic. An example of a digital phase-shifter of the Nth order is given in Fig. 13.7(a). For $N = 20$ and $g = 0.75$ the associated impulse response and frequency characteristics are shown in Figs 13.7(b) and (c). It can be seen that this type of network also creates a whole series of pulses of decreasing amplitude from a single incident pulse. On this basis the density of the 'reflections' in the artificial reverberation can be increased further without colouring.

By appropriate use of several delay elements and filter coefficients a more general type of phase-shifter is obtained (Fig. 13.8). This gives more scope for defining the frequency dependence of the phase shift, and hence also the decay of each individual frequency component.

The system function $H(z)$ of the comb filter shown in Fig. 13.4(a) is given by

$$H(z) = \frac{1}{1 - gz^{-N}} \qquad (13.1)$$

and hence the associated frequency response is given by

$$H(e^{j\omega T}) = \frac{1}{1 - ge^{-j\omega TN}} \qquad (13.2)$$

It can easily be seen that this frequency response is a periodic function of ω with a period $2\pi/NT$, because it gives exactly the same value for ω and $(\omega + i2\pi/NT)$ for any integer number i.

The system function of the digital filter of Fig. 13.5 is:

$$H(z) = \frac{a + bz^{-1}}{1 - cz^{-1}} \qquad (13.3)$$

with the associated frequency response

$$H(e^{j\omega T}) = \frac{a + be^{-j\omega T}}{1 - ce^{-j\omega T}} \qquad (13.4)$$

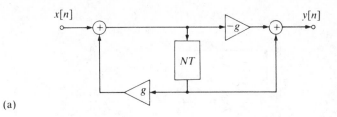

(a)

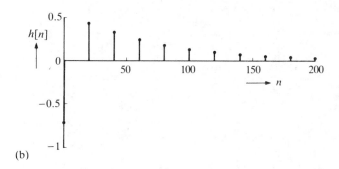

(b)

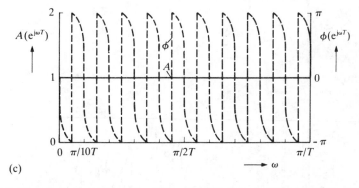

(c)

FIGURE 13.7 Digital phase-shifter of the *N*th order with only two filter coefficients ($|g| < 1$). (a) Block diagram. (b) Impulse response $h[n]$ for $N = 20$ and $g = 0.75$. It can be seen that the 'main pulse' $h[0]$ is the only sample with a negative value. (c) Associated phase and amplitude characteristics.

This function is also periodic in ω; now, however, the period is exactly equal to the sampling rate $\omega_s = 2\pi/T$, as indeed it is for any digital filter. With appropriate choice of a, b and c, (13.4) represents a lowpass filter; with other choices it can be a highpass filter.

The system function and frequency response of the phase-shifter in Fig. 13.7 are given by ($|g| < 1$):

$$H(z) = \frac{-g + z^{-N}}{1 - gz^{-N}} \qquad (13.5)$$

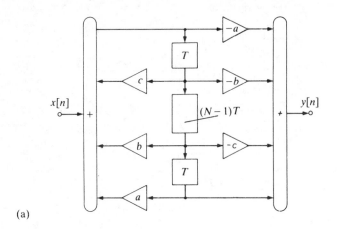

(a)

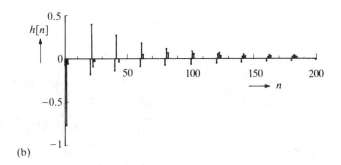

(b)

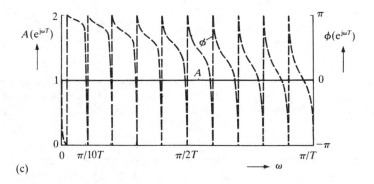

(c)

FIGURE 13.8 Digital phase-shifter of the $(N+1)$th order with six filter coefficients. If a, b and c are selected appropriately, this network will give the desired variation of phase shift and hence decay time with frequency; the high-frequency components could for example be made to decay faster than low-frequency components. (a) Block diagram. (b) Impulse response $h[n]$ for $N = 20$, $a = 0.1$, $b = 0.75$ and $c = 0.1$. (c) Associated phase and amplitude characteristics.

and

$$H(e^{j\omega T}) = \frac{-g + e^{-j\omega TN}}{1 - ge^{-j\omega TN}} \qquad (13.6)$$

or

$$H(e^{j\omega T}) = e^{-j\omega TN} \cdot \frac{1 - ge^{j\omega TN}}{1 - ge^{-j\omega TN}} \qquad (13.7)$$

From this last expression we can easily see that we are concerned with a pure phase-shifter, because the modulus of the first factor is 1, while the numerator and denominator of the fraction are complex conjugates. We can reach the same conclusion by determining the poles and zeros of $H(z)$ in the z-plane in (13.5). It then turns out that poles and zeros appear in pairs that are reflected in the unit circle $|z| = 1$. (In this case there are N zeros on the circle $|z| = g^{-1/N}$ and N poles on the circle $|z| = g^{1/N}$.) This reflection property is characteristic of phase-shifters. It can be shown that this property always applies if the coefficients of the denominator of $H(z)$ have the opposite order to those of the numerator,[98] i.e.

$$H(z) = \frac{a_0 + a_1 z^{-1} + a_2 z^{-2} + \ldots a_{M-1} z^{-M+1} + a_M z^{-M}}{a_M + a_{M-1} z^{-1} + a_{M-2} z^{-2} + \ldots a_1 z^{-M+1} + a_0 z^{-M}} \qquad (13.8)$$

Taking this as the starting point it can easily be seen that the system function $H(z)$ of Fig. 13.8, which is

$$H(z) = \frac{-a - bz^{-1} - cz^{-N} + z^{-N-1}}{1 - cz^{-1} - bz^{-N} - az^{-N-1}} \qquad (13.9)$$

also represents a phase-shifter.

Both comb filters and phase-shifters carry out linear processing on the signal. This means that the designer of an electronic reverberation system can select the order of these processes arbitrarily.

Figure 13.9 shows the block diagram of a (stereo) reverberation system that can be produced with the ASP. It includes two phase-shifters F_1 and F_2 in cascade and six modified comb filters $K_1, \ldots, K_6$ in parallel. In this system two *different* (uncorrelated) reverberation signals for the left-hand and right-hand sound channels can be formed from a single mono signal equal to the sum of the two original stereo subsignals. This approach gives a considerable saving in the number of operations required per unit time and in the amount of memory required to effect the different delays. The input signal IN of the reverberation system is first filtered in a digital highpass filter H_h that suppresses frequencies below 55 Hz. This filter has the same structure as that in Fig. 13.5(a), but its coefficients are modified. The delay circuit with a delay of $N_i T$ provides the initial delay of the first reflections with respect to the direct sound (τ_{in} in Fig. 13.3). At various points

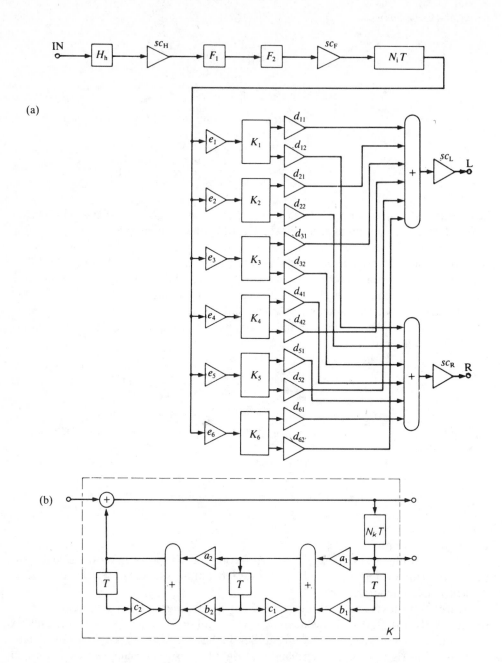

FIGURE 13.9 (a) Block diagram of a complete digital stereo reverberation system. IN mono input. L, R stereo outputs for left-hand and right-hand channels. The most important components are: a highpass filter H_h for suppressing frequencies below 55 Hz, two phase-shifters F of the type shown in Fig. 13.8, a delay $N_i T$ and a parallel arrangement of six modified comb filter $K_1, \ldots, K_6$. The delay $N_i T$ corresponds to the initial delay τ_{in} between the direct sound and the first 'reflection'. sc scaling factor. d, e gains. (b) Detailed block diagram of one of the modified comb filters K. The feedback loop includes a combination of two lowpass filters as in Fig. 13.5(a). This makes the reverberation time frequency-dependent, a, b, c filter coefficients.

in the circuit gains d and e and scaling factors sc are incorporated. The gains determine the relative magnitudes of the subsignals; the scaling factors ensure that the signals occurring 'match' the numerical values allowed in the system as well as possible. Without scaling there might be overflow, or alternatively it could happen that some of the most-significant bits were never used at all.

All six modified comb filters have the same structure K (Fig. 13.9(b)), but the values of N_k and the filter coefficients vary. In addition each comb filter has two outputs, one contributing to the left-hand reverberation signal L and the other to the right-hand reverberation signal R. The signals L and R are ultimately added to the two original stereo subsignals that formed the starting point.

During the design of this reverberation system extensive use was made of listening tests (see also ref. [99]) which in this case were largely carried out with the help of prototypes ('breadboard' models). This was done both for selecting the number of phase-shifters and comb filters and also for determining the delays and coefficients. These tests showed that a set of four comb filters could not provide a sufficiently colour-free reverberation. They also showed that the bandwidth of the reverberation signals can be limited to 10 kHz, for example, without reducing the subjective quality. This makes it possible to operate the whole reverberation system at a lower sampling rate than the original stereo signal (e.g. 22.05 kHz instead of 44.1 kHz). To do this it is necessary that the reverberation circuit is preceded and followed by sampling-rate converters, which decrease or increase the sampling rate. A further considerable saving in the number of operations and memory capacity is achieved in this way.

To give an impression of the operation of the reverberation system described above, Fig. 13.10 shows its impulse response and amplitude characteristic as obtained after digital-to-analog conversion. Both the curves shown have the desired irregular structure and are free from periodicity. The impulse response also reproduces the initial delay τ_{in} of the first reflection with respect to the direct sound. This quantity will in general be made adjustable, so that the listener can simulate the reverberation of a small room or of a large one.

The length of the delay incorporated in the filters is related to the volume of the room being simulated: the larger the room, the longer the time between two reflections of a sound signal from the walls. In the same way the coefficients in the feedback circuits represent the absorption of sound at a reflection. By making the delay periods and coefficients adjustable, the reverberation of enclosures of very different size and nature can be simulated – from a bathroom to a sports hall!

13.4 EQUALIZATION OF THE FREQUENCY CHARACTERISTIC

In quality reproduction the audio equipment is required to reproduce all audio frequencies, both low and high, at equal strength. This is a flat frequency response. Audio frequencies are usually considered as the band from 20 Hz to 20 000 Hz.

The components that require most attention in such applications are the transducers: the microphone, the pick-up stylus and the loudspeaker. The design of the loudspeaker is usually based on measurements carried out in an anechoic chamber. In such an enclosure

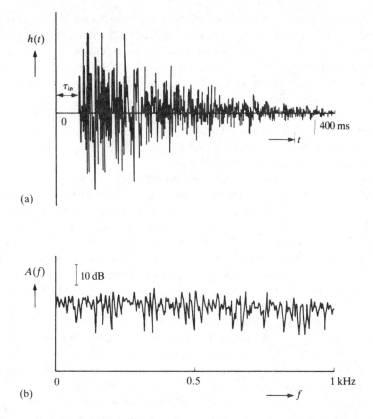

FIGURE 13.10 (a) Impulse response $h(t)$ (over 400 ms), and (b) amplitude characteristic $A(f)$ of a reverberation system made with the ASP. Both curves were obtained after digital-to-analog conversion and relate to a frequency band deliberately limited to 1000 Hz to avoid too much detail in the figures. They are largely free from regularity or periodicity, giving reverberation without 'colouring'.

there are no sound reflections at all from the walls or other surfaces, so that there is an unambiguous result that is the same for every anechoic chamber.

In ordinary use the situation is quite different. The room where the loudspeaker is used does not have completely absorbing walls, and standing waves (resonances) occur in it. The same applies to sound reproduction in a car. These resonances have a considerable effect on the final result, particularly at low frequencies where their ratios are fairly widely spaced, and they enhance certain notes and suppress others. In spite of efforts to approach the desired flat characteristic of loudspeakers as closely as possible in the anechoic chamber, in a practical 'listening situation' differences in level of more than 10 dB can easily occur. However, by adjusting the gain as a function of the frequency and depending on the actual situation ('equalization') it is nevertheless possible to obtain a flat characteristic.

For this purpose we have used adjustable octave-bandfilters, i.e. a series of filters whose centre frequencies and bandwidths increase regularly by a factor of 2. The complete frequency range is therefore divided into 10 octave bands. Each octave band can be amplified or attenuated separately. The 10 octave-band filters are connected in cascade (Fig. 13.11). The transmission at the midpoint of each octave can be adjusted between +12 dB and −12 dB in steps of 1 dB. See Fig. 13.12.

Figure 13.13(a) shows how the sound level in different frequency bands can vary on reproduction in a room. Correction with the 10 octave-band filters gives a much more even distribution of the sound level over the frequency scale (Fig. 13.13(b)).

The peaks and troughs that appear in the frequency characteristic because of the characteristics of the room are about an octave wide at low frequencies. Octave-band filters are therefore suitable for correcting them. The location of the centre frequency of the octave-band filters is not always optimum, of course. The digital audio system in which the ASP is used is therefore programmed in such a way that the centre frequency for the octave-band filters below 1000 Hz can be raised or lowered by a third of an octave. This can be done independently in each stereo channel.

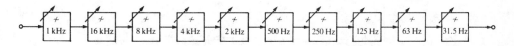

FIGURE 13.11 Cascade arrangement of 10 octave-band filters for equalization of a frequency characteristic. The gain at the centre of each octave band can be adjusted in steps of 1 dB between −12 dB and +12 dB. For the realization with the ASP the sequence of octave bands was chosen so that the noise caused by the finite precision of the digital processing was kept to a minimum at the end of the chain (see subsection 13.4.3).

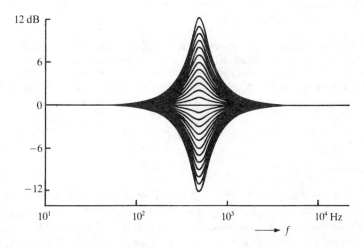

FIGURE 13.12 The 25 possible response curves of one of the octave-band filters (the filter centred on 500 Hz).

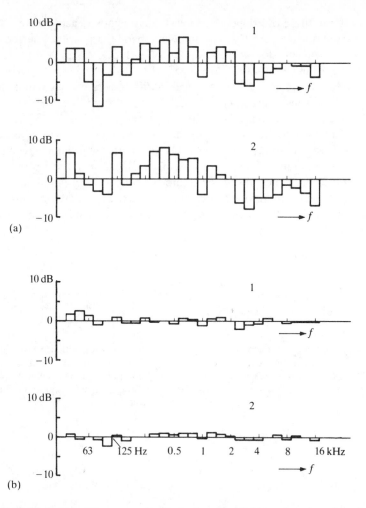

FIGURE 13.13 Variation of the power density in the sound field as a function of the frequency f at a certain point in a room on reproduction of an electrical noise signal with a uniform power spectral density, (a) before and (b) after equalization with the aid of octave-band filters. Recordings 1 and 2 were made by driving the two channels of a stereo system separately. In the figure the frequency is divided into bands of one-third of an octave.

13.4.1 Filter structure

In the design of the digital octave-band filters it was not possible to use the same filter structure in all cases. The normal direct-form-1 recursive filters (Fig. 13.14), for example, are less suitable for the lower frequency ranges. This is because only a limited number of bits (in our case 12) are available for each coefficient. This means that only a limited number of frequency responses are obtainable for each filter, and also that they are dependent on the filter structure selected. In the case of the adjustable octave-band filters which we wanted, the requirements with 12-bit coefficients were satisfied best by using the direct-form-1 structure (Fig. 13.14) for the higher octaves (centred on 1, 2, 4, 8 and 16 kHz) and the coupled form (Fig. 13.15) for the lower octaves (centred on 31.5, 63, 125, 250 and 500 Hz). Furthermore, the desired filter curves can be approached more closely by quantizing the coefficients in each filter in a well-considered coordinated manner during the design. In realizing the two coefficients a in Fig. 13.15, which would have exactly the same value without quantization, as 12-bit numbers it might for example be advantageous to round one of them up and the other one down.

Quite different problems (which nevertheless fall under the same general heading of 'finite-word-length effects') arise because results of calculations in the filter must constantly be quantized, because the number of bits available for storing the signal obtained (the word length) is limited. Thus Fig. 13.14 includes a quantizer Q that limits the word length of the sum of the five products (in our case to 24 bits).

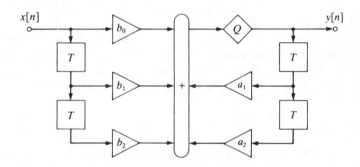

FIGURE 13.14 Recursive digital filter of the second order (direct-form-1 structure). a_1, a_2, b_0, b_1, b_2 filter coefficients. Q quantizer, in which the word length of the signal samples is reduced to 24 bits.

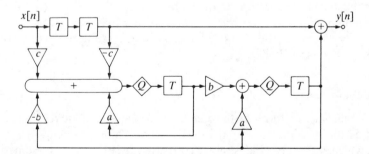

FIGURE 13.15 Recursive digital filter of the second order (coupled-form structure). This type has been selected for the 500-Hz, 250-Hz, 125-Hz, 63-Hz and 31.5-Hz octave-band filters, because the desired filter curves can be obtained more accurately with it for these lower frequencies, in spite of the quantization of the filter coefficients to no more than 12 bits. a, b, c filter coefficients. Q quantizer in which the word length of the signal samples is reduced to 24 bits.

This is necessary because the delay elements in the feedback loops have only a fixed number of bit places available. However, the execution of operations (mainly multiplications) generally causes an increase in the word length. The required quantization of intermediate results has two main undesirable consequences: the possible occurrence of 'limit cycles' and degradation of the signal-to-noise ratio. We shall look more closely at these effects in the next subsections.

If we neglect the quantizers Q, for a direct-form-1 filter (Fig. 13.14) we find the system function

$$H(z) = \frac{b_0 + b_1 z^{-1} + b_2 z^{-2}}{1 - a_1 z^{-1} - a_2 z^{-2}}. \qquad (13.10)$$

and for a filter of the coupled form (Fig. 13.15)

$$H(z) = z^{-2} \left\{ 1 + cb \frac{1 - z^{-2}}{1 - 2az^{-1} + (a^2 + b^2)z^{-2}} \right\} \qquad (13.11)$$

The octave-band filters that we require can be realized with both types of system function. The most important differences only become apparent if we limit the word lengths of filter coefficients and intermediate results, particularly if attention is paid to limit cycles, quantization noise and realizability of certain frequency responses.

13.4.2 Limit cycles

Limit cycles are oscillations of small amplitude that can be maintained by the filter if the input signal has a constant value (including zero) for some time, or is periodic. In spite of their low level these oscillations can be particularly troublesome in audio systems. The first countermeasure that can be used here is to use 'magnitude truncation' in all the filters as a method of quantizing the intermediate results. This was sufficient for all our octave-band filters except the 1-kHz filter, in which limit cycles could still occur. A double word length was therefore used for the signals in this filter (i.e. 46 bits instead of 24 bits), making the quantization step smaller by a factor of 2^{22}. This meant that any limit cycles also had an amplitude that was 2^{22} times smaller than before. Subsequent limitation of the output signal of the filter to 24 (most-significant) bits makes these oscillations undetectable outside the filter.

13.4.3 Signal-to-noise ratio

If the input signal varies sufficiently the quantization of intermediate results has about the same effect as adding some noise to the useful signal. This could degrade the signal-to-noise ratio, which is of course undesirable. To analyse these effects it is assumed that each quantizer can be considered as an additional noise source of noise with a flat frequency spectrum, located at the quantizer.

Quantization by rounding off with a quantization step q is accompanied by a rounding-off error between $-q/2$ and $q/2$; the corresponding noise source has a total noise power of $q^2/12$. Magnitude truncation with the same quantization step gives a quantization error between $-q$ and 0 for positive signal values and between 0 and q for negative signal values. The power of the corresponding noise source can be taken to be $q^2/3$. The direct-form-1 filter (Fig. 13.14) requires only one quantizer, but the coupled form (Fig. 13.15) requires two. This alone gives different noise characteristics, although it is not apparent beforehand which structure produces most noise.

As the quantizers are located inside the filter sections, the noise no longer has a flat spectrum at the output of the section in which it arises. Before the noise finally reaches the output of the complete equalizer it must first pass through all the following sections and acquire further 'colouring'. Meanwhile new noise contributions are constantly added by the successive quantization operations. The important feature is the total amount of noise power that appears at the output and the shape of its spectrum (as a result of colouring). In view of this the sequence chosen for the cascade arrangement of octave-band filters is the one shown in Fig. 13.11. The smallest noise contribution comes from the 1-kHz filter because it is the only one with a double word length for its internal operation. This filter is therefore made the first one. Next come the sections with the direct-form-1 structure and then the coupled-form sections, in order of descending centre frequency. The coupled-form sections make the largest contributions to the total noise. To illustrate this Fig. 13.16 shows the different noise contributions from four octave-band filters as they appear at the output of each of these filters and as calculated by a computer. The level of 0 dB corresponds to the noise level at the quantizer; the noise spectrum is initially assumed to be flat there. For frequencies below about 6 kHz the internal noise in most of the sections is therefore amplified. The dashed line indicates the sum of the noise

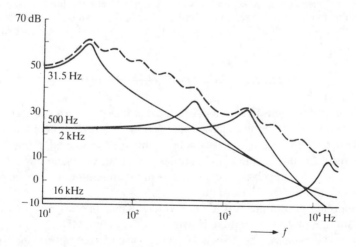

FIGURE 13.16 Continuous curves: frequency spectra of the individual noise contributions from four octave-band filters from Fig. 13.11 as they appear at the outputs of each of these filters. The 16-kHz filter and the 2-kHz filter have the structure of Fig. 13.14. The 500-Hz filter and the 31.5-Hz filter have the structure of Fig. 13.15. Dashed line: spectrum of the total noise from all ten octave-band filters together measured at the output of the equalizer. These curves are for the worst case, i.e. for the gain of all the filters set to $+12$ dB. 0 dB on the vertical axis corresponds to the level at which quantization noise arises in each quantizer. It can also be seen from the dashed line that the 1-kHz filter contributes hardly any noise itself (as a consequence of the 'double precision').

contributions from all ten filter sections as it appears at the output of the equalizer if each filter is set to maximum gain (+12 dB). Averaged over the whole audio band from 0 to 20 kHz the total noise level is about 31 dB higher than the level of each individual noise source. To ensure that the equalizer does not noticeably degrade the original good signal-to-noise ratio (maximum value about 96 dB) of the 16-bit input signal, the internal noise sources must each be at least 31 dB below the input noise. This requires a smaller internal quantization step and hence a greater word length than that of the input signal. Each 6 dB requires 1 additional bit. We have therefore added 6 bits.

With a maximum gain of 12 dB for each octave-band filter there may be appreciable amplification for certain input signals. To prevent this from causing overflow the internal word length must again be increased. We therefore added 2 more bits, so that our total internal word length is $16 + 6 + 2 = 24$ bits (Fig. 13.17).

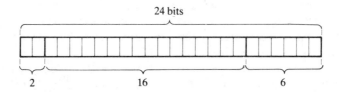

FIGURE 13.17 The word length of the original input signal is normally 16 bits. In the internal processing, however, the signals consist of words with a length of 24 bits. This can be explained as follows. The possible gain of +12 dB (a factor of 4 in amplitude) in the equalization filters requires an extension of 2 bits at the most-significant value end, to prevent overflow. On the other hand an internal refinement of the quantization by 6 bits relative to the input signal is necessary to prevent the equalizer from degrading the signal-to-noise ratio of the input signal.

13.5 THE ARCHITECTURE OF THE ASP

The ASP is an integrated digital signal processor that can perform all the operations arising in the applications described above, and more. It is used either alone or with other ASPs and with ancillary circuits such as standard microprocessors. It is highly suitable for the processing of 44 100 discrete signal samples per second (the sampling rate of 44.1 kHz is then the same as for the Compact Disc). The input and output signals take the form of words with a maximum length of 24 bits.

In one sampling period the ASP can execute 128 machine instructions, which brings the maximum repetition rate of these instructions to about 6 MHz.

There are two versions of the ASP, a ROM version and a RAM version. The difference relates to the program memory. If this is a random-access memory (RAM), the same ASP can be programmed for all sorts of different operations, which makes it very versatile. If the program memory is a read-only memory (ROM) then the program is fixed permanently in the IC. This also fixes the application. This version is simpler than the other; it requires a smaller area of silicon and is intended for use in the applications such as the digital audio system that we shall describe later.

In the first part of this chapter on digital audio signal processing we frequently encountered the three basic operations: multiplication, addition and delay. Probably the most common operation is the addition of a number of multiplication results. There are also operations such as rectification and the quantization of intermediate results. High-precision computation also occupies a special place. In a practical system the transfer of signal samples (which we shall call 'data' from now on for brevity), coefficients and results plays an important part. The ASP is designed to carry out all these operations as efficiently as possible.

13.5.1 Block diagram of the ASP

Figure 13.18 is a block diagram of the ASP. Many of the blocks are interconnected. Some connections are common to several blocks; these are the 'buses'. There are three buses: the data bus, the coefficient bus and the control bus ('C Bus'). The most important blocks are the multiplier Mpy and the ALU (arithmetic and logic unit). The multiplier is asymmetrical: it multiplies words of 12 bits by words of 24 bits. The 24-bit words represent signal data, the 12-bit words represent filter coefficients when an ASP is used to carry out filtering processes. The result of the multiplication is given as 36 bits.

If the task being carried out by the ASP is one of automatic control, as in dynamic-range compression, the 12-bit words include control information derived from the signal itself. For dynamic range compression quantization to 12 bits is too coarse; the steps in the adjustment of the sound level would be audible. Therefore two words of 12 bits are always taken together to represent one control value ('double precision'). The signals can also be handled with 'double precision', which is used in the 1000-Hz octave-band filter in the equalizer mentioned above. The ASP can in fact operate at better than double precision. However, this facility has not been used in the audio signal processing described so far.

For the multiplication of a signal value at double precision by a coefficient at double precision, four consecutive multiplications are performed as shown in the diagram of Fig. 13.19. The signal value is represented by two words of 24 bits. The first word, which we will call d, contains the 24 most-significant bits; the second word d' contains two zeros in the first two places and then the 22 least-significant bits. The signal value, which is thus represented by 46 bits, takes up two memory locations; it can be expressed as $d + d' \times 2^{-22}$. In a similar way a 12-bit coefficient acquires double precision by the addition of 11 less-significant bits; we can then write the coefficient as $c + c' \times 2^{-11}$. In the execution of the product

$$(d + d' \times 2^{-22})(c + c' \times 2^{-11}) \approx p + p' \times 2^{-22} \qquad (13.12)$$

the least-significant parts are first multiplied together. The resulting product $c' \times d'$ is then shifted by 11 places. The product $c \times d'$ is then formed and added to it. The sum is again shifted by 11 places, and then $c' \times d$ is added to it. After a further shift $c \times d$ is finally added. Of the total thus obtained the 24 most-significant bits (the word p) are stored in a 'high' register, the following 22 bits (the word p') in a 'low' register. Both registers transfer their contents to a selector, a wide register, which transfers parts of the long words stored in it to the data bus when it receives a control signal. These registers are collectively indicated by Res Reg in the block diagram (Fig. 13.18).

The unit ALU is an adder with a width of 40 bits. This adds the results of multi-plications carried out in the multiplier Mpy to earlier results stored in the accumulator register Accu, which also has a width of 40 bits. The output of this register is therefore connected to a second input of ALU.

The results of the addition in ALU are sent back to the data bus via the Res Reg register. They are in fact too long for this bus and are therefore quantized to 24 bits in the circuit Ovf Clip. Various types of quantization and overflow are possible here. In cases where calculations must be made with double precision the least-significant bits and the most-significant bits are handled separately.

The coefficients are supplied via the coefficient bus Coef Bus. They come from the memory Coef RAM, which is loaded from outside the ASP with the coefficients appropriate to the task in hand (e.g. by a microprocessor via the IIC circuit). When the

---

FIGURE 13.18 Block diagram of the ASP. The central section surrounded by a dashed line is the arithmetic unit in which the actual calculations take place. The multiplier Mpy multiplies a 24-bit word (data) by a 12-bit word (often a filter coefficient); if desired the product can be added in the arithmetic logic unit ALU to the sum obtained from the previous calculation and stored in the accumulator register Accu. The data are brought from the random-access memory Data RAM via the data bus (Data Bus); the coefficients are brought via the coefficient bus (Coef Bus) from a memory coef RAM, which has been supplied with the appropriate coefficients, e.g. via the connection unit IIC Interface from an external microprocessor; the latter is the link between the controls that are set by the user and the ASP. The significance of the abbreviations is as follows:

Accu	accumulator register for intermediate results
Addr Sel	address control for data memory
ALU	arithmetic and logic unit
C Bus	control bus
Clock	clock signal
Coef Bus	coefficient bus
Coef RAM	random access memory for coefficients
Coef Reg	coefficient register
Data Bus	data bus
Data Bus Monitor	unit that monitors the traffic on the data bus for control purposes
Data RAM	random-access memory for data
Data Reg	data register
IC	integrated circuit
IIC	standardized control signal ('Inter-IC Control')
IIC Interface	connection with the (external) IIC bus that carries the signals to and from other ICs in accordance with a standard protocol
IIS	standardized digital signal ('Inter-IC Signal')
Mpy	multiplier (24×12 bits)
Ovf Clip	quantization and overflow circuit
Par I/O	parallel input and output unit
Pr C	program counter
Prog ROM	read-only memory containing the program
Res Reg	results register
Ser In, Out	serial input and output units
Sgn	register for the sign bit
S/P Conv	serial/parallel converter
Sync	synchronization circuit
μC	microcode

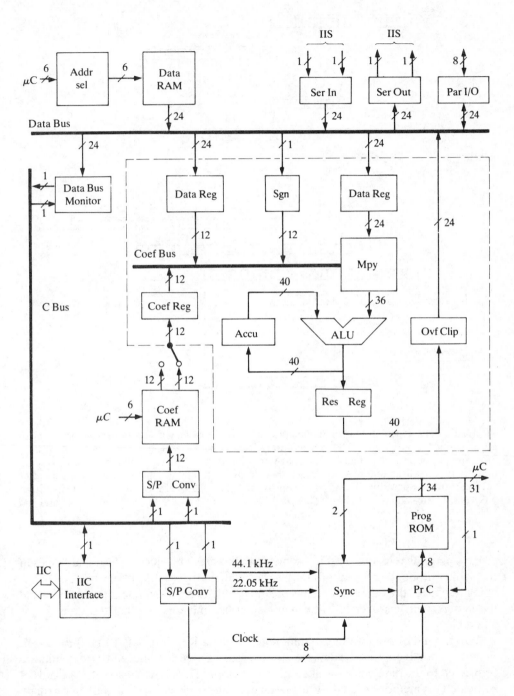

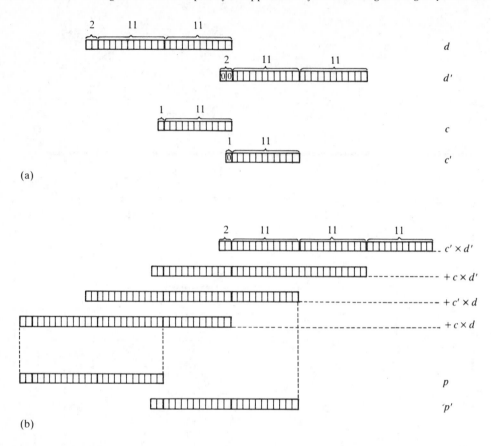

(a)

(b)

FIGURE 13.19 Calculation with 'double precision'. (a) Obtaining a data word of 46 bits from 24 most-significant bits d and 22 least-significant bits d', and a coefficient word of 23 bits from a most-significant part of 12 bits c and a least-significant part c'. (b) The creation of the product in double precision from four partial products that are given the right weight by shifting by 11 bits after each part of the operation. The product consists of 46 bits, which are split into 24 most-significant bits p and 22 least-significant bits p'. Thus $p + p' \times 2^{-22} = (d + d' \times 2^{-22})(c + c' \times 2^{-11})$, apart from quantization effects.

task changes the reloading requires some time; this may introduce undesirable signals in the audio output. To prevent this the coefficient memory is duplicated; while one half is being reloaded, the other half retains the old coefficients. When reloading is completed the system switches over to the newly loaded half-memory, so that the transition is inaudible.

With the aid of one of the two data registers Data Reg (capacity 24 bits) the 24-bit signals are obtained from the data bus. Both parallel and series input and output units are connected to the data bus, as well as the data memory Data RAM (capacity 64 words of 24 bits). This memory is used for temporary storage of received data and intermediate results and final results that are awaiting further processing or transfer to the output.

The data memory Data RAM therefore also performs the elementary delay operation, i.e. delaying a signal by a sampling period (at the usual audio sampling rate of 44.1 kHz this is $22.7\,\mu s$). This operation is a regular part of the filtering process, for which the signals are held temporarily in Data RAM. Once they arrive there they are not shifted along (as they would have to be if a shift register was used as the delay element) but put in fixed memory locations. The unit Addr Sel records this and computes the addresses from which the signals must be fetched when required.

If a series of second-order filters are to be realized with an ASP, then a large number of multiplications must be performed in each sampling period rather than a single multiplication. These multiplications occur in sequence. Acceleration of the process is obtained by 'pipelining': the operations are prepared by bringing in the necessary data and feeding it through any intermediate registers at the same time as the operations on previous data are completed. To make this possible intermediate registers are necessary.

Routing the data streams and addressing the memories correctly require extensive control facilities, of course. An important part is played by the program read-only memory Prog ROM, which contains the complete program of consecutive operations that the ASP must perform. Prog ROM is directly connected to all the appropriate subsidiary units and controls them via 'microcode' μC. For clarity only a few of the μC connections are shown in Fig. 13.18. For the control of the ASP the control bus C Bus is also available, with a number of special components such as a program counter, a synchronization and clock-signal unit and an IIC unit for connection to an external IIC bus that can carry data traffic with other ICs to a standard protocol ('IIC' stands for 'Inter-IC Control').

Let us now consider the input and output units for a moment. The parallel input/output unit Par I/O consists of a whole complex of registers, one of whose tasks is to separate the internal synchronization of the ASP from that of the incident signals. In addition, Par I/O does not just transfer data but is also used in addressing the external memories that are used in dynamic-range compression and the excitation of reverberation. (The internal memories of the ASP are much too small for these applications.) The external memory addresses required are computed by Par I/O.

The serial input and output units are not simple shift registers but separately controllable 'latches' in which the most significant bit of the received word always arrives at the same place whatever the length of the word. This means that traffic with other ICs can follow the IIS protocol, a standard in which a word-selection signal is provided to indicate the start of each new word ('IIS' stands for 'Inter-IC Signal'). This protocol is also used in all kinds of other modern ICs for digital audio, such as digital filters and digital-to-analog converters.

It is characteristic of the ASP that many operations run in parallel; the term 'pipelining' has already been mentioned. This means that the control is complicated. The complete IC was therefore simulated on a computer in the design phase. A full set of software was written for this simulation. An assembler language was defined in which the commands could be expressed, and an assembler program was written to translate these commands into machine instructions for the ASP. In addition a simulation program was written, and during the design phase prototypes ('breadboard' models) were used for carrying out listening tests.

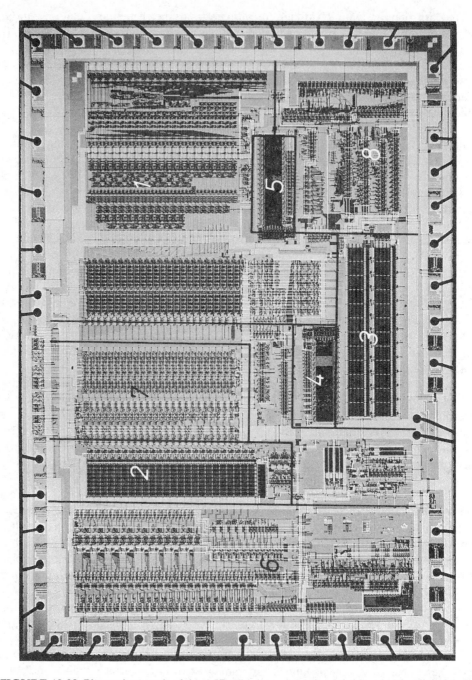

FIGURE 13.20 Photomicrograph of the ASP. This integrated circuit is intended for digital audio signal processing. It is made in 2.5-μm NMOS technology, with a chip area of 46 mm^2. The most important components are: *1* the part of Fig. 13.18 inside the dashed line; *2* data RAM; *3* coefficient RAM; *4, 5* program ROM; *6* parallel input and output units; *7* serial input and output units; *8* IIC circuit.

Finally, we should mention that the current ASP is an integrated NMOS circuit with an area of $46 \, mm^2$ (see Fig. 13.20). It has 40 connection pins.

13.6 A DIGITAL AUDIO SYSTEM MADE UP OF ASPs

As mentioned earlier the ASP IC does not carry out the tasks we have described on its own. Figure 13.21 represents the block diagram of a digital audio system that contains five ASPs, two standard microprocessors and a number of ancillary ICs. These are used mainly as input and output circuits and memories.

One of the microprocessors (μP1) provides the main control for the system. All the ASPs are connected to this microprocessor via the IIC bus, and the instructions resulting from the commands that the user gives are supplied to the ASP in this way. The second microprocessor (μP2) is provided to control the display panel, which shows information such as the equalizer setting. (The same panel can also display the frequency spectrum of the sound, split into ten octave bands.) To simplify the application of the system, the practical version is combined with two power amplifiers ($2 \times 100 \, W$) in a single unit.

Serial transfer of signals between the different ICs follows the IIS standard protocol. This requires three lines. One of these is for the serial clock signal, which tells the receiving IC how fast the bits are arriving; this rate is independent of the internal operating rate of the IC. A second line carries a word-selection signal. This is a symmetrical square wave whose edges indicate the start of a new data word (a trailing edge indicates a word for the left-hand audio channel, a rising edge indicates a word for the right-hand channel), so that the length of the words does not need to be known in advance by the receiving IC. The third line carries the actual signals (in 'two's complement' representation, with the most significant bit first).

The digital audio system works almost exclusively (see below) with a sampling rate of $f_{s1} = 44.1 \, kHz$. Either analog sources (record player, radio tuner, tape player) or digital sources can be connected to it. With analog sources analog-to-digital conversion of both the stereo channels is necessary. Digital sources would include the Compact Disc player, a digital cassette player, a video recorder adapted for digital audio or a receiver for the proposed digital radio broadcast from future broadcasting satellites. The signals from the digital sources must be put into the required serial form; this is done in the IC ADIC.

First the dynamic range is compressed or expanded in ASP1. The time required for preparing this process after peak detection is obtained by temporarily storing the signal in a 64-kbit memory. This time is not inconsiderable, because of the slow nature of the control signals. After processing in the first ASP the signal words have the standard length of 24 bits. They are sent to ASP2.

To introduce reverberation ASP2 splits off a mono signal with half the bandwidth (up to 10 kHz), which is sampled at $f_{s2} = 22.05 \, kHz$ and used as the starting point (see also page 371). This signal is applied to ASP3, which is also connected by the parallel input/output unit to a 256-kbit memory consisting of 4 ICs each with a capacity of 64 kbits. The addresses of these memories are sent via the parallel input/output lines of ASP2. The storage time in these memories gives the signals the delay necessary for the operation of the reverberation circuits. In ASP3 a reverberation signal with a sampling rate of 22.05 kHz is calculated for each of the two stereo channels as indicated in Fig.

13.9. These signals are fed back to ASP2. In ASP2 new reverberation signals are then calculated at twice the sampling rate and these are finally added to the two original stereo subsignals.

So much processing is required for the equalization of the frequency characteristic by adjusting the 10 octave-band filters that a separate ASP has to be used for each of the two stereo channels (ASP4 and ASP5). For the moment the equalization is set up by hand.

After the processing the 24-bit audio signal is scaled in such a way that it does not overload the digital-to-analog converter. The converter is intended for 16-bit signals and where possible it is driven to full scale to obtain the best possible signal-to-noise ratio. It supplies an analog signal to the gain control of the power amplifier. Two 16-bit output signals are also available at two digital outputs.

FIGURE 13.21 Block diagram of a digital signal-processing system for audio, based on ASPs. There are analog inputs for various signal sources and analog outputs for the processed or unprocessed signals; there are also digital inputs and outputs. Five ASP ICs perform the operations. The significance of the abbreviations is as follows:

A/D	analog-to-digital converter
ADIC	converter to IIS format (for incoming digital signal)
ASP	audio signal processor
AOut	analog output
Aux	auxiliary analog input
CD-A	Compact Disc player (analog signal)
CD-D	Compact Disc player (digital signal)
CD2B	converter to external signal format (for outgoing IIS signal)
Comp/Exp	compression/expansion
D/A	digital-to-analog converter
DOut	digital output
DT	digital tape player
EEPROM	electrically erasable programmable read-only memory for preset states and for final state on switching off
Eq	equalization
f_s	sampling rate ($f_{s1} = 44.1$ kHz, $f_{s2} = 22.05$ kHz)
IIC, IIS	(see Fig. 13.18)
L	left-hand channel
PA	power amplifier
Phono	record player
R	right-hand channel
RAM	random-access memory
Reverb	reverberation generation
ROM	read-only memory for program and filter coefficients
S	digitally-controlled selector
Tape	tape player
Tuner	radio tuner
VC	volume control
μP	microprocessor

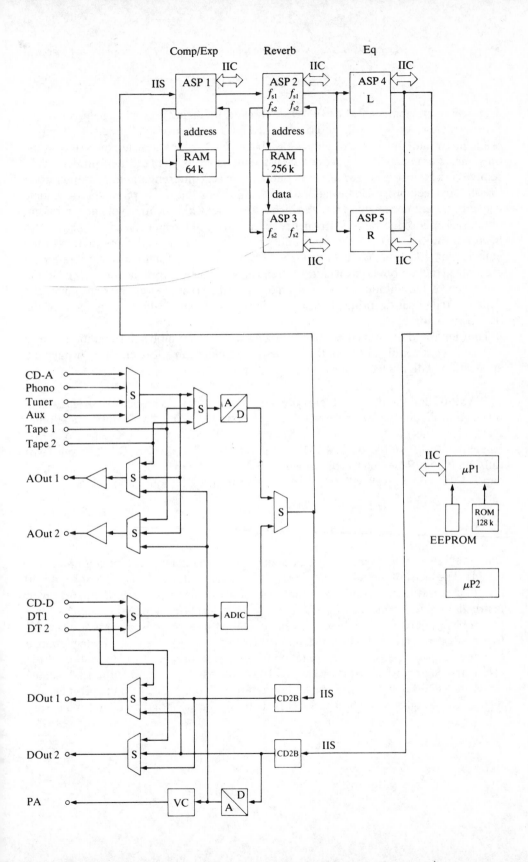

13.7 RETROSPECT AND PROSPECT

The ASP is one of the first ICs to be 'custom-made' for digital processing in hi-fi audio. It is nevertheless very versatile and can be used for a number of applications with appropriate programming. Three examples have been given above; these are by no means the only ones. For instance, we have not discussed the suppression of the 'tick' produced by a scratch on a gramophone record, or of similar defects of different origin. This application has already been proved in the laboratory. Another possible use is 'physiological volume control': the enhancement of low and high frequencies when the total volume is reduced, to take account of the variable frequency characteristic of the human ear.[99] This has also been put into practice. A third possibility now close at hand is the automation of the equalization function. By means of the signal from a microphone, incorporated for example in the remote-control unit, the frequency characteristic of the reproduced sound can be corrected automatically. In a similar way the compression or expansion of the dynamic range can in principle also be controlled automatically as a function of the background noise.

The digital audio system described here shows how such complicated operations can be carried out with equipment that the 'ordinary' consumer can afford and use, through the application of digital techniques.

The ASP IC and the digital signal-processing system for audio based on it originated in the Digital Audio Signal Processor Project Group led by the two authors. The other members of this group were H. W. A. Begas, P. J. Berkhout, M. H. Geelen, W. J. W. Kitzen, E. A. M. Odijk, A. C. A. M. van der Steen, E. F. Stikvoort, R. A. H. van Twist (all with Philips Research Laboratories), P. Bakker, F. J. Op de Beek, J. M. Rijnsburger and J. J. H. Verspay (all with the Consumer Electronics Division, Philips NPB).

13.8 SUMMARY

The ASP integrated circuit has been developed for the digital processing of hi-fi audio signals. The heart of the IC is a multiplier (24×12 bits) followed by an adder of width 40 bits that can add the result of a multiplication to earlier results. About 6 million instructions per second can be executed. With external programming the ASP can perform various digital filter processes. In a digital audio system described as an example, five ASPs are combined with ancillary circuits to perform three tasks: dynamic-range compression or expansion, the excitation of reverberation and the equalization of the frequency characteristic with the aid of 10 octave-band filters, which are adjustable between $+12\,dB$ and $-12\,dB$. The first part of this chapter deals with the practical and theoretical background to these tasks and with the implications of their digital execution.

14

A digital 'decimating' filter for analog-to-digital conversion of hi-fi audio signals

J. J. van der Kam[†]

14.1 INTRODUCTION

The digital processing of audio signals offers many new prospects. Transmission, recording, adaptation of the frequency response to the characteristics of the human ear, compression and expansion of the dynamic range, the addition of reverberation to music[12] – all this can be done by digital methods with less distortion, with a better signal-to-noise ratio and often more cheaply than by analog methods.

In 'professional' applications there is nothing new about this. In telephony, speech signals in digital form have been transmitted over long distances for many years. Now, however, digital techniques are increasingly penetrating the more consumer-oriented areas of hi-fi audio technology; a typical example is the Compact Disc.[14]

Before audio signals can be processed digitally they have to be converted from the analog form, in which they originated, into a digital form, by means of an analog-to-digital (A/D) converter. This conversion must not introduce any distortion or noise. The specification for hi-fi audio here are stricter than for telephony, since the frequency bandwidth for hi-fi audio is much larger: 20 kHz as compared with only 3.4 kHz in telephony, and the signal-to-noise ratio required is very much higher than the 33 dB that is sufficient for telephony.

It was only with the advent of fast and very complex integrated circuits that it became possible to meet the special requirements that audio imposes on A/D conversion at an economic price. However, it is still necessary to select the best technique. This chapter

†Formerly with Philips Research Laboratories, Eindhoven, now with the Elcoma Division, Philips NPB, Eindhoven. This chapter is reprinted from *Philips Tech. Rev.* **42** No. 6/7: 230–8, Apr. 1986.

gives prominence to one of the many different choices. It deals with the application to hi-fi audio of a concept that has already been applied successfully in telephony: A/D conversion via the intermediate stage of 1-bit coding[100,101,102] at a very high sampling rate. The 1-bit coding stage is followed by a complex digital filtering operation in which the 1-bit coding is replaced, in our case, by a 16-bit coding and in which the sampling rate is reduced to a more manageable lower value.

Digital filters in which the sampling rate is reduced are known as 'decimating filters', even if the reduction factor has no clear relation to the number ten. The design of such a filter for hi-fi audio is the particular subject of this chapter. However, we should first look at the requirements to be met by an A/D converter and at the arguments in favour of the intermediate stage of a 1-bit coding.

14.2 ANALOG-TO-DIGITAL CONVERSION

Any form of A/D conversion consists of at least two operations:

- Sampling, i.e. taking samples of the analog signal at regular intervals.
- Quantization, i.e. limiting the number of different values that the samples can assume so that each sample can be expressed in a digital word of a finite number of bits.

In sampling, the sampling theorem has an important part to play. This theorem states that the minimum sampling rate is equal to twice the highest frequency present in the analog signal. Conversely, at a given sampling rate the signal frequencies may not exceed half the sampling frequency to prevent the formation of distortion products ('aliasing').

14.2.1 Pulse-code modulation

The kind of digital signal most commonly encountered in audio technology is 16-bit pulse-code modulation (16-bit PCM). An audio signal then consists of a series of words (or pairs of words in the case of stereo) with each word consisting of 16 bits. This means that $2^{16} = 65\,536$ different signal values are available, making high-quality audio possible. The highest signal frequency to be reproduced is usually 20 kHz. For the Compact Disc a sampling rate of 44.1 kHz was accordingly adopted, and in this chapter we shall adopt the same value.

The conversion of an analog audio signal into a PCM signal usually takes place in a circuit as represented in Fig. 14.1. First, the bandwidth of the input signal is limited to about half the sampling rate. This is followed by sampling and quantization. The quantization circuit calculates the 16-bit word that most closely reproduces the value of each sample, e.g. by comparing each sample value with a set of reference voltages.

The three successive steps of the operations in Fig. 14.1 seem simple in principle, but are far more difficult in practice. For a start let us consider the lowpass input filter. If, as is usual in digital technology, the audio band is allowed to extend to 20 kHz at the usual sampling rate of 44.1 kHz, then the filter must have a flat characteristic up to 20 kHz and it must attenuate strongly above 24.1 kHz to prevent aliasing. (An original frequency component of say 24.5 kHz is 'aliased' as a result of sampling to $44.1 - 24.5 = 19.6$ kHz, i.e. to a position within the audio band where it appears as distortion.) This filter characteristic has to be realized with a filter for analog signals, i.e. in analog technology. This implies a high-order filter that requires accurate trimming, which will make it relatively expensive. A further difficulty is that such a filter will inevitably cause considerable phase shifts at some frequencies in the audio signal, so that certain waveforms will be distorted.

Very high precision is also required for the operations in the quantizer in Fig. 14.1. Reproducible discrimination between the 65 536 voltage levels mentioned above, in a range of say -5 V to $+5$ V (corresponding to a quantization-step size of $153\,\mu$V) requires an exceptionally stable circuit and very high precision in the voltage dividers.

An analog-to-digital converter of the kind we have described is very largely an *analog* circuit, and very complex. It would be much better if more of the operations required could be performed *digitally*.

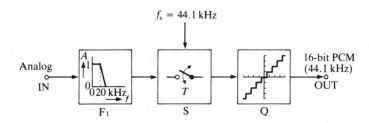

FIGURE 14.1 Schematic diagram illustrating analog-to-digital conversion by 16-bit pulse-code modulation (16-bit PCM) with a sampling rate of $f_s = 44.1$ kHz. In the analog prefilter F_1 with amplitude characteristic A the frequency spectrum of the input signal is limited to about half the sampling rate, leaving the audio band, which extends up to 20 kHz, largely unaffected. In the sampling circuit S one sample is taken in every sampling interval $T = 1/f_s$. This sample is then quantized in the quantization circuit Q and translated into a 16-bit word.

The need for an input filter with a sharp cut-off (i.e. a narrow transition band) can be avoided by using a higher sampling rate. If, however, a digital signal at a given sampling rate is ultimately required – perhaps at 44.1 kHz as mentioned above – then an extra digital conversion must be provided to give this lower sampling rate. This then requires a filter that meets much the same specifications as the original analog input filter. Now, however, we are dealing with a digital filter, so that completely different rules apply: the high precision required can now be obtained without accurate trimming of the filter. What is more, a digital filter can have a linear phase characteristic, so that there is no signal distortion, and it can be made as an integrated circuit.

14.2.2 1-bit coding

Another advantage of using a higher sampling rate is that fewer bits per sample are necessary in the quantization, especially in a circuit arrangement where the quantizer is included in a feedback loop.[20] Indeed, it is even possible for each sample to be represented by only 1 bit. This is referred to as 1-bit coding: the most familiar example of this is delta modulation. The basic circuit diagram of a delta modulator is shown in Fig.

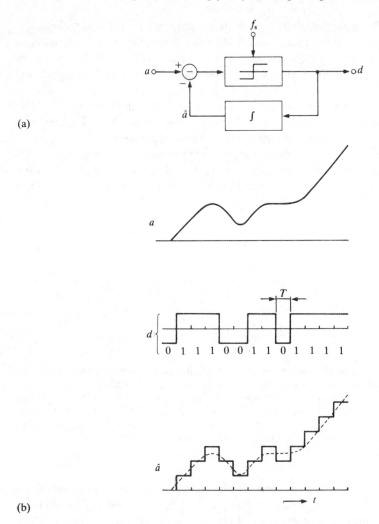

FIGURE 14.2 (a) Schematic diagram of a delta modulator, consisting of a subtracter, decision circuit and integrator. (b) The analog signal a is converted into a bit stream d in the delta modulator; d represents a digital signal consisting of 1-bit words with a sampling rate of $f_s = 1/T$. Integrating d gives a stepped signal $\hat{a}$ that is a good approximation to a. The decision circuit operates on the difference signal $a - \hat{a}$.

14.2. Here the 1-bit coder consists simply of a subtracter, a clocked two-level quantizer (a 'decision circuit') and an integrator. The operation of the circuit can be understood in general terms from the signals a, d and $\hat{a}$ shown in the diagram. The attraction of this coding is mainly due to its simplicity and the small number of critical components.

With 1-bit coding as an intermediate step an analog audio signal can be converted into a 16-bit PCM signal with a sampling rate of 44.1 kHz in a circuit as shown in Fig. 14.3. The analog prefilter F_2 is very much simpler than F_1 in Fig. 14.1, since the filter specification is now far less strict, mainly because the transition band can now be much wider. The 1-bit coding with a sampling rate of $R \times 44.1$ kHz is followed by a digital filter F_3 and a 'sampling-rate decreaser' SRD. F_3 and SRD form a decimating digital filter that supplies the required PCM signal. In this chapter we shall confine ourselves to the case $R = 72$, which means that the 1-bit coding takes place at a sampling rate of 72×44.1 kHz $= 3175.2$ kHz.

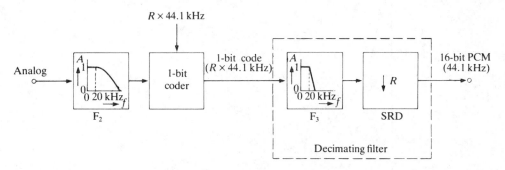

FIGURE 14.3 Schematic diagram illustrating analog-to-digital conversion with 1-bit coding at a high sampling rate as an intermediate stage. The specification for the analog prefilter F_2 is far less strict than that for F_1 in Fig. 14.1. The output signal of the 1-bit coder can now be converted by digital methods alone into a 16-bit PCM signal with a sampling rate of 44.1 kHz. This is done by means of a decimating filter, which consists in principle of a lowpass digital filter F_3 and a sampling-rate decreaser SRD.

14.2.3 Sigma-delta modulation

Depending on the situation, the 1-bit coding in Fig. 14.2 can be improved in various respect. In our application an important aspect is the inevitable quantization noise that is always produced in any A/D conversion, including 1-bit coding. We would prefer the quantization noise to be least in the audio band and immediately above it, because such noise cannot be removed, while quantization noise at higher frequencies (up to $0.5 \times 3.1752 = 1.5876$ MHz) can in principle be filtered out by F_3 (Fig. 14.3). This 'noise' shaping' can be achieved by sigma-delta modulation ($\Sigma\Delta$M or ΣDM).[103]

A block diagram of a sigma-delta modulator is shown in Fig. 14.4(a). As can be seen, it closely resembles the 1-bit coder of Fig. 14.2, but its exact operation is rather more difficult to illustrate. The noise shaping is determined by the integrating filter LF. In the choice of this filter care must be taken to ensure the stability of the complete circuit. To illustrate the noise shaping Fig. 14.4(b) shows the amplitude characteristic of a possible filter LF and Fig. 14.4(c) gives the frequency spectrum of the resultant quantization noise. The low-pass characteristic of LF gives the noise a 'high-pass' character.

Further details of the sigma-delta modulator will not be given here. It is merely assumed that it delivers a series of bits at the repetition rate mentioned above. We now have to consider how a digital decimating filter uses it to produce a series of 16-bit words with a repetition rate of 44.1 kHz, yet without adding impermissible noise or causing distortion during the transformation process.

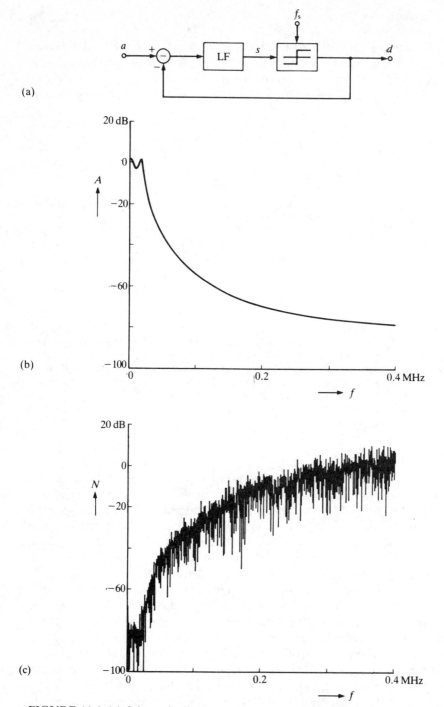

FIGURE 14.4 (a) Schematic diagram of a sigma-delta modulator consisting of a subtracter, a lowpass filter LF and a decision circuit. The rectangular output signal d is now directly subtracted from the analog input signal a (see Fig. 14.2). s input signal of the decision circuit. (b) Typical amplitude characteristic A of the filter LF. (c) Corresponding power-density spectrum N of the quantization noise in the signal d, when a is a sinusoidal signal at 990 Hz and the sampling rate f_s is 3.1752 MHz. (For clarity the frequency scale is shown only from 0 to 0.4 MHz.)

14.3 THE DECIMATING FILTER

The first task of the digital decimating filter, which follows the sigma-delta modulator, is to suppress frequencies above the audio band, i.e. higher than 20 kHz. In the context of Fig. 14.1 we have already discussed the necessity of suppressing these frequencies in direct A/D conversion. It is just as necessary in the present arrangement, where the ultimate sampling rate of 44.1 kHz is achieved in a roundabout way. Ideally, the digital filter F_3 should have the amplitude characteristic shown in Fig. 14.5: a passband with a gain of 1 from 0 Hz to 20 kHz, an infinitely narrow transition band at 20 kHz and complete suppression in the stopband between 20 kHz and 1.5876 MHz (half the sampling rate).

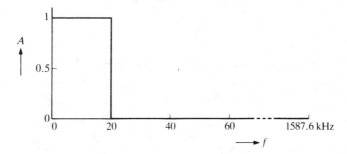

FIGURE 14.5 Ideal amplitude characteristic A of the low-pass digital filter F_3 in Fig. 14.3.

14.3.1 The stop band

Apart from the fact that the filter in Fig. 14.5 cannot be realized (since it would require an infinitely large number of filter coefficients and infinitely high accuracy for the coefficient values), we do not strictly speaking need it, since the measure for the required quality of the ultimate 16-bit PCM signal is the direct conversion in Fig. 14.1. This means for one thing that the maximum signal-to-noise ratio for a digitized sinusoidal signal need never be more than 98 dB. From this we may deduce a specification for the attenuation in the stop band of the digital filter F_3. To start with we shall use a number of approximations. In the first place we assume that the noise shaping of the 1-bit coding is perfect, i.e. all the quantization noise is located in the frequency range above the audio band. In the second place we consider the largest sinusoidal input signal that does not saturate the decision circuit in the sigma-delta modulator (Fig. 14.4), so that the peak value of the input signal s is equal to the peak value of the signal d (Fig. 14.6). For the time being we put both at 1 V. Applied across a hypothetical load resistance of 1 ohm the signal s then represents a power of 0.5 W and the signal d a power of 1 W. Of this 1 W, 0.5 W represents signal power and 0.5 W represents noise power. The signal d therefore now has a signal-to-noise ratio of 0 dB (where the noise in the full frequency band up to half the sampling rate is taken into account).

With these simplifying assumptions we can achieve the ultimately required maximum signal-to-noise ratio of 98 dB by specifying for the filter F_3 a stop-band attenuation of 98 dB over the complete bandwidth of 20 kHz up to 1.5876 MHz (Fig. 14.7(a)).

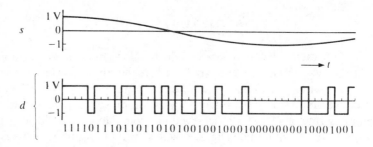

FIGURE 14.6 Determining the signal-to-noise ratio in the output signal of a sigma-delta modulator with the decision circuit just below saturation. The sinusoidal input signal s of the decision circuit then has the same peak value as the output signal d. The total power of d is twice that of s. This means that the signal power and the noise power in d are identical. The signal-to-noise ratio is thus 0 dB.

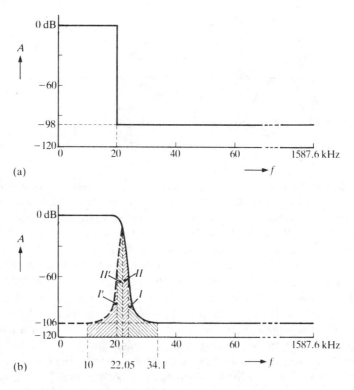

FIGURE 14.7 (a) The attenuation in the stop band of the filter F_3 in Fig. 14.3 does not have to be infinite. With an infinitely narrow transition band and 'perfect' noise shaping an attenuation of 98 dB is sufficient. (b) Amplitude characteristic used for our filter F_3. The transition band extends from 20 kHz to 34.1 kHz, so that aliasing can occur in the 16-bit PCM signal, which has a sampling rate of 44.1 kHz. This is indicated by the hatching. A distinction is made here between aliasing products that appear in the audio band (I–I') and aliasing products that are outside it (II–II'). To avoid an increase in the total noise power, the required stop band attenuation is increased to 106 dB.

14.3.2 The transition band

The filter characteristic in Fig. 14.7(a) also cannot be realised because of the sharp transition from pass band to stop band. In a practical filter a certain bandwidth will always have to be reserved for this. This has two consequences for the 16-bit PCM output signal:

- the quantization noise in the transition band is less strongly attenuated, and
- frequency components of the input signal of the A/D converter that fall within the part of the transition band above 22.05 kHz will be aliased to frequencies below 22.05 kHz.

Input frequencies between 22.05 kHz and 24.1 kHz are aliased to between 22.05 kHz and 20 kHz, i.e. *above* the audio band (Fig. 14.7(b)). As long as they are not too strong they will cause no interference. But even a cut-off band of 20 kHz to 24.1 kHz would give an impractically complicated filter, since both the number of filter coefficients and the accuracy required (i.e. the number of bits) would be excessively high. To obtain a realizable filter (although still a very complex one) yet another concession to the original specificaton is required: input frequencies above 24.1 kHz are indeed aliased *into* the audio band, but to a limited extent this is acceptable because the sensitivity of the ear to frequencies between 10 kHz and 20 kHz is weak at low levels.

On the basis of all these considerations we opted for a filter with a transition band from 20 kHz to 34.1 kHz (Fig. 14.7(b)). Above about 22 kHz the attenuation first increases rapidly and then rather less rapidly up to 106 dB. The increased stopband attenuation compared with Fig. 14.7(a) serves to compensate for the quantization noise in the transition band and for the noise shaping, assumed to be perfect, in the 1-bit coding. Another factor that we have not yet taken into account is that the necessary limitation of the word length of the output signal of the decimating filter to 16 bits itself limits the maximum signal-to-noise ratio to 98 dB. For this reason also a greater attenuation is required in the stop band than is suggested in Fig. 14.7(a).

14.3.3 The pass band

In the pass band from 0 Hz to 20 kHz some ripple in the amplitude characteristic is acceptable; this is also a relaxation of the filter specifications. We decided to limit the ripple to ±0.2 dB. To make the phase characteristic linear as well we decided to use a transversal digital filter (we shall return to this in the next section). In view of all the requirements that the filter characteristic had to satisfy, as many as 2304 filter coefficients were required. It was mainly because we represented the input signal by a 1-bit code and used a decimating filter that we were able to realize this digital filter with only two ICs, as will appear later.

14.4 PRACTICAL DESIGN OF THE FILTER

14.4.1 Structure of the filter

In the practical design of a digital filter it is always necessary to choose between a recursive structure and a non-recursive structure. In a recursive filter the signal samples

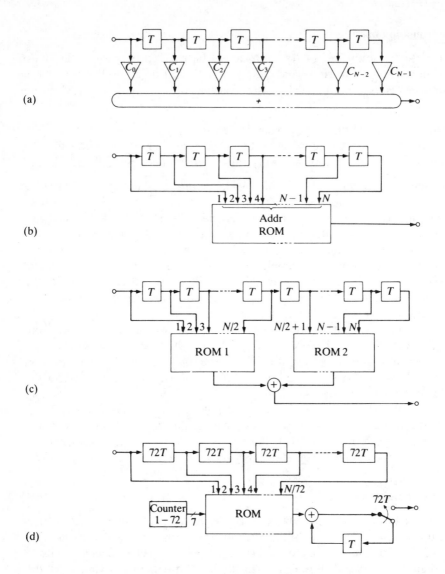

(a)

(b)

(c)

(d)

FIGURE 14.8 Derivation of the structure of the decimating filter. (a) Transversal filter in its basic form. T unit-delay element. $c_0, c_1, \ldots, c_{N-1}$ filter coefficients. (b) The 1-bit words at the taps of the delay elements can be treated as one large N-bit address word addr. This can be used to address a read-only memory (ROM) in which all the possible (i.e. 2^N) values of the output signal are already preprogrammed and stored. No further calculation is therefore required. (c) Combination of the two previous structures: the total number of ROM addresses is $2 \times 2^{N/2}$, and one addition is required per output sample. (d) Since the sampling rate has been decreased by a factor of 72, the calculation of each output sample can take place in 72 successive steps. These are counted by a '72-counter'; the 72 results are added together in an adder circuit with a feedback loop. The number of ROM addresses is now $72 \times 2^{N/72}$. (e) The filter has a symmetrical impulse response and hence a symmetrical set of coefficients. By 'folding in two' the row of delay elements, the 1-bit words that have to be multiplied by the same coefficient come close to one another. In this diagram the sampling rate is reduced by a factor of 72 and the ROM is divided into smaller units that are each addressed by only four 1-bit words. This produces filter cells, two of which are shown here.

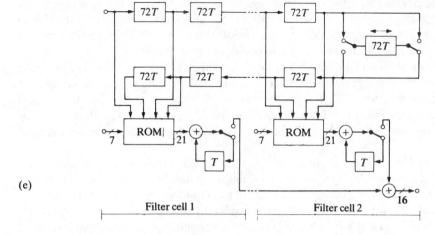

(e)

Filter cell 1   Filter cell 2

FIGURE 14.8 *contd.*

are fed back from at least one point in the filter, via delays and possible multiplications, to a point closer to the input. In our case the input samples consist of only one bit. All that is necessary to delay them are 1-bit memory cells (bistables or flip-flops). However, the signal samples that appear as intermediate results and at the output of our filter have a much greater word length (16 bits and more) and therefore require much more elaborate memory cells for producing delays. It is therefore better to choose a filter structure in which only input samples are delayed. Such a structure is found in non-recursive filters of the transversal type. With filters of this type it is also easy to take advantage of the fact that the output samples have a lower sampling rate than the input samples.

The structure that our digital filter took will be explained in a number of steps, which are illustrated in Fig. 14.8. Figure 14.8(a) shows a transversal filter in its most familiar form. After each unit-delay element T, which corresponds to one sampling interval for the input signal, the delayed signal sample is tapped off and applied to a multiplier, where it is multiplied by a coefficient. The products are then added to form a total sum. In our case, however, each input sample is only 1 bit long and represents either $+1$ or -1. The multiplication is therefore simplified to an operation on the sign and the total sum is arrived at by the addition and subtraction of coefficients. It will be obvious that this speeds up the processing procedure very considerably and that it is a definite advantage of using the 1-bit code as the first phase of the A/D conversion.

All the possible combinations that can be obtained by adding together N coefficients or subtracting them from each other in a transversal filter together form a finite set of numbers with at most 2^N different values. Each output sample of the filter is an element of this set, but always a different one, depending on the last N input samples. The filter can now be arranged in such a way that this sum does not have to be recalculated every time, because all the possible results are already stored in a read-only memory (ROM). The last N bits of the input signal are then treated as the address of the memory location where the relevant result is stored (Fig. 14.8(b)).

It is of course also possible to make combined forms of Fig. 14.8(a) and (b), drastically reducing the number of additions, although not to zero, while at the same time substantially reducing the number of memory locations required in the ROM. Figure 14.8(c) shows such an arrangement, which requires two ROMs each with $2^{N/2}$ memory locations (instead of 2^N in Fig. 14.8(b)) and one addition per output sample.

The transversal structures outlined are not the complete answer, however. We have still taken no advantage of the fact that what we are finally aiming at is a decimating filter with a decimation factor of 72. This means that we do not have to generate a new output sample after every input bit, but only after every 72 input bits. We can use this for calculating the output sample in 72 steps and then adding the 72 results together successively.

This can be done in the way shown in Fig. 14.8(d). We now have delay circuits (shift registers) that can contain 72 input bits. The number of taps has been reduced from N to $N/72$. With a '72-counter' a cycle of 72 subsets of ROM addresses is counted. The counter uses a 7-bit address for this. The total number of ROM addresses is $72 \times 2^{N/72}$. An adder with a feedback loop and a switch that substitutes a connection to the output instead of the feedback once every $72T$ provide the output samples at the reduced sampling rate. But even with the arrangement in Fig. 14.8(d) we have still not arrived at our final structure.

We have already mentioned that the transversal digital filter can have a linear phase characteristic – a feature that gives it an advantage over a conventional filter in analog technology. It can be shown that a transversal filter with a linear phase characteristic and a low-pass amplitude characteristic has a symmetrical impulse response (Fig. 14.9). This implies that the row of coefficients $c_0, c_1, \ldots, c_{N-1}$ is symmetrical: the first coefficient and the last are identical, the second and the next-to-last are identical, and so on. Consequently the number of coefficients with a different value is halved. We take advantage of this feature to obtain a further improvement in the design of our filter. We do this by 'folding the filter in two'. This is illustrated in Fig. 14.8(e), which also shows the division into sub-ROMs in accordance with Fig. 14.8(c) and the decimation as in Fig. 14.8(d). The two input bits, which always have to be multiplied by the same coefficient, therefore come close together and serve as address bits for the same sub-ROM.

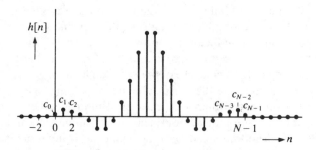

FIGURE 14.9 Example of a symmetrical impulse response $h[n]$ of a digital lowpass filter with a linear phase characteristic. Since the successive coefficients of a transversal filter correspond exactly to the successive sample values of the impulse response, the coefficients are also symmetrical, i.e.
$$c_i = c_{N-1-i} \text{ for } i = 0, 1, 2, \ldots, N-1.$$

Halfway through the filter, i.e. 'on the fold', a special step has to be taken. The groups of 72 bits have to reverse direction at this point to preserve the symmetry of the operation. They are therefore entered into a shift register and extracted backwards ('last in, first out'), like a train reversing out of a terminus. In the final version of our filter we use eight sub-ROMs, which together with four shift registers each with a delay of $72T$, and a few other subsidiary circuits, form a unit or 'filter cell'. Seven of these filter cells are identical, and the eighth is slightly different because of the 'last in, first out' effect. We shall now describe one filter cell in rather more detail.

14.4.2 The filter cell

The operations in a filter cell (Fig. 14.10) involve four input bits b_1, b_2, b_3, b_4, each of which may represent the value $+1$ or -1. There can therefore be a total of $2^4 = 16$ different combinations of bits. The four bits address the ROM, which is required to give the output signal $r = c_1(b_1 + b_2) + c_2(b_3 + b_4)$; c_1 and c_2 are two filter coefficients from a linear-phase filter. The remarkable thing here is that there are only nine possible different results for r. If we look at its absolute value, we are concerned with only four numbers, different from zero, which must be available to sufficient accuracy and must therefore be present in the memory. These have the values $|2c_1|$, $|2c_2|$, $|2c_1 + 2c_2|$ and $|2c_1 - 2c_2|$. In this way valuable memory space can be saved. Each of the numbers still has to be given the correct sign, and it is also possible for the result to be zero. This can all be taken as an extra multiplication by $g = -1, 0$ or 1. The value of g is separately derived from the four bits b_1, b_2, b_3, b_4 by a logic circuit.

As we have seen in Fig. 14.8(d), our decimating filter uses a 72-counter that selects one of the 72 subsets in the ROM by means of a 7-bit address. Altogether the ROM of each

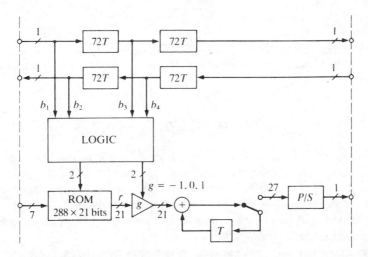

FIGURE 14.10 More detailed diagram of a filter cell. Although the four bits b_1, b_2, b_3, b_4 can form sixteen different combinations, the corresponding ROM contents r can assume only nine different values. One of these is zero and the others are in pairs differing only in sign. For every 4-bit combination the LOGIC circuit therefore calculates a 2-bit address for the ROM plus a multiplication constant $g = -1, 0$ or 1. *P/S* parallel/series converter.

filter cell must therefore contain $72 \times 4 = 288$ numbers r. Each number r in the ROM consists of a word of 21 bits. The ROM of the filter cell therefore has a capacity of $288 \times 21 = 6048$ bits. In the calculation of each output sample 72 of the 21-bit words are always added together in an adder circuit with a feedback loop, which, in this case, makes an extra addition width of 6 bits necessary to prevent overflow. The adder is thus 27 bits wide. After every period of $72T$ the total is passed in parallel to a shift register and the adder is reset to zero. The same signal also resets the memory counter to its initial state, and a new cycle starts. Meanwhile the output register is read out serially.

The same thing happens in the other cells of the filter and the results are added up two by two (Fig. 14.11). Because of the serial read-out serial adders can be used, which require relatively little hardware. In three steps of addition a final result is thus obtained, which is then rounded off to 16 bits.

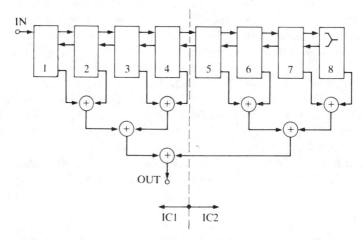

FIGURE 14.11 Our decimating filter consists of eight filter cells arranged on two chips IC1 and IC2. The outputs of the filter cells are added in pairs to give a final result OUT.

14.5 USE OF THE COMPUTER FOR DESIGN AND SIMULATION

After the choice of the filter structure – in our case the transversal filter – an important task is to calculate the coefficients. Computer programs for this are available.[104] We have made use of these and we have also written some supplementary programs for showing the impulse response of a filter that has been calculated and for determining the frequency response. A special program does the same thing for decimating filters, with choice of value for the decimation factor.

The resulting filter-software package would not have been complete without a program that calculates the effect of the finite word length of the coefficients on the filter characteristics. We have added such a program; the number of bits in which the coefficients are expressed can be input as required.

Figure 14.12 shows the amplitude characteristic of the filter F_3 in Fig. 14.3 as calculated

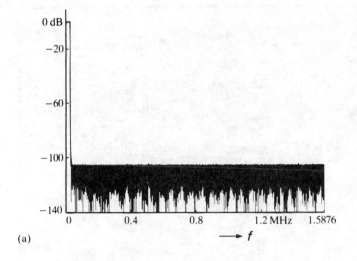

(a)

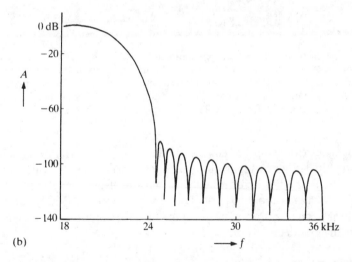

(b)

FIGURE 14.12 Computer-calculated amplitude characteristic of the filter F_3 from Fig. 14.3, with 2304 coefficients and the coefficient values rounded off to 21 bits, (a) between 0 Hz and 1.5876 MHz, and (b) on an expanded scale between 18 kHz and 36 kHz (see also Fig. 14.7(b)).

by the computer; Fig. 14.12(a) shows the entire frequency range up to half the sampling rate, and Fig. 14.12(b) shows the transition band and the closely adjacent range with an expanded frequency scale. The 2304 coefficient values are rounded off to 21 bits here.

14.6 INTEGRATION OF THE COMPLETE FILTER ON TWO CHIPS

Experimental ICs of the decimating filter were made. Because of its size, the filter divided between two integrated circuits (chips). Each chip contains four of the eight filter

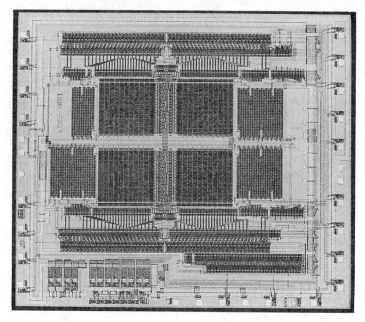

FIGURE 14.13 Experimental chip of the decimating filter in NMOS technology. The chip shown here contains the second half of the filter (the second four filter cells, including the reversal cell). The four central fields are the ROMs in which the required combinations of filter coefficients are stored. The first half is contained on another chip; each chip has an area of 50 mm². They were made by the 'N 700' process. The minimum detail size is 4 μm.

cells. The signal goes from chip 1 to chip 2, changes direction, goes through chip 2 again and then on to chip 1. Figure 14.13 shows a photograph of chip 2, which contains the reversal stage. The four fields in the middle are four ROMs.

The area of each chip is about 50 mm². They are produced in NMOS technology with smallest details of 4 μm in a process in which the thickness of the oxide beneath the gate electrode of the transistors is 70 nm.

14.7 SUMMARY

For analog-to-digital conversion of audio signals with no loss of hi-fi quality, it can be advantageous first to use sigma-delta modulation to convert the audio signal into a 1-bit signal with a high sampling rate, and then to convert the result into the required 16-bit signal by means of a 'decimating' transversal digital filter. The requirements for the analog prefiltering of the audio signal then become far less stringent and the problems are mainly shifted to the digital domain. A large digital filter that meets the requirements has been integrated on two chips; it has 2304 coefficients and decreases the sampling rate from 3.1752 MHz to 44.1 kHz. Computer programs were used for calculating the filter coefficients, the resulting frequency response and the irregularities that occur on rounding off the coefficients to a limited number of bits (21).

15

Developments in integrated digital signal processors, and the PCB 5010

J. L. van Meerbergen[†]

Since computers first appeared on the scene, more than 40 years ago, their dimensions and energy consumption have rapidly diminished, while their performance and capabilities have steadily increased. Even the vast numbers of calculations per second required for processing digital signals can nowadays be performed on a very small number of chips, or sometimes just one. This is done with a special type of 'computer': the digital signal processor. Some time ago the author of this chapter presented a paper[105] on the developments in this field. At Philips, one recent result of these developments is the PCB 5010 integrated digital signal processor. A general picture of these developments and the PCB 5010 itself are the main topics of this chapter.

15.1 INTRODUCTION

For many people, the word 'computer' used to conjure up the standard image of a number of metal cabinets set up in a special room and usually fitted with the well-known magnetic tape units. This picture is changing, however, now that so many of us have home computers and personal computers.

And there seems to be no end to the developments: many users of a Compact Disc player, for example, will not always realize that it contains a special kind of computer for processing the signals. In the near future rather similar devices will also be found in telephones, telephone exchanges, television receivers and all kinds of audio equipment. This will mean the large-scale use of special integrated circuits ('application-specific integrated circuits' or ASICs) designed for digital signal processing. These will include chips designed for one particular application (as in the Compact Disc player, for example)

†Philips Research Laboratories, Eindhoven. This chapter is reprinted from *Philips Tech. Rev.* **44** No. 1: 1–14, Mar. 1988.

as well as more generally applicable chips, usually known as *digital signal processors* (Fig. 15.1).

The first part of this chapter gives the general picture of the gradual evolution of the digital signal processor. This is followed by a discussion of the PCB 5010 digital signal processor now available from Philips, with attention to its architecture and also to programming facilities and supporting accessories. This signal processor is primarily intended for a wide range of applications in telecommunications, audio and speech-processing.

FIGURE 15.1 A digital signal processor is a 'computer on a chip' designed for processing digital signals. These chips contain many tens of thousands of transistors and are designed so that several million typical signal-processing instructions can be executed per second.

15.2 THE ARCHITECTURE OF COMPUTERS

Every computing process can be broken down into four elementary operations:

- The input and output of data.
- Storage in a memory (of data, intermediate results, final results, and computation procedures or algorithms).
- The execution of the computations.
- The control of the entire process.

This functional division has been of great importance in the design of computers from the earliest days. The four different functions have been performed more or less independently, by:

- Input and output (I/O) devices
- A memory
- An arithmetic unit
- A control unit

The interconnection of the individual devices or units in a specific pattern determined the ultimate architecture of the computer. The combination of the arithmetic unit and the

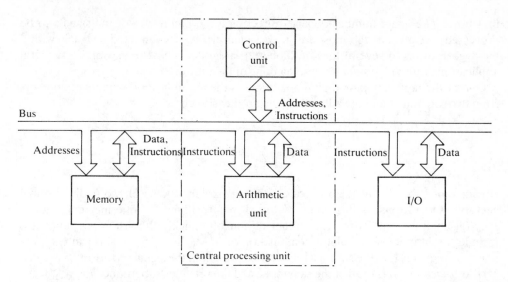

FIGURE 15.2 One of the oldest and best-established forms of computer architecture is the von Neumann architecture shown here. All communication between the main parts of the computer (memory, arithmetic unit, input and output devices I/O, control unit) goes via a single common route: the bus.

control unit is often called the central processing unit. Originally a single common provision was made for the interconnections: a *bus*. All communication between the different parts of the computer was made via the bus in successive steps. Data, memory addresses and instructions were exchanged in turn via this 'main route'. This is generally referred to as a *von Neumann* architecture (Fig. 15.2). It was the standard computer design for many years, first for the large 'main-frame' computers, and later for mini-computers and microcomputers too. In microcomputers the central processing unit consisted of only a few chips, and was soon to consist of only one – the microprocessor.

Signal processing was performed on main-frame computers right from the start, and efforts were soon made to use microprocessors in the same way. Hopes were expressed that it would be possible to process signals 'in real time', and ultimately on a single chip, so that all the advantages of digital signal processing would become available for countless applications.

It was soon realized, however, that in some ways the ordinary microprocessor was not so suitable for this purpose and that it was necessary to adapt the design more specifically to signal processing. The chips that resulted from this development are referred to as digital signal processors. Almost without exception, the first types commercially available had the original von Neumann architecture.

15.3 DIGITAL SIGNAL PROCESSORS

So just how do digital signal processors differ from microprocessors? In the first place, a signal processor has to be capable of performing very large numbers of operations per unit time. The exact number depends directly on the bandwidth of the signals to be processed. At present the most advanced signal processors can handle signals with a bandwidth of up to several tens of kilohertz; integrated signal processors for general applications with video signals are still a thing of the future.

One of the most common digital operations is the multiplication of pairs of numbers from two sequences and the addition of the products, such as

$$y = \sum_{i=1}^{N} a_i x_i$$

(Incidentally, this is equivalent to calculating the scalar product of two N-dimensional vectors.) Operations of this type are found in algorithms for filtering, correlation, spectral analysis, etc. To perform them reasonably quickly the arithmetic unit must include a multiplier/accumulator combination or MAC (Fig. 15.3). This can compute one subproduct $a_i x_i$ in the smallest unit of time present in the signal processor (one *clock cycle* or *machine cycle*) and at the same time add the previous subproduct $a_{i-1} x_{i-1}$ to the sum already calculated from all the earlier subproducts. The presence of a MAC is in itself a distinctive difference as compared with the original microprocessors. Much more radical changes in the architecture are required, however, if we are to obtain signal processors that are fast enough; this aspect will be dealt with at some length in the following sections.

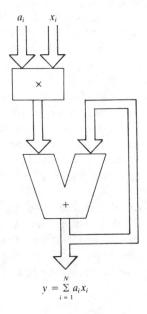

FIGURE 15.3 In digital signal processing it is often necessary to sum N products of the type $a_i x_i$. This takes a disproportionate amount of time in the von Neumann architecture shown in Fig. 15.2. The situation can be improved by providing the arithmetic unit with a multiplier/accumulator combination as shown here.

The maximum processing rate of a signal processor is also affected by the number of chips used; as the number of chips diminishes less time will be lost in transferring the signals. As well as an arithmetic unit and a control unit, signal processors must therefore also contain memory facilities. These consist of a ROM (Read-Only Memory) for storing unchanging quantities, such as the program and constant values, and a RAM (Random-Access Memory) for storing intermediate and final results. The versatility of application is considerably increased if these on-chip memories can be supplemented by external storage as required.

Since the input signals will often be analog in origin and signal processors are digital devices, analog/digital (A/D) conversion will frequently be encountered as a preliminary operation and D/A conversion as a final operation. In one type of signal processor (the Intel 2920) the converters are present on the chip as part of the input/output devices. In general, however, it seems preferable (at least in the present state of the technology) to add the converters to the signal processor as separate components. The specific requirements of individual applications, which can vary considerably, are then more easily taken into account.

15.4 THE ARCHITECTURE OF SIGNAL PROCESSORS

The basic von Neumann architecture has one serious disadvantage: everything happens *consecutively*. Before any one operation is completed, many steps (often very many) have to be completed. For example:

- The location (the 'address') where the next instruction is stored in the memory is determined (e.g. by adding 1 to the previous address).
- The instruction is read from the memory and transferred to the control unit.
- The instruction is interpreted ('decoded').
- The address of data necessary for executing the instruction is sent to the memory.
- The data are sent from the memory to the arithmetic unit.
- The arithmetic unit then executes the instruction.
- The result of the instruction is stored in the memory.

Then the complete cycle (the *instruction cycle*), which clearly requires more than one *machine cycle*, may be repeated.

The component that restricts the speed most of all is the one most characteristic of the von Neumann structure – the common signal bus, which handles every exchange of information between the various parts of the processor. It acts as a bottleneck. For higher speeds it is necessary to change to a 'non-von Neumann' architecture (sometimes just called a 'non-von' architecture).

The most common alternative is the Harvard architecture, in which data and instructions are stored in separate memories and which therefore has to have separate connections for data and control information. This gives the architecture shown in Fig. 15.4, where the signal processor is divided into two parts, called the controller and the data path. The exact nature of the connections *between* these two parts is mostly of less importance.

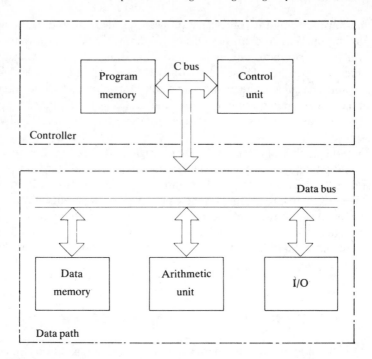

FIGURE 15.4 A modern alternative to the von Neumann architecture is the Harvard architecture shown here. This has separate memories for the program and the data, and separate connections for the control information and the data. These connections are called the control bus (or C bus) and the data bus. The total architecture can now be divided into two parts, called the controller and the data path.

15.5 THE ARCHITECTURE OF THE PCB 5010

The design of our PCB 5010 signal processor is based on the Harvard architecture of Fig. 15.4. It has been modified in some essential aspects, however, to obtain sufficient versatility and signal-processing capacity.

One of the main modifications is the duplication of the data bus, resulting in an X bus and a Y bus, each 16 bits wide.[13] The reason for this is that most signal-processing operations have two operands and can only be performed efficiently with a duplicated data bus. There is also the advantage that complex numbers can be more easily manipulated with a duplicated bus. As a direct consequence of this duplication, the RAM for the data is divided into two parts. A separate ROM is also available for data.

As we have seen, every signal processor has to have a multiplier/accumulator combination MAC to reach the speed required in the many 'vector-like' operations that arise. High-speed processing is also required for other types of operation, such as logic AND, NOT, OR etc., for operations on absolute values and for operations on individual bits ('masks'). The PCB 5010 has a separate arithmetic and logic unit ALU for these activities.

Finally, to facilitate data input and output our signal processor has two serial input devices and two serial output devices, as well as a combined parallel input/output device. This brings us to the block diagram[106,107,108] of the PCB 5010 in Fig. 15.5.

A computer drawing of the actual plan of the PCB 5010 is shown in Fig. 15.6; the numbers 1 to 5 refer to the main components in the previous figure. There are some 135 000 transistors in all in this drawing, yet this IC only occupies an area of 61 mm^2 when fabricated in 1.5-μm CMOS technology.

In the following sections of this chapter we shall take a closer look at the structure of the controller and of the two principal components of the data path: the data memory and the arithmetic unit. We shall see that in each of these components every effort is made to decentralize as many activities as possible to prevent bottlenecks, e.g. by providing a separate address-computation unit for each memory.

Besides the architecture, various other aspects are important in assessing the performance of a signal processor. They include:

- The time (in seconds and in number of machine cycles) required for executing an instruction.
- The number of (sub)operations that can be performed simultaneously.
- Facilities for interaction with the outside world.
- The possible degree of overlap in time of executions of successive instructions in different parts of the signal processor ('pipelining').

These all depend greatly on the way in which the signal processor can be programmed (the 'microcode'). We shall return to this point later.

The ultimate critical factor in comparing signal processors is the time required for performing a number of standard operations at a given accuracy. Typical operations might be a 128-point FFT ('fast Fourier transform'), a complex multiplication or a particular elementary filtering operation. These operations are often called 'bench-marks'.

15.6 THE ARITHMETIC UNIT

15.6.1 The multiplier/accumulator combination MAC

Figure 15.7 shows a block diagram of the multiplier/accumulator with its various supporting devices. The operation of multiplier MPY is based on the 'modified Booth's algorithm'[109] and in one machine cycle it can compute the product of two 16-bit words P and Q. The product is stored temporarily as a 32-bit word in the product register PR. At the same time the accumulator ACC can add the previous contents of PR to the contents of the accumulator register ACR. The contents of ACR are then multiplied, under the control of the block S/SD, either by 1, -1, -2^{-15}, 2^{-15} or by 0. The output of the accumulator has a width of 40 bits, so that even if large numbers of products are added together, there are no overflow errors.

The 'barrel shifter', BS, the corresponding barrel-shift register BSR and the format adjuster FA derive one group or two groups of 16 bits from the 40-bit contents of ACR to form the output signal of the MAC unit.

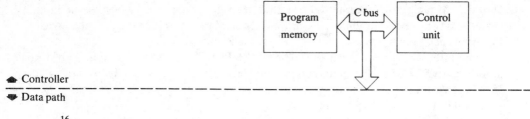

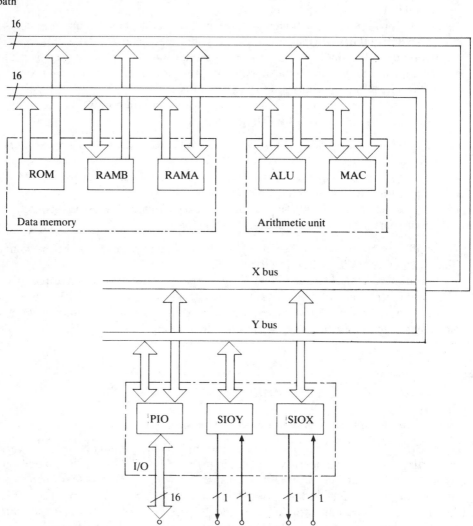

FIGURE 15.5 The PCB 5010 digital signal processor has a Harvard architecture with two independent 16-bit data buses: the X bus and the Y bus. The data memory consists of three parts: a ROM (read-only memory) for unchanging data and two RAMs (random-access memories) RAMA and RAMB. The arithmetic unit consists of two parts: a multiplier/accumulator combination MAC and an arithmetic and logic unit ALU. The input and output devices I/O consist of a parallel unit PIO, which has a 16-bit connection to the outside world, and two serial units SIOX and SIOY, each with a 1-bit input and a 1-bit output.

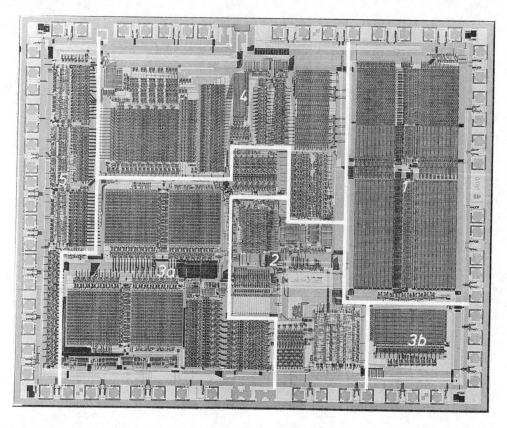

FIGURE 15.6 Computer drawing of the integrated digital signal processor PCB 5010. The most important components are: *1* program memory; *2* control unit; *3a*, *3b* data memory; *4* arithmetic unit; *5* input and output devices I/O. The complete circuit contains some 135 000 transistors and is fabricated in 1.5-μm CMOS technology. It occupies an area of 61 mm^2.

The operands P and Q presented to MPY may come directly from the X bus and the Y bus via the X and Y input selectors ILX and ILY. It is also possible to select the previous X and Y information, which is automatically stored in the latches MXL and MYL. In addition, P can take the value −1, and Q the logically inverted value of the current Y information.

Certain special occurrences, such as overflow in ACC, are reported directly to the control unit by means of a 'flag' or 'flag signal'.

15.6.2 Multiplication with greater precision

Blocks S/SD, BS, BSR and FA are also important because they permit calculations to be made at a greater precision than 16 bits, though at the expense of more processing time. For example, the product of a 46-bit operand and a 31-bit operand can be computed in 7

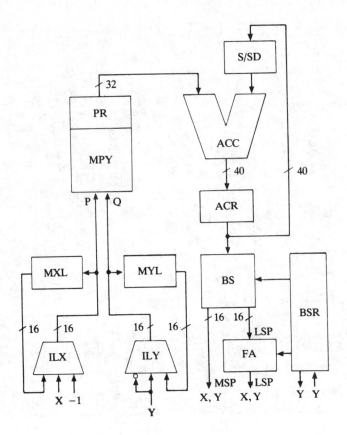

FIGURE 15.7 Block diagram of the multiplier/accumulator combination MAC of the PCB 5010. The number of operations that can be performed in each machine cycle has been greatly increased by the addition of extra registers and selectors. The operations required are specified by a 7-bit code via the C bus (not shown explicitly). The numbers beside the oblique strokes in the connections indicate how many bits are transferred in parallel. The significance of the designations is:

ACC	accumulator
ACR	accumulator register
BS	barrel shifter
BSR	barrel-shift register
FA	format adjuster
ILX	X-input selector
ILY	Y-input selector
LSP	least-significant part
MPY	multiplier
MSP	most-significant part
MXL	X latch
MYL	Y latch
P	operand 1
PR	product register
Q	operand 2
S/SD	sign/scale-down block
X	16-bit connection to the X bus
Y	16-bit connection to the Y bus

machine cycles at most. This is done in much the same way as multiplying two large numbers together conventionally: subproducts are determined first, then shifted appropriately and added. The length of the operands is not quite an integer multiple of 16, because one bit in each group of 16 is always reserved as a sign bit, and the length can only increase effectively by multiples of 15, thus: 16, 31, 46, ...

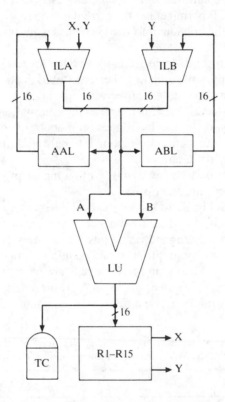

FIGURE 15.8 Block diagram of the arithmetic and logic unit ALU of the PCB 5010. As well as two selectors and two latches, this unit contains a file of 15 independent 16-bit registers. The operations required from ALU are specified in each machine cycle by a 7-bit code via the C bus. Also, 4 bits are reserved on the C bus for selecting one of the registers from the register file. The significance of the designations is:

A	operand 1
AAL	A latch
ABL	B latch
B	operand 2
ILA	A-input selector
ILB	B-input selector
LU	logic unit
R1–R15	register file
TC	trash can
X	16-bit connection to the X bus
Y	16-bit connection to the Y bus

15.6.3 The arithmetic and logic unit ALU

A block diagram of the arithmetic and logic unit ALU of the PCB 5010 is shown in Fig. 15.8. Grouped around the logic unit LU, which can perform operations on a single operand A or on two operands A and B, there are a number of registers and other supporting devices.

A total of 31 different arithmetic and logic operations can be executed, including addition, subtraction, absolute-value determination, AND, OR, NOT and shift operations. The selectors ILA and ILB permit either the current information on the X and Y buses to be used for A and B, or the previous information, which is always automatically stored in the latches AAL and ABL.

At the output of LU a set of 15 registers is available for temporary storage. During each machine cycle the present result of LU can be stored in one of these registers, and the contents of any other two registers can be read out via the X bus and the Y bus.

The arithmetic and logic unit ALU can send status information directly to the control unit of the signal processor by means of flags (e.g. about overflow or the sign of the computed result). If just this status information is to be stored, the rest of the computed result is consigned to the 'trash can' TC.

If required, ALU can perform computations to a higher precision than 16 bits, but again it will take longer for the processing.

The units MAC and ALU are to some extent complementary. MAC, for instance, is designed for processing 16-bit words (and hence for vector operations), while ALU is particularly suited for processing individual bits. There are seven types of instructions for MAC, all concerned with multiplication and addition, while ALU has 31, some very different from the others. Finally, in MAC there is always room for intermediate and final results with a maximum length of 40 bits; in ALU the length is always 16 bits, unless special arrangements are made. These differences are summarized in Table 15.1.

TABLE 15.1 Comparison of some complementary features of the multiplier/accumulator combination MAC and the arithmetic and logic unit ALU

MAC	ALU
Designed for vector operations (16-bit word level)	Designed for other operations (bit level)
7 slightly different instructions (all connected with adding and multiplying)	31 sometimes very different instructions (such as AND, OR, NOT, EXOR, shifting, addition, subtraction, incrementing)
Gives 40-bit intermediate result and 40-bit final result (can be extended)	Gives 16-bit result (can be extended)

15.7 THE DATA MEMORY

As already mentioned, three memories are provided for storing all the various data required during the operation of the signal processor. One memory with 512 locations, each of 16 bits, stores unchanging data, such as constants and filter coefficients, and is therefore a ROM. These data are entered into the memory once only, during manufacture.

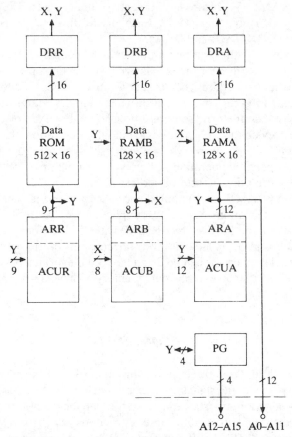

FIGURE 15.9 Block diagram of the three data memories of the PCB 5010. Each memory has its own address-computation unit and its own output-data register. The 12-bit address of one of the two RAMs is also available externally. If this address is extended by an extra 4 bits from a page register, an external data memory with 64k ($=2^{16}$) memory locations can be used. In each machine cycle for each of the address-computation units three bits of information are reserved on the C bus. The significance of the designations is:

ACUA	address-computation unit A
ACUB	address-computation unit B
ACUR	address-computation unit R
ARA	address register A
ARB	address register B
ARR	address register R
A0–A15	address bits 0–15
DRA	output-data register A
DRB	output-data register B
DRR	output-data register R
PG	page register
RAMA	random-access memory A
RAMB	random-access memory B
ROM	read-only memory
X	16-bit connection to the X bus
Y	16-bit connection to the Y bus

The other two are RAMs; these store intermediate or final results. They each have 128 memory locations of 16 bits (Fig. 15.9). Each of the three memories has its own output-data register (DRR, DRB and DRA).

To keep the capacity of the arithmetic unit and the X and Y buses of the signal processor free as far as possible for the actual signal-processing operations, each of the three memories has its own address-computation unit, or ACU, denoted by ACUR, ACUB and ACUA. During each machine cycle a new memory address can be computed in an ACU from the old address and some additional data. For example, a fixed number can be added to the previous address. If this is a 'modulo-N' addition, the data memory is scanned repeatedly in a fixed pattern, which is very useful for repetitive look-up of filter coefficients. Another possible operation is the reversal of the order of the address bits, which facilitates the calculation of fast Fourier transforms.

In both RAMA and RAMB the computed addresses are longer than is necessary for addressing 128 different locations. As ACUB supplies an 8-bit address, a later version of this signal processor can thus have a larger RAMB with 256 locations. ACUA supplies a 12-bit address, which is also available at the connector pins of the signal processor for addressing an external memory. It can be extended by 4 bits from a page register PG to form a 16-bit address, so that an external data memory with 64k ($=65\,536$) locations can be used.

15.8 THE CONTROLLER

The controller is the 'brain' that directs the entire operation of the signal processor. It determines what happens in the signal processor in each individual machine cycle. The program memory is of major importance here. It stores all the possible instructions for all the components of the signal processor. These instructions have a fixed length of 40 bits. The program memory of the PCB 5010 can contain 1024 such instructions (992 in a ROM and 32 in a RAM). The instructions stored in the ROM are entered permanently when the signal processor is manufactured; the instructions in the RAM can be changed at any time. During each machine cycle a 40-bit instruction is entered into the instruction register IR (Fig. 15.10), and the various sub-instructions are then distributed over the entire chip. A part of some instructions goes straight to the X bus and the Y bus ('load immediate', see the section on the microcode); the rest of the information in the instructions goes to the various parts of the signal processor via connections that we shall not consider further here (the C bus). At the same time the program counter PC determines the next address for the program memory.

For certain frequently occurring types of program special provisions have been made. A particular instruction may have to be repeated N times, and a separate instruction-repeat register (RPR) is available for such operations. While this is in use, the contents of the program counter PC remain unchanged. Other regularly occurring events include interruptions and the execution of subroutines. The current program is then stopped for a moment, and this is noted in a 'stack register', so that the program can be resumed later. The stack register has five levels, but if required it can be extended by a part of one of the data memories described in the previous section. The stack register enables other subroutines and interruptions to be processed inside a subroutine.

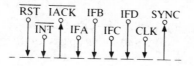

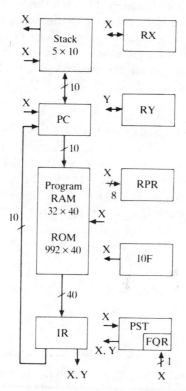

FIGURE 15.10 Block diagram of the controller of the PCB 5010, consisting of the program memory and the control unit. The program memory is mostly read-only, with a small random-access section. The memory address is supplied by the program counter. The contents of the program memory are written to the 40-bit instruction register, and from there the information is distributed via the C bus (not shown) and some of it possibly also via the X bus and the Y bus. The control unit consists mainly of a number of registers that serve a variety of purposes. The arrows at the top indicate that the control unit can exchange certain information with the outside world. The significance of the designation is:

CLK	clock	PST	processor-status register
FQR	mode bit (P mode/NP mode)	RAM	random-access memory
IACK	acknowledgement	ROM	read-only memory
IFA	user flag A	RPR	instruction-repeat register
IFB	user flag B	RST	reset signal
IFC	user flag C	RX	X register
IFD	user flag D	RY	Y register
INT	interrupt signal	STACK	stack register
IOF	input/output status and user flag register	SYNC	synchronization signal
		X	16-bit connection to the X bus
IR	instruction register	Y	16-bit connection to the Y bus
PC	program counter		

Since the controller represents the 'nerve centre' of the signal processor, it has many connections to the outside world and other parts of the chip. We have already encountered several examples, such as the capability for external interruption of a current program. Clock-synchronization signals and any reset signals also reach the controller from outside. Status information about the various components of the signal processor is always stored in the 16-bit processor-status register PST in the form of 1-bit flag signals. Similarly the register IOF contains input/output flags that indicate the status of the input and output circuits of the processor, and four user flags that originate outside the chip. Finally, the controller contains two registers, RX and RY, that save the signals on the X bus and the Y bus if there is an interruption.

15.9 GENERAL DIAGRAM

A block diagram giving a general picture of the component parts of the PCB 5010 signal processor discussed above is shown in Fig. 15.11. The numbers beside the inter-connections indicate the number of parallel bits. The symbols X and Y indicate a direct connection to the X bus and the Y bus. Some connections to the outside world are also indicated (the chip is in an encapsulation with 68 connector pins). The internal connections for control purposes (the C bus) are not shown, however.

FIGURE 15.11 General diagram of the PCB 5010, produced by combining the block diagrams from the four previous figures and adding the data buses and the input and output devices. The significance of the designations not given earlier is:

CIX, CIY	X, Y input clock
COX, COY	X, Y output clock
DIX, DIY	serial X, Y input
DOX, DOY	serial X, Y output
$\overline{\text{DS}}$	data strobe
D0–D15	parallel input/output
GND	ground
PI	parallel-input latch
PO	parallel-output latch
R/$\overline{\text{W}}$	read/write
SIOST	serial I/O control register
SIX, SIY	serial X, Y input latch
$\overline{\text{SIXEN}}$, $\overline{\text{SIYEN}}$	X, Y input enable
$\overline{\text{SIXRQ}}$, $\overline{\text{SIYRQ}}$	X, Y input request
SOX, SOY	serial X, Y output latch
$\overline{\text{SOXEN}}$, $\overline{\text{SOYEN}}$	X, Y output enable
$\overline{\text{SOXRQ}}$, $\overline{\text{SOYRQ}}$	X, Y output request
VDD	supply voltage
$\overline{\text{WAIT}}$	wait signal

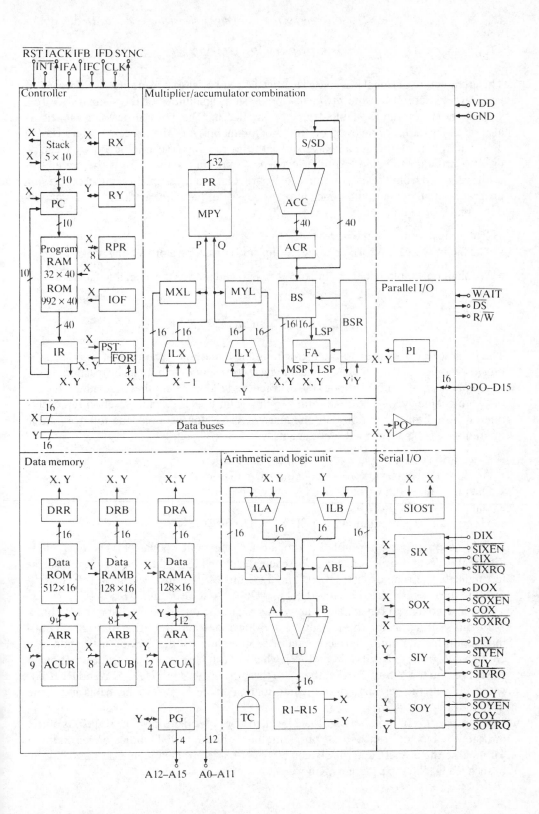

15.10 PROGRAMMING

The architecture of signal processors has been discussed at some length above, because it very largely determines the *theoretical* processing capabilities of the signal processor, such as the maximum available degree of parallelism. The *actual* processing capacity is also very dependent, however, on the 'programmability' of the processor. This can be deduced from the structure and diversity of the instructions (the 'microcode') that can be used to program the chip. Another extremely important point here is the amount of effort required from the user to translate a particular required function into a processing program (an algorithm) and hence into a sequence of basic instructions. This is mainly determined by two factors:

- The 'transparency' of the microcode.
- The facilities available for creating, testing and correcting a processing program.

We shall now look first at the microcode itself, and then at the available facilities.

15.11 THE MICROCODE

The PCB 5010 works with 40-bit instructions and a machine cycle time of 125 ns. In the *pipeline mode* (the P mode) each machine cycle corresponds to a single instruction; in the *non-pipeline mode* (the NP mode) the instructions follow one another at regular intervals of two machine cycles (we shall return to this point later). Each instruction represents one or more basic operations. In one machine cycle, for example, the following basic operations can be performed simultaneously:

- Calculation of the product of two 16-bit numbers.
- Addition of the previous product in the accumulator.
- Data transfer via the X bus.
- Data transfer via the Y bus.
- Three address computations in ACUA, ACUB or ACUR.

The basic operations available are not always the same, however. There are four different types of instruction, as shown schematically in Fig. 15.12. The two bits on the far left indicate the type of instruction. The 3×3 bits of each instruction on the far right indicate the operations to be carried out in ACUA, ACUB and ACUR.

The instructions of types 0 and 1 are very similar; the difference is that one type has a 7-bit sub-instruction for the arithmetic and logic unit and the other has a 7-bit sub-instruction for the multiplier/accumulator combination. Both types of instruction also contain two groups of 5 bits (SX and SY), which indicate the source of the information on the X bus and the Y bus. Finally, there are three groups of 4 bits (DX, DY and RFILE), which indicate the destinations of the information on the X bus, on the Y bus and at the output of LU.

An instruction of type 2 in Fig. 15.12 is called a branch operation. The three bits of BR indicate the type of branch in the program; the six bits of COND indicate the condition for making the branch, and the sixteen bits of NAP determine the new address that the branch leads to in the program memory.

Instruction type 0:

39	38	37 36 35 34 33	32 31	30 29 28 27 26	25 24 23 22 21	20 19 18 17	16 15 14 13	12 11 10 9	8 7 6	5 4 3	2 1 0
0	0	ALU	AOPS	SX	SY	DX	DY	RFILE	ACUA	ACUR	ACUB

Instruction type 1:

39	38	37 36 35	34 33 32 31	30 29 28 27 26	25 24 23 22 21	20 19 18 17	16 15 14 13	12 11 10 9	8 7 6	5 4 3	2 1 0
0	1	MPY	MOPS	SX	SY	DX	DY	RFILE	ACUA	ACUR	ACUB

Instruction type 2:

39	38	37 36 35 34 33 32 31 30 29 28 27 26 25 24 23 22	21 20 19	18 17 16 15 14 13	12 11 10 9	8 7 6	5 4 3	2 1 0
1	0	NAP	BR	COND	–	ACUA	ACUR	ACUB

Instruction type 3:

39	38	37	36 35 34 33 32 31 30 29 28 27 26 25 24 23 22 21	20 19 18 17	16 15 14 13	12 11 10 9	8 7 6	5 4 3	2 1 0
1	1	–	DATA	DX	DY	RFILE	ACUA	ACUR	ACUB

FIGURE 15.12 The PCB 5010 operates with instructions of four types. Each instruction consists of 40 bits, numbered here from 0 to 39. The bits numbered 38 and 39 indicate the type. The instructions are individually divided into segments of from 2 to 16 bits. Each segment represents a sub-instruction and is indicated by one of the letter combinations listed here:

ACUA	type of ACUA operation
ACUB	type of ACUB operation
ACUR	type of ACUR operation
ALU	type of ALU operation
AOPS	ALU operands
BR	type of branch operation
COND	branch condition
DATA	16-bit data word transmitted on X bus and Y bus
DX	destination on X bus
DY	destination on Y bus
MOPS	multiply operands
MPY	type of multiplier/accumulator operation
NAP	address of next instruction if COND is true
RFILE	destination in register file
SX	source on X bus
SY	source on Y bus

Instruction type 3 can be used to feed a group of 16 bits directly to the X bus and the Y bus as new data; this is called a 'load immediate' instruction.

Since the sub-instructions always have a fixed location, programming is greatly simplified. For example, the entire data stream of a program can be selected and then the corresponding address computations can be determined.

15.12 ACCESSORIES

Besides the PCB 5010, there is a very similar processor, the PCB 5011. This has no program memory, however, and no on-chip data ROM; the chip has connector pins for these, so that external memories can be connected to it. The PCB 5011 contains 70 000 transistors, while the PCB 5010 has about 135 000. Figure 15.13 is a photograph of the PCB 5011. The left-hand two-thirds of Fig. 15.6 can clearly be identified here. The large

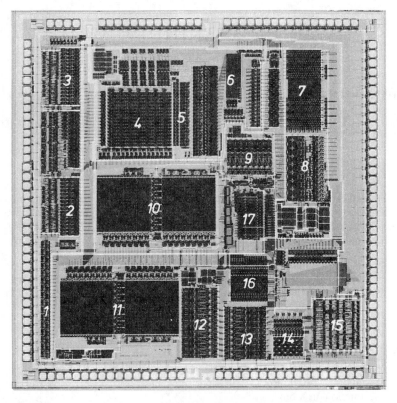

FIGURE 15.13 Photograph of the PCB 5011 digital signal processor. This processor is identical to the PCB 5010 in Fig. 15.6 and Fig. 15.11, except that it does not have the program memory and the data memory ROM. The connections for these go to connector pins, so that external memories can be used instead. The following components are indicated by numbers (the abbreviations are the same as in the previous figures): *1* PI and PO; *2* SIX and SOX; *3* SIY and SOY; *4* multiplier; *5* accumulator; *6* barrel shifter; *7* register file; *8* logic unit; *9* ACUB; *10* RAMB; *11* RAMA; *12* ACUA; *13* ACUR; *14* STACK and PC; *15, 16* various components of the control unit; *17* PST.

number of external connections is also immediately obvious: the PCB 5011 has 144, the PCB 5010 'only' 68. (The integrated circuit shown in Fig. 15.1, incidentally, is also of type PCB 5011.)

The PCB 5011 can be used in the design or development phase of a system (before any final decision has been made about the contents of the ROM in the PCB 5010), in applications where it is not worth making a special version of the PCB 5010 with ROMs specified by the user, and in applications where a very large program memory is required.

To facilitate the use of the PCB 5010/PCB 5011 special software has been written in the programming language PASCAL. This can be used on several widely used computers (VAX, IBM-PC). In the first place there is a *simulator program*, which can simulate the entire operation of a signal processor programmed for a specific application. There is also an *assembly program* that makes it unnecessary to specify the contents of the instructions bit by bit, requiring only symbolic indications that can be handled more readily (i.e. groups of letters – 'mnemonics' – that look like abbreviations). In this program these are automatically translated into bit sequences. The program also contains a *macro library*, in which frequently occurring algorithms, such as certain kinds of filtering and FFT operations, are stored in 'macrocode' as ready-to-use subroutines for the signal processor.

To test a system in which the PCB 5010/PCB 5011 is used under realistic conditions, e.g. in real time, the 'Stand-alone Debug System' (SDS) can be used. The SDS is an *emulator*, i.e. a device that functions in exactly the same way as a later definitive version of the PCB 5010, but also has a variety of facilities for interrupting a program being run in the signal processor at any moment and for investigating the internal status of the signal processor at that moment. Program modifications are also easily made.

Finally, there is a *prototype board* containing the PCB 5011 and all the external memories and circuits required for loading these memories. This board can be used, for example, for making a prototype of a system that will later include one or more PCB 5010 chips.

15.13 APPLICATIONS

The PCB 5010 has its greatest signal-processing capacity when it is used in the pipeline mode: in pipelining a new operation starts while the last part of the previous operation is still being performed at another location on the chip during the same machine cycle. This is done in product accumulation, for example. A single product accumulation really takes two machine cycles, but by pipelining the multiplication part and the addition part the effective duration is only about one machine cycle in long sequences of product accumulations. Programming in the pipeline mode is rather more difficult than in the non-pipeline mode. It is therefore possible to select one of the two modes and even to switch from one to the other within the same signal-processing program.

The PCB 5010 can be used in a variety of system configurations; in a minimum configuration only A/D and D/A converters have to be added. In many communication applications incoming and outgoing signals can even be processed effectively simultaneously (Fig. 15.14). It is also possible to use an external microprocessor to control the signal processor; this makes it even more versatile (Fig. 15.15). Also, since the PCB 5010

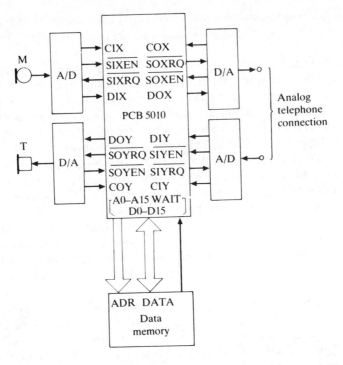

FIGURE 15.14 Example of the application of the PCB 5010 in telecommunications. The figure shows how the signal processor can be used in an analog telephone connection: after addition of a telephone T, a microphone M and A/D and D/A converters, a single PCB 5010 can perform all the processing operations required for transmitting and receiving. It can also be seen how the internal data memory RAMA can be replaced by an external memory.

chips have extensive input/output facilities, they can easily be combined to form a multiprocessor system suitable for more demanding applications (Fig. 15.16).

    Just how powerful the PCB 5010/PCB 5011 signal processors are can be seen most clearly from the time required to execute a number of characteristic processing operations. A summary of these is given in Table 15.2. Unless otherwise stated, the information in this Table relates to 16-bit quantities, for both signal samples and filter coefficients. In the examples relating to the fast Fourier transform (FFT) 'looped code' processing programs were used. This reduces the number of steps in the program to only about 40. It is also possible to halve the time required by using 'straight-line code', which avoids program loops. However, this requires about 100 times the number of program steps, and therefore about 100 times the memory capacity.[107] With the processing times given in Table 15.2, the processors can be used for many applications in the fields for which they were originally developed: telecommunications (especially in telephony),[110] audio and many kinds of speech-processing (such as speech coding, voice recognition and speaker identification).

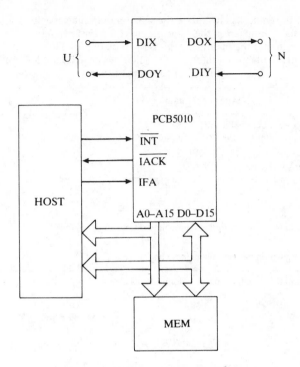

FIGURE 15.15 A PCB 5010 can be controlled from an external microprocessor HOST in combination with an external data memory MEM. The example shown here relates to a terminal in a telecommunication system; the PCB 5010 links the user U to the rest of the network N.

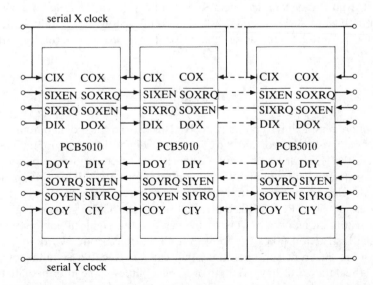

FIGURE 15.16 Even greater versatility in processing can be obtained by combining several PCB 5010 chips. One method of combining the chips is illustrated here.

TABLE 15.2 Some benchmarks illustrating the signal-processing capacity of the PCB 5010/PCB 5011. The right-hand column gives the time required for performing a number of frequently occurring operations; in filtering operations this is the time required for calculating one sample of the output signal. Unless otherwise stated, all the quantities (signal samples, coefficients) have the standard word length of 16 bits

Type of operation	Processing time (μs)
Non-recursive filtering (per filter coefficient)	0.125
Non-recursive filtering (64 filter coefficients; including input and output)	9.250
Recursive filtering (2nd-order section)	0.625
Recursive filtering (2nd-order section)[†]	1.875
Recursive filtering (2nd-order section; including input and output)[†]	3.375
Complex multiplication	1
128-point FFT	927
128-point FFT (including window function and input and output)	1100
256-point FFT	2112
256-point FFT (including window function and input and output)	2300

†Both signal samples and filter coefficients have double word length

15.14 TEAMWORK

The PCB 5010 is one of the results of the 'SIGMAPI' project. The team included staff from Philips Research Laboratories (Eindhoven), Valvo (Hamburg) and TeKaDe (Nuremberg). Besides the author, F. J. A. van Wijk and F. P. J. M. Welten also shared the responsibility for the development of the chip architecture. They received considerable support from the system designers R. J. Sluijter, P. Vary and K. Hellwig. Others who contributed were A. Delaruelle, J. A. Huisken, J. Stoter, W. Gubbels, J. Schmid, K. Rinner and J. Wittek (in the design), and K. J. E. van Eerdewijk (in the testing).

15.15 SUMMARY

Digital signal processors have gradually evolved away from the older computer concepts to become a separate class of large to very large digital integrated circuits. Modern versions have the 'Harvard architecture', which is characterized by separate arrangements for transfer and storage of data and control information. This also applies to the PCB 5010, developed primarily for applications in telecommunications, audio and speech-processing. The PCB 5010, fabricated in 1.5-μm CMOS technology, contains 135 000 transistors on an area of 61 mm^2 and can execute eight million instructions per second. Each instruction takes the form of a 40-bit 'microcode' word and specifies a maximum of six different sub-operations that can be executed simultaneously. As a general rule the data words have a length of 16 bits, but for some intermediate results 40 bits are available and if required, computations can be carried out with greater precision.

The PCB 5010 has three data memories (a 512×16-bit ROM and two 128×16-bit RAMs) and a program memory (1024×16 bits, mostly in a ROM). Various items of supporting software and hardware are available to facilitate the application of the PCB 5010.

Appendix I
Summarizing the Fourier integral

I.1 MOST IMPORTANT PROPERTIES

(a)	$x(t)$	$\circ\!\!-\!\!\!-\!\!\circ X(\omega); y(t) \circ\!\!-\!\!\!-\!\!\circ Y(\omega)$	Elementary notation for Fourier pairs
(b)	$ax(t) + by(t)$	$\circ\!\!-\!\!\!-\!\!\circ aX(\omega) + bY(\omega)$	Linearity
(c)†	$X(t)$	$\circ\!\!-\!\!\!-\!\!\circ 2\pi \cdot x(-\omega)$	Time/frequency symmetry
(d)	$x(kt)$	$\circ\!\!-\!\!\!-\!\!\circ \dfrac{1}{|k|}X(\omega/k)$	Time scaling
(e)	$\dfrac{1}{|k|}x(t/k)$	$\circ\!\!-\!\!\!-\!\!\circ X(k\omega)$	Frequency scaling
(f)	$x(t-\tau)$	$\circ\!\!-\!\!\!-\!\!\circ X(\omega)e^{-j\omega\tau}$	Time shift
(g)	$e^{jvt}x(t)$	$\circ\!\!-\!\!\!-\!\!\circ X(\omega - v)$	Frequency shift
(h)	$2\cos(vt)\cdot x(t)$	$\circ\!\!-\!\!\!-\!\!\circ X(\omega - v) + X(\omega + v)$	Double frequency shift
(i)	$\dfrac{d^n x(t)}{dt^n}$	$\circ\!\!-\!\!\!-\!\!\circ (j\omega)^n X(\omega)$	Differentiation with respect to time
(j)	$(-jt)^n x(t)$	$\circ\!\!-\!\!\!-\!\!\circ \dfrac{d^n X(\omega)}{d\omega^n}$	Differentiation with respect to frequency
(k)	$x(t) * y(t)$	$\circ\!\!-\!\!\!-\!\!\circ X(\omega)Y(\omega)$	Convolution in the time domain
(l)	$x(t)y(t)$	$\circ\!\!-\!\!\!-\!\!\circ \dfrac{1}{2\pi}X(\omega) * Y(\omega)$	Convolution in the frequency domain
(m)	$\displaystyle\int_{-\infty}^{\infty} |x(t)|^2 \, dt = \dfrac{1}{2\pi} \int_{-\infty}^{\infty} |X(\omega)|^2 \, d\omega$		Parseval's theorem

(n) For real $x(t)$: $X(-\omega) = X^*(\omega)$ Note: * = complex conjugate

(o) For real even $x(t)$, i.e. $x(t) = x(-t)$:
$\text{Im}\{X(\omega)\} = 0$ Note: Im = imaginary part

(p) For real odd $x(t)$, i.e. $x(t) = -x(-t)$:
$\text{Re}\{X(\omega)\} = 0$ Note: Re = real part

---

†Be careful! In this exceptional case the time function is indicated by an upper-case letter and the frequency function by a lower-case letter.

I.2 IMPORTANT FOURIER PAIRS

Time domain	Frequency domain

(a)

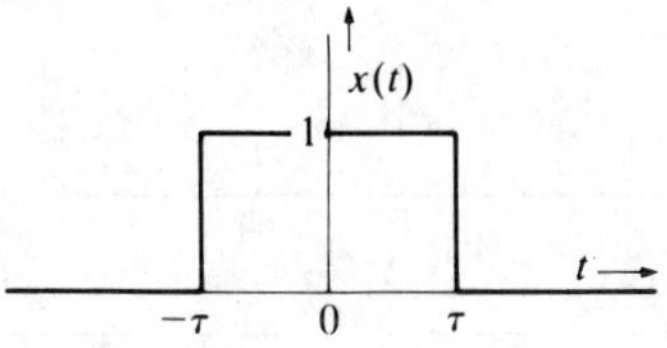

$$x(t) = \begin{cases} 1 & \text{for } |t| < \tau \\ 0 & \text{for } |t| > \tau \end{cases}$$

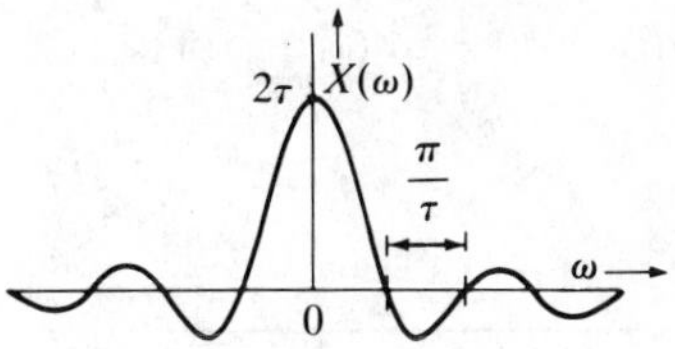

$$X(\omega) = \frac{2\sin(\omega\tau)}{\omega}$$

(b)

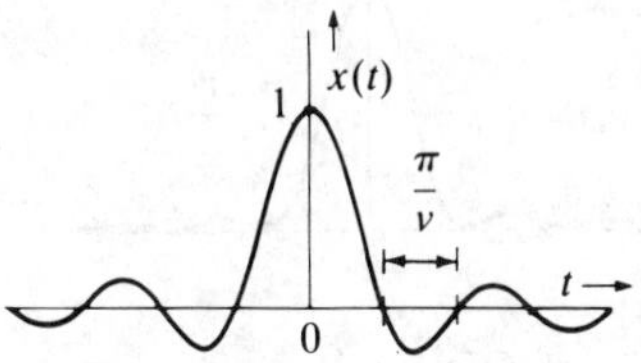

$$x(t) = \frac{\sin(vt)}{vt}$$

$$X(\omega) = \begin{cases} \pi/v & \text{for } |\omega| < v \\ 0 & \text{for } |\omega| > v \end{cases}$$

(c)

$$x(t) = \delta(t)$$

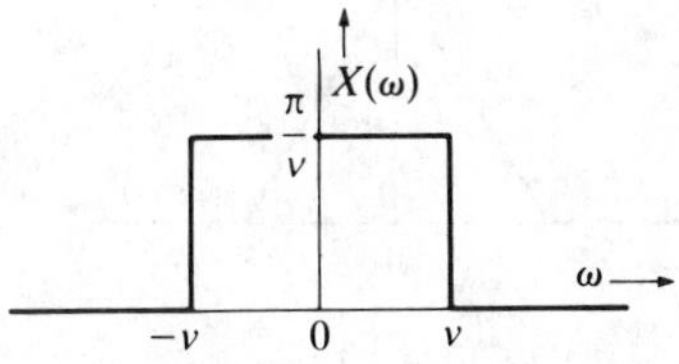

$$X(\omega) = 1$$

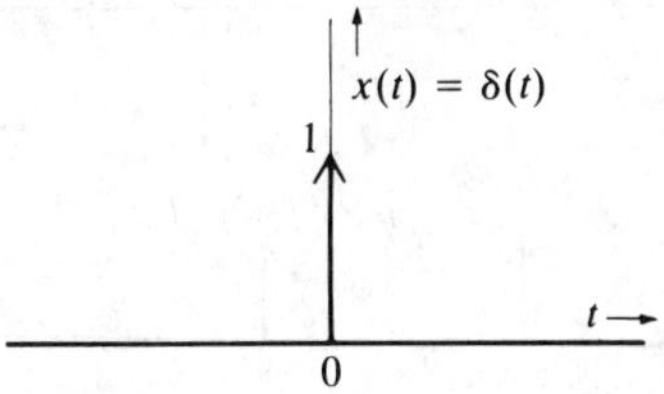

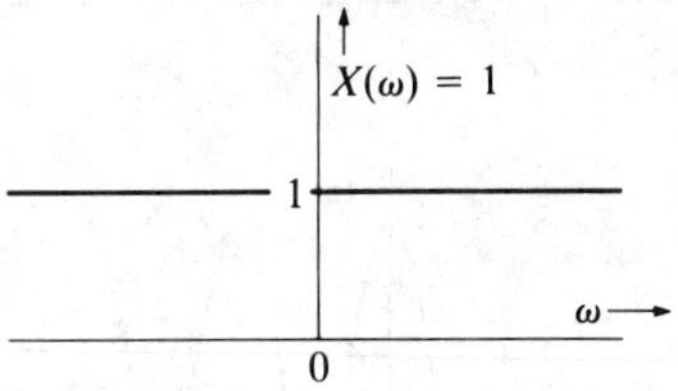

(d)

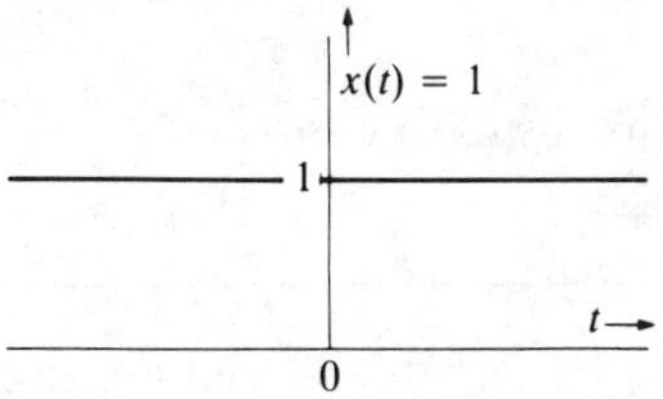

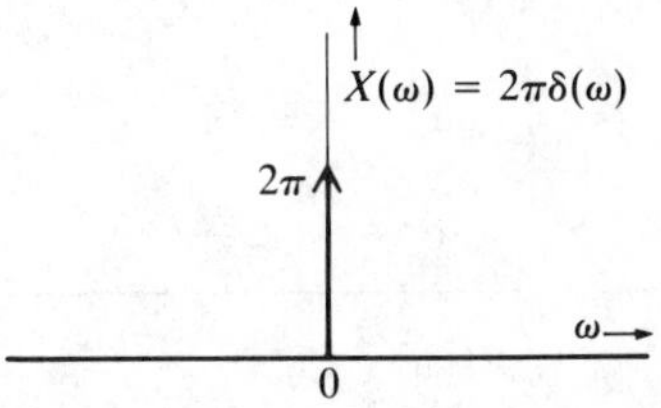

$$x(t) = 1$$

$$X(\omega) = 2\pi\delta(\omega)$$

(e)

$$x(t) = \delta(t - t_1)$$

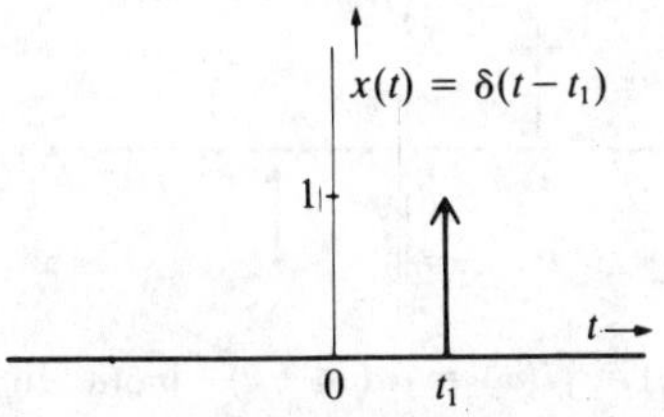

$$X(\omega) = e^{-j\omega t_1} = \cos(\omega t_1) - j\sin(\omega t_1)$$

	Time domain	Frequency domain

(f)

$$x(t) = e^{j\omega_1 t} = \cos(\omega_1 t) + j\sin(\omega_1 t)$$

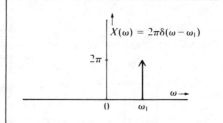

$$X(\omega) = 2\pi\delta(\omega - \omega_1)$$

(g)

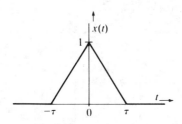

$$x(t) = \begin{cases} 1 - |t|/\tau & \text{for } |t| < \tau \\ 0 & \text{for } |t| \geq \tau \end{cases}$$

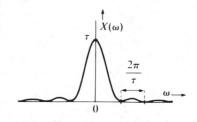

$$X(\omega) = \frac{4\sin^2(\omega\tau/2)}{\omega^2\tau}$$

(h)

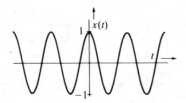

$$x(t) = \cos(vt)$$

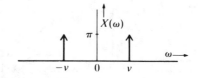

$$X(\omega) = \pi\delta(\omega + v) + \pi\delta(\omega - v)$$

(i)

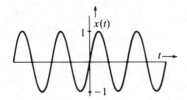

$$x(t) = \sin(vt)$$

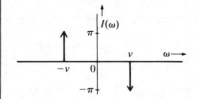

$$X(\omega) = jI(\omega) = j\pi\delta(\omega + v) - j\pi\delta(\omega - v)$$

Time domain	Frequency domain

(j)

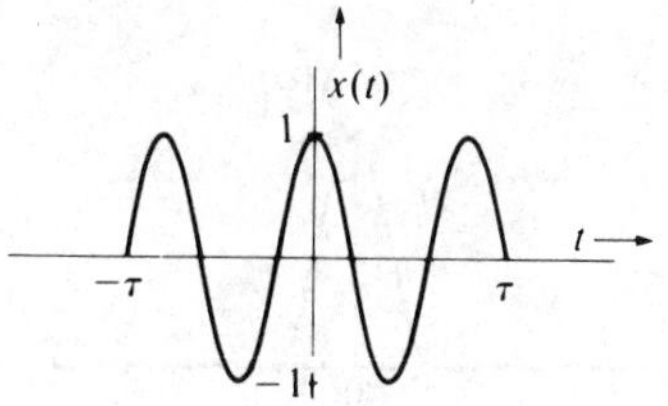

$$x(t) = \begin{cases} \cos(vt) & \text{for } |t| < \tau \\ 0 & \text{for } |t| \geq \tau \end{cases}$$

$$X(\omega) = \frac{\sin\{(\omega + v)\tau\}}{\omega + v} + \frac{\sin\{(\omega - v)\tau\}}{\omega + v}$$

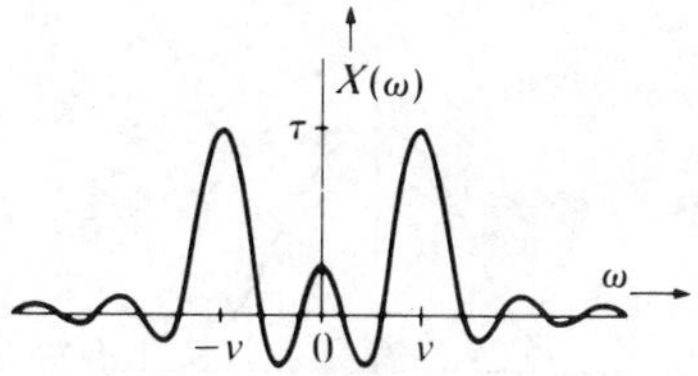

(k)

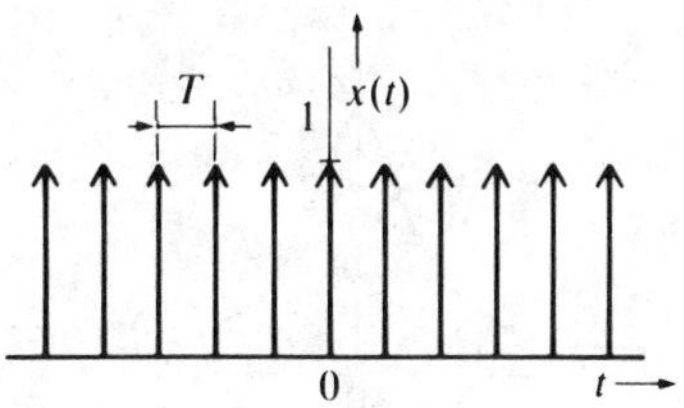

$$x(t) = \delta_T(t) = \sum_{n=-\infty}^{\infty} \delta(t - nT)$$

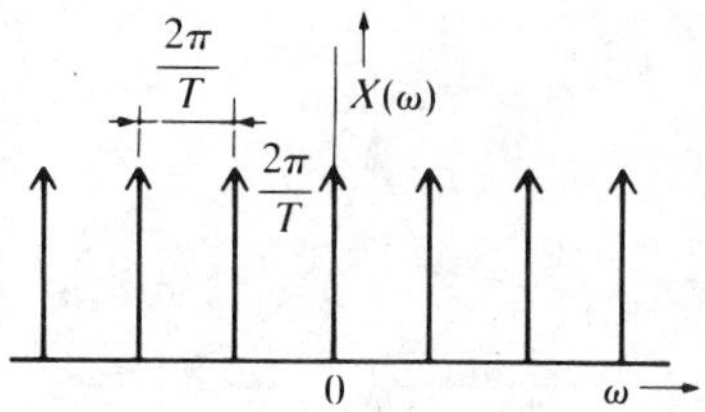

$$X(\omega) = \frac{2\pi}{T} \sum_{k=-\infty}^{\infty} \delta\left(\omega - \frac{2\pi k}{T}\right)$$

(l)

$$x(t) = \sqrt{\frac{\alpha}{\pi}} \cdot e^{-\alpha t^2}$$

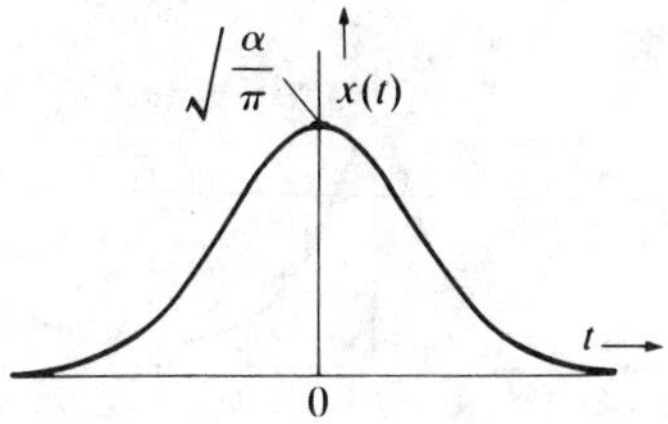

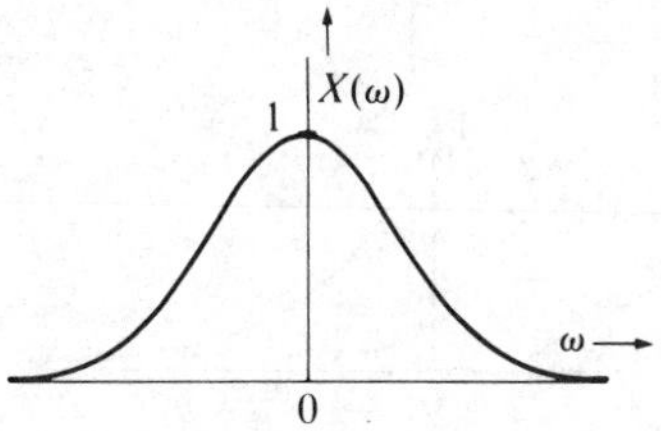

$$X(\omega) = e^{-\omega^2/4\alpha}$$

(m)

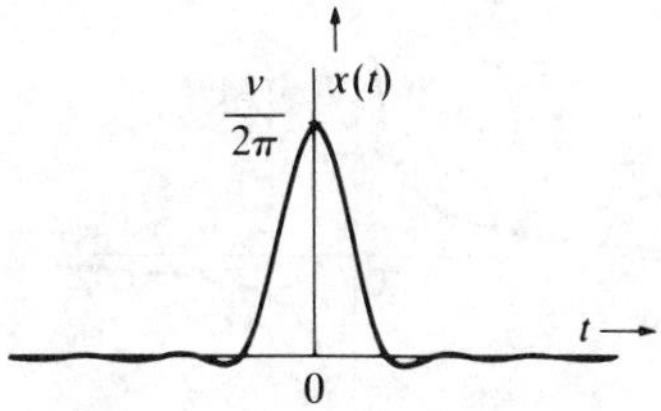

$$x(t) = \frac{\sin(vt)}{2\pi t} - \frac{\sin(vt)}{4\pi(t + \pi/v)} - \frac{\sin(vt)}{4\pi(t - \pi/v)}$$

$$X(\omega) = \begin{cases} \dfrac{1 + \cos(\pi\omega/v)}{2} & \text{for } |\omega| < v \\ 0 & \text{for } |\omega| \geq v \end{cases}$$

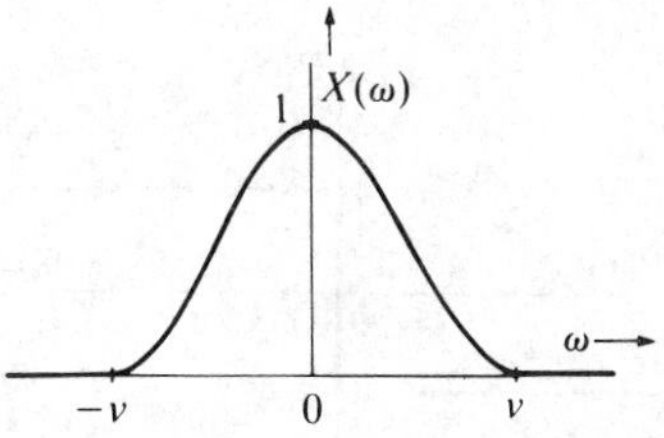

Time domain	Frequency domain
(n)	

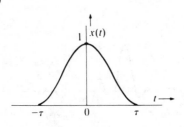

$$x(t) = \begin{cases} \dfrac{1 + \cos(\pi t/\tau)}{2} & \text{for } |t| < \tau \\ 0 & \text{for } |t| \ge \tau \end{cases}$$

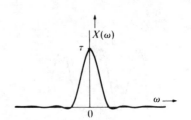

$$X(\omega) = \frac{\sin(\omega\tau)}{\omega} - \frac{\sin(\omega\tau)}{2(\omega + \pi/\tau)} - \frac{\sin(\omega\tau)}{2(\omega - \pi/\tau)}$$

(o)

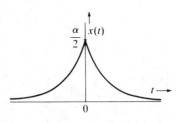

$$x(t) = \frac{\alpha}{2} e^{-\alpha|t|}$$

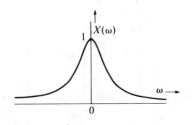

$$X(\omega) = \frac{\alpha^2}{\alpha^2 + \omega^2}$$

(p)

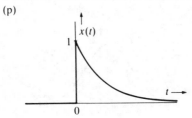

$$x(t) = \begin{cases} e^{-\alpha t} & \text{for } t > 0 \\ 0 & \text{for } t < 0 \end{cases}$$

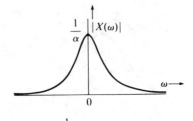

$$X(\omega) = \frac{1}{\alpha + j\omega}$$

(q)

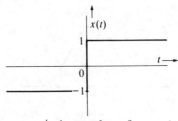

$$x(t) = \begin{cases} 1 & \text{for } t > 0 \\ -1 & \text{for } t < 0 \end{cases}$$

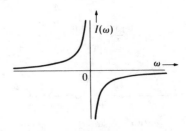

$$X(\omega) = j I(\omega) = -2j/\omega$$

Time domain	Frequency domain
(r) $$x(t) = \begin{cases} 1 & \text{for } t>0 \\ 0 & \text{for } t<0 \end{cases}$$	$$X(\omega) = \pi\delta(\omega) + \frac{1}{j\omega}$$
(s) $$x(t) = \begin{cases} \cos(vt) & \text{for } t>0 \\ 0 & \text{for } t<0 \end{cases}$$	$$X(\omega) = \frac{\pi}{2}\delta(\omega-v) + \frac{\pi}{2}\delta(\omega+v) + \frac{j\omega}{v^2-\omega^2}$$
(t) $$x(t) = \begin{cases} \sin(vt) & \text{for } t>0 \\ 0 & \text{for } t\leq0 \end{cases}$$	$$X(\omega) = \frac{j\pi}{2}\delta(\omega+v) - \frac{j\pi}{2}\delta(\omega-v) + \frac{v}{v^2-\omega^2}$$
(u) $$x(t) = \begin{cases} 1 & \text{for } t\geq\tau/2 \\ \frac{1}{2}+t/\tau & \text{for } |t|<\tau/2 \\ 0 & \text{for } t\leq-\tau/2 \end{cases}$$	$$X(\omega) = \frac{2\sin(\omega\tau/2)}{j\omega^2\tau} + \pi\delta(\omega)$$

Appendix II
Expansion in partial fractions

II.1 INTRODUCTION

In the inverse transformations that are necessary for working with Fourier transforms, Laplace transforms and z-transforms it can often be helpful to use the technique of expansion in partial fractions. The idea is to convert a ratio of two polynomials in the (arbitrary) complex variable x into a sum of simple fractions, i.e. to convert:

$$F(x) = \frac{N(x)}{D(x)} = \frac{b_N x^N + b_{N-1} x^{N-1} + b_{N-2} x^{N-2} + \ldots + b_1 x + b_0}{x^M + a_{M-1} x^{M-1} + a_{M-2} x^{M-2} + \ldots + a_1 x + a_0} \tag{II.1}$$

(with $N < M$) into, say:

$$F(x) = \frac{c_1}{x - x_1} + \frac{c_2}{x - x_2} + \ldots + \frac{c_i}{x - x_i} + \ldots + \frac{c_M}{x - x_M} \tag{II.2}$$

The most important reason for this is that the separate inverse transforms of the various terms of (II.2) are usually very simple. The conversion from (II.1) to (II.2) can always be made if the values $x_1, x_2, \ldots, x_i, \ldots, x_M$ are all different or, in other words, if $F(x)$ has M simple poles.

In principle there are two different methods of calculation:

1. Multiplying out.
2. Using residues.

First we should like to give a simple example.

Example 1

$$F(x) = \frac{N(x)}{D(x)} = \frac{3x - 10}{x^2 - 7x + 12} \tag{II.3}$$

Factorizing the denominator, we have $D(x) = (x - 3)(x - 4)$, so that $x_1 = 3$ and $x_2 = 4$. We want to find c_1 and c_2 such that:

$$\frac{3x - 10}{(x - 3)(x - 4)} = \frac{c_1}{x - 3} + \frac{c_2}{x - 4} \tag{II.4}$$

• *Method 1 (multiplying out)*
Multiplying out the right-hand side of (II.4) gives:

$$\frac{3x-10}{(x-3)(x-4)} = \frac{(c_1+c_2)x - (4c_1+3c_2)}{(x-3)(x-4)} \tag{II.5}$$

Since the numerators on the left-hand side and the right-hand side must be the same, we must have:

$$c_1+c_2 = 3 \quad \text{and} \quad 4c_1+3c_2 = 10 \tag{II.6a}$$

or $c_1 = 1$ and $c_2 = 2$, so that:

$$F(x) = \frac{3x-10}{x^2-7x+12} = \frac{1}{x-3} + \frac{2}{x-4} \tag{II.6b}$$

• *Method 2 (residues)*
We can multiply both sides of (II.4) by $(x-3)$. We then find:

$$\frac{3x-10}{x-4} = c_1 + \frac{(x-3)c_2}{x-4} \tag{II.7}$$

To find c_1 we put $x = 3$ in (II.7), so that:

$$c_1 = \frac{3 \cdot 3 - 10}{3-4} = 1 \tag{II.8}$$

To find c_2 we multiply both sides of (II.4) by $(x-4)$:

$$\frac{3x-10}{x-3} = \frac{(x-4)c_1}{x-3} + c_2 \tag{II.9}$$

Now by putting $x = 4$ we find $c_2 = 2$, and we see that this method also gives (II.6b).

In general Method 2 requires less calculation than Method 1. In formal terms we can say that the values obtained for c_i in (II.2) by using Method 2 are given by:

$$c_i = (x-x_i) \cdot F(x)|_{x=x_i} \quad \text{for } i = 1, 2, 3, \ldots, M \tag{II.10}$$

Mathematically, the value of c_i found from (II.10) for the complex function $F(x)$ at the pole $x = x_i$ is called *the residue of $F(x)$ at x_i*, hence the name of the method.

II.2 COMPLEX CONJUGATE POLES

The same methods can also be applied if $F(x)$ in (II.1) has complex conjugate poles.

Example 2

$$F(x) = \frac{N(x)}{D(x)} = \frac{2x^2+6x-26}{x(x^2+4x+13)} \tag{II.11}$$

Factorizing the denominator, we have $D(x) = x(x + 2 + 3j)(x + 2 - 3j)$, so that $x_1 = 0$, $x_2 = -2 - 3j$ and $x_3 = -2 + 3j$. We now want c_1, c_2 and c_3 such that:

$$\frac{2x^2 + 6x - 26}{x(x + 2 + 3j)(x + 2 - 3j)} = \frac{c_1}{x} + \frac{c_2}{x + 2 + 3j} + \frac{c_3}{x + 2 - 3j} \tag{II.12}$$

- *Method 1 (multiplying out)*

From (II.12) we find:

$$\frac{2x^2 + 6x - 26}{x(x + 2 + 3j)(x + 2 - 3j)} = \frac{(c_1 + c_2 + c_3)x^2 + \{4c_1 + (2 - 3j)c_2 + (2 + 3j)c_3\}x + 13c_1}{x(x + 2 + 3j)(x + 2 - 3j)} \tag{II.13a}$$

Therefore:

$$\begin{cases} c_1 + c_2 + c_3 = 2 \\ 4c_1 + (2 - 3j)c_2 + (2 + 3j)c_3 = 6 \\ 13c_1 = -26 \end{cases} \tag{II.13b}$$

From this it follows that:

$$c_1 = -2, \; c_2 = 2 + j \text{ and } c_3 = 2 - j \tag{II.13c}$$

- *Method 2 (residues)*

Applying (II.10) to the left-hand side of (II.12) gives:

$$c_1 = (x - 0)F(x)\Big|_{x = 0} = \frac{2 \cdot 0^2 + 6 \cdot 0 - 26}{(0 + 2 + 3j)(0 + 2 - 3j)} = -2 \tag{II.14a}$$

$$c_2 = (x + 2 + 3j)F(x)\Big|_{x = -2 - 3j} = \frac{2(-2 - 3j)^2 + 6(-2 - 3j) - 26}{(-2 - 3j)(-2 - 3j + 2 - 3j)} = 2 + j \tag{II.14b}$$

$$c_3 = (x + 2 - 3j)F(x)\Big|_{x = -2 + 3j} = \frac{2(-2 + 3j)^2 + 6(-2 + 3j) - 26}{(-2 + 3j)(-2 + 3j + 2 + 3j)} = 2 - j \tag{II.14c}$$

- *Comment.* Complex poles are often paired together in partial-fraction expansion; in this example we then find the result:

$$\frac{2x^2 + 6x - 26}{x(x^2 + 4x + 13)} = \frac{-2}{x} + \frac{4x + 14}{x^2 + 4x + 13} \tag{II.15}$$

Verify this!

II.3 DEGREE OF NUMERATOR AND DENOMINATOR

In the examples so far given for $F(x)$ the degree N of the numerator $N(x)$ has always been lower than the degree M of the denominator $D(x)$; see equation (II.1). If $N = M$ a constant term d_0 appears in the partial-fraction expansion; if $N > M$, terms of the type $d_1 x$, $d_2 x^2$, $d_3 x^3$, ..., $d_{N-M} x^{N-M}$ also appear. The values of d_0, d_1, d_2, ..., c_1, c_2, ... can again be found by using method 1. Method 2 can be used only to determine the values of c_1, c_2, ...; d_0, d_1, d_2, ... have to be found by dividing $N(x)$ into $D(x)$ until the degree of the numerator of the residual fraction is lower than the degree of the denominator.

Example 3

$$F(x) = \frac{3x^2 - 19x + 14}{x^2 - 7x + 10} \tag{II.16}$$

We now expand this as follows:

$$\frac{3x^2 - 19x + 14}{(x-2)(x-5)} = d_0 + \frac{c_1}{x-2} + \frac{c_2}{x-5} \tag{II.17}$$

- *Method 1 (multiplying out)*
We find in the usual way $d_0 = 3$, $c_1 = 4$ and $c_2 = -2$ (verify this).

- *Method 2 (residues)*
Dividing out the right-hand side of (II.16) gives:

$$\frac{3x^2 - 19x + 14}{x^2 - 7x + 10} = 3 + \frac{2x - 16}{x^2 - 7x + 10} \tag{II.18}$$

We apply the ordinary residue method to the new fraction in the right-hand side of (II.18). This gives $c_1 = 4$ and $c_2 = -2$.

Example 4

$$F(x) = \frac{2x^3 + 11x^2 + 19x + 2}{x^2 + 4x + 3} \tag{II.19}$$

We expand this as follows:

$$\frac{2x^3 + 11x^2 + 19x + 2}{(x+1)(x+3)} = d_0 + d_1 x + \frac{c_1}{x+1} + \frac{c_2}{x+3} \tag{II.20}$$

- *Method 1 (multiplying out)*
We now find $d_0 = 3$; $d_1 = 2$; $c_1 = -4$ and $c_2 = 5$.

- *Method 2 (residues)*
Dividing out the right-hand side of (II.19) gives:

$$\frac{2x^3 + 11x^2 + 19x + 2}{x^2 + 4x + 3} = 2x + \frac{3x^2 + 13x + 2}{x^2 + 4x + 3} = 2x + 3 + \frac{x - 7}{(x+1)(x+3)} \tag{II.21}$$

Using the ordinary residue method we now find $c_1 = -4$ and $c_2 = 5$.

II.4 MULTIPLE POLES

If $F(x)$ has multiple poles, we cannot make the conversion from (II.1) to (II.2). We can see this from the following example:

Example 5

$$F(x) = \frac{2x - 2}{(x - 3)(x - 5)^2} \tag{II.22}$$

The function $F(x)$ has a double pole at $x = 5$. It turns out that we can in fact expand as follows:

$$\frac{2x - 2}{(x - 3)(x - 5)^2} = \frac{c_1}{x - 3} + \frac{e_1}{x - 5} + \frac{e_2}{(x - 5)^2} \tag{II.23}$$

- *Method 1 (multiplying out)*

With this method we find fairly easily $c_1 = 1$, $e_1 = -1$ and $e_2 = 4$.

- *Method 2 (residues)*

Method 2 cannot be simply applied directly; it must first be extended for use with multiple poles. It can be shown that we find e_2 and e_1 from:

$$e_2 = (x - 5)^2 F(x)_{x = 5} \tag{II.24a}$$

and:

$$e_1 = \frac{\mathrm{d}}{\mathrm{d}x} \left\{ (x - 5)^2 F(x) \right\} \Bigg|_{x = 5} \tag{II.24b}$$

Calculating the values from the last two equations again gives $e_1 = -1$ and $e_2 = 4$ (verify this!).

In general, for a K-fold pole at $x = x_r$, we must introduce K terms of the form:

$$\frac{e_1}{x - x_r} + \frac{e_2}{(x - x_r)^2} + \ldots + \frac{e_i}{(x - x_r)^i} + \ldots + \frac{e_K}{(x - x_r)^K} \tag{II.25}$$

in the partial-fraction expansion. The value of e_i ($i = 1, 2, \ldots, K$) can be found from the relation

$$e_i = \frac{1}{(K - i)!} \cdot \frac{\mathrm{d}^{K-i}}{\mathrm{d}x^{K-i}} \left\{ (x - x_r)^K F(x) \right\} \Bigg|_{x = x_r} \tag{II.26}$$

Appendix III
The inverse z-transform

III.1 INTRODUCTION

We have already encountered the general equation for the inverse z-transform (IZT) in (4.80). We now recapitulate the equation[†] here:

$$x[n] = \frac{1}{2\pi j} \oint_C X(z) z^{n-1} dz \qquad \text{(IZT)} \qquad \text{(III.1)}$$

We do not want to *prove* the complete validity of this equation here, but we do want to show that it is *reasonable* to assume that it holds at least for all the discrete signals in which we are most interested in practice. These are the discrete signals for which an FTD exists or, in other words, for which the ZT always converges on the unit circle in the z-plane. For these signals the IFTD is given by (4.19):

$$x[n] = \frac{1}{2\pi} \int_{-\pi}^{\pi} X(e^{j\theta}) e^{jn\theta} d\theta \qquad \text{(IFTD)} \qquad \text{(III.2)}$$

If we now substitute $e^{j\theta} = z$ on the right-hand side, then the integration path of $-\pi \le \theta \le \pi$ becomes a counterclockwise path around the unit circle in the z-plane. We also have:

$$dz = j e^{j\theta} d\theta = j z d\theta \qquad \text{(III.3a)}$$

so that:

$$d\theta = \frac{dz}{jz} \qquad \text{(III.3b)}$$

Using (III.3b) and substituting $e^{j\theta} = z$, we can change (III.2) into (III.1), where the integration contour coincides with the unit circle.

[†]The contour C must lie entirely inside the region of convergence of the function $X(z)$. Very often we are interested in functions that are 'right-sided'; this is true, for example, for all *causal* functions $x[n]$. We must then be sure that the contour we take for C *encloses all of the poles* (see Table 4.2). If the pole furthest from the origin is at a distance $|a|$ from the origin, we can take any arbitrary circle for which $|z| > |a|$ for C, for example.

III.2 METHOD OF CALCULATION

For the actual calculation of the IZT from (III.1) we have to make use of the theory of complex functions. From this theory it is known that for a contour integral of the complex function $F(z)$ over a closed contour (counterclockwise) in the z-plane:

$$\frac{1}{2\pi j} \oint_C F(z)\,dz = \sum \begin{array}{l} \text{residues of the poles of } F(z) \\ \text{inside the contour C} \end{array} \tag{III.4}$$

The residue of a K-fold pole at $z = z_0$ is given by:

$$\text{Res}\{F(z) \text{ at } z = z_0\} = \frac{1}{(K-1)!} \cdot \frac{d^{K-1}}{dz^{K-1}}\left\{(z-z_0)^K F(z)\right\}\bigg|_{z=z_0} \tag{III.5a}$$

For a simple pole this is equal to:

$$\text{Res}\{F(z) \text{ at } z = z_0\} = (z-z_0)F(z)|_{z=z_0} \tag{III.5b}$$

(see also (II.10) and (II.26) in Appendix II.

Example
Given $X(z) = z/(z-a)$, find $x[n]$. This is done as follows: $\qquad$ (III.6)

$$x[n] = \frac{1}{2\pi j} \oint_C \frac{z}{z-a} z^{n-1}\,dz = \frac{1}{2\pi j} \oint_C \frac{z^n}{z-a}\,dz \tag{III.7}$$

For $n \geq 0$, $F(z) = z^n/(z-a)$ has only a single simple pole at $z = a$. The inverse transform is therefore:

$$x[n] = \text{Res}\{F(z) \text{ at } z = a\} = (z-a)\frac{z^n}{z-a}\bigg|_{z=a} = a^n \text{ for } n \geq 0 \tag{III.8}$$

For $n < 0$, $F(z)$ has an $|n|$-fold pole at $z = 0$. This means that a different number of poles has to be considered for each negative value of n, which does not make the calculation of $x(n]$ for $n < 0$ any easier. There are again certain stratagems that can help in tackling this problem (replacing z by $1/z$), but we shall not go into the matter further here. We think that we have already made it sufficiently plain that the direct application of the IZT via (III.1) is best avoided whenever possible.

III.3 CALCULATION OF DISCRETE NOISE

A completely different situation, which is nevertheless closely related to the previous one, should not be left unmentioned. We are thinking here of the calculation of noise powers, as in the analysis of quantization effects in digital filters (see the chapter 10, 'Finite word length in digital signals and systems'). In (10.14) we encountered the following integral, which is characteristic of such calculations:

$$P = \frac{1}{2\pi} \int_{-\pi}^{\pi} |H(e^{j\theta})|^2 \, d\theta \tag{III.9}$$

An integral of this kind is in fact very conveniently calculated by converting it to a contour integral in the z-plane and calculating its value there by residues. This is done in the following way. Provided we are dealing with a real $h[n]$, $H(e^{j\theta})$ is always a complex function with an even real part and an odd imaginary part; see (4.34). Consequently:

$$|H(e^{j\theta})|^2 = H(e^{j\theta})H^*(e^{j\theta}) = H(e^{j\theta})H(e^{-j\theta}) \tag{III.10}$$

Equation (III.9) therefore becomes:

$$P = \frac{1}{2\pi} \int_{-\pi}^{\pi} H(e^{j\theta})H(e^{-j\theta}) \, d\theta \tag{III.11}$$

Making use of (III.3b) we obtain in the same way as in section (III.1):

$$P = \frac{1}{2\pi j} \oint_C H(z)H(z^{-1})z^{-1} \, dz = \frac{1}{2\pi j} \oint_C F(z) \, dz \tag{III.12}$$

where the contour C again represents the unit circle. We can generally calculate this contour integral very easily by using (III.4). The difficulties we encountered in the example in the previous section in calculating (III.7) with the aid of (III.4) originated with the factor z^{n-1}. Equation (III.12) does not contain a factor of this type, in which n occurs as an exponent, and in many cases it is therefore a very practical aid in calculations on discrete noise.

Example

Given

$$H(e^{j\theta}) = \frac{15e^{j\theta}}{(e^{j\theta} - \frac{1}{2})(e^{j\theta} + \frac{1}{2})} \tag{III.13}$$

calculate P from (III.9). From (III.11) and (III.12) we find:

$$P = \frac{1}{2\pi j} \oint_C \frac{15z}{(z - \frac{1}{2})(z + \frac{1}{2})} \cdot \frac{15z^{-1}}{(z^{-1} - \frac{1}{2})(z^{-1} + \frac{1}{2})} z^{-1} \, dz$$

$$= \frac{1}{2\pi j} \oint_C \frac{900z}{(z - \frac{1}{2})(z + \frac{1}{2})(2 - z)(2 + z)} \, dz \tag{III.14}$$

We solve this by using (III.4). Only the simple poles at $z = \frac{1}{2}$ and $z = -\frac{1}{2}$ are contained within the integration contour C (the unit circle) and make a contribution to the integral. Therefore:

$$P = \frac{900z}{(z+\frac{1}{2})(2-z)(2+z)}\bigg|_{z=1/2} + \frac{900z}{(z-\frac{1}{2})(2-z)(2+z)}\bigg|_{z=-1/2}$$

$$= 120 + 120 = 240 \tag{III.15}$$

Appendix IV
Answers to the exercises

IV.1 ANSWERS TO CHAPTER 2

2.1 (a) $X(\omega) = \dfrac{4\sin^2(\omega\tau/2)}{\omega^2\tau}$

(b) $X(\omega) = \dfrac{2\alpha}{\alpha^2 + \omega^2}$

2.2 $x(t) = \dfrac{\sin(\omega_0 t)}{\pi t}$

2.3 (a) $a_0 = 4\tau/T;\ a_k = \dfrac{2}{\pi k}\sin(2k\pi\tau/T)$ for $k>0;\ b_k = 0;$

$\quad \alpha_0 = 2\tau/T;\ \alpha_k = \dfrac{1}{\pi k}\sin(2k\pi\tau/T)$ for $k \neq 0;$

(b) $a_k = \dfrac{4(-1)^k}{\pi(1-4k^2)};\ b_k = 0;\ \alpha_k = \dfrac{2(-1)^k}{\pi(1-4k^2)}$

(c) $a_k = \dfrac{4\sqrt{2}(-1)^k}{\pi(1-16k^2)};\ b_k = 0;\ \alpha_k = \dfrac{2\sqrt{2}(-1)^k}{\pi(1-16k^2)}$

(d) $a_k = 0;\ b_k = 0$ for k even; $b_k = \dfrac{2T}{\pi^2 k^2 \tau}\sin\left(\dfrac{2\pi k\tau}{T}\right)$ for k odd;

$\quad \alpha_k = 0$ for k even; $\alpha_k = \dfrac{-jT}{\pi^2 k^2 \tau}\sin\left(\dfrac{2\pi k\tau}{T}\right)$ for k odd

(e) $a_k = 0$ for k even; $a_k = \dfrac{4}{\pi k}\sin(2\pi k\tau/T)$ for k odd; $b_k = 0$;

$\alpha_k = 0$ for k even; $\alpha_k = \dfrac{2}{\pi k}\sin(2\pi k\tau/T)$ for k odd

2.4 $x(t) = \dfrac{1}{2\pi}\displaystyle\int_{-\infty}^{\infty} 2\pi \sum_{k=-\infty}^{\infty} \alpha_k\delta(\omega-k\omega_0)e^{j\omega t}\,d\omega$

$\qquad = \displaystyle\sum_{k=-\infty}^{\infty}\alpha_k\int_{-\infty}^{\infty}\delta(\omega-k\omega_0)e^{j\omega t}\,d\omega = \sum_{k=-\infty}^{\infty}\alpha_k\,e^{jk\omega_0 t}$

2.5 (a) $\alpha_k = 1$ for $k = 0$ and $\alpha_k = 0$ for $k \ne 0$
     $X(\omega) = 2\pi\delta(\omega)$
 (b) $\alpha_k = 0.5$ for $k = 1$ and $k = -1$; $\alpha_k = 0$ for all other k
     $X(\omega) = \pi\delta(\omega-\omega_0)+\pi\delta(\omega+\omega_0)$
 (c) $\alpha_k = -j/2$ for $k = 1$; $\alpha_k = j/2$ for $k = -1$; $\alpha_k = 0$ for all other k
     $X(\omega) = -j\pi\delta(\omega-\omega_0)+j\pi\delta(\omega+\omega_0)$
 (d) $\alpha_k = 1$ for $k = 1$; $\alpha_k = 0$ for all other k
     $X(\omega) = 2\pi\delta(\omega-\omega_0)$

2.6 The function $x(t)$ is a periodic sequence of Dirac pulses of area 1 and period T.

$\alpha_k = 1/T$ for all k; $X(\omega) = \dfrac{2\pi}{T}\displaystyle\sum_{k=-\infty}^{\infty}\delta(\omega-2\pi k/T)$

2.7 (a) $x(t) = y(t)+RC\dfrac{dy(t)}{dt}$

 (b) $H(\omega) = 1/(1+j\omega RC)$

 (c) $h(t) = \dfrac{1}{RC}e^{-t/RC}u(t);\ u(t) = 0.5$ at $t = 0$

 (d) $y(t) = (1-e^{-t/RC})u(t)$
2.8 (a) $y(t) = 0$ for $t<-1$; $y(t) = 1+t$ for $|t|<1$; $y(t) = 2$ for $t>1$
 (b) $y(t) = 2-|t|$ for $|t|<2$; $y(t) = 0$ for $|t|>2$
 (c) $y(t) = [1+\cos(\pi t)]/\pi$ for $|t|<1$; $y(t) = 0$ for $|t|>1$
2.9 See Fig. IV.1

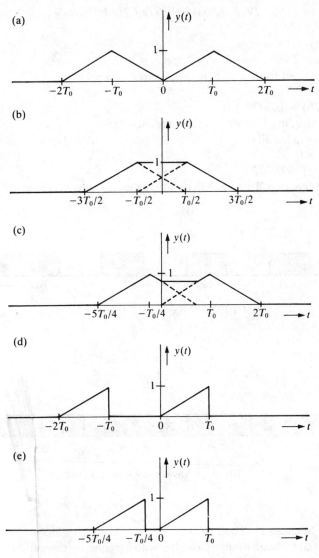

FIGURE IV.1 Answer to Exercise 2.9.

2.10

	$H_1(p)$	$H_2(p)$	$H_1(p)H_2(p)$
Poles	2 and $-3 \pm 4j$	$\pm 5j$	$-3 \pm 4j$
Zeros	$\pm 5j$	2 and $3 \pm 4j$	$3 \pm 4j$
Comments	Not stable $H_1 = 0$ for $|\omega| = 5$ Minimum-phase Not all-pass Physically realizable	Not stable $H_2 = \infty$ for $|\omega| = 5$ Non-minimum-phase Not all-pass Not physically realizable	Stable Never 0 or ∞ Non-minimum-phase All-pass Physically realizable

IV.2 ANSWERS TO CHAPTER 3

3.1 (a) $x(nT_1) = 0$ for n even
        $x(nT_1) = 1$ for $n = -3, n = 1$ and $n = 5$
        $x(nT_1) = -1$ for $n = -5, n = -1$ and $n = 3$
   (b) $x(1/8) = 0.71$ (A)
        $\sin(2\pi/8) = 0.707$ (B)
        Difference: in (A) only 15 values are included
   (c) $x(nT_2) = 0$ for all n
   (d) $x(1/8) = 0$ (C)
        $\sin(2\pi/8) = 0.707$ (D)
        Difference: in (C) we just fail to satisfy the sampling theorem

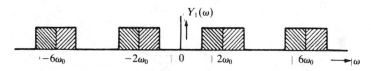

FIGURE IV.2 Answer to Exercise 3.2a.

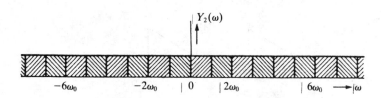

FIGURE IV.3 Answer to Exercise 3.2b.

3.2 (a) and (b) See Figs IV.2 and IV.3
   (c) Pass $Y_2(\omega)$ through a band-pass filter with a frequency response $H(\omega)$ such that:

$$H(\omega) = \begin{cases} T & \text{for } \omega_0 < |\omega| < 2\omega_0 \\ 0 & \text{elsewhere} \end{cases}$$

3.3 (a) $\omega_{min,1} = 2\omega_0$ and $\omega_{min,2} = 5\omega_0$
   (b) See Fig. IV.4
   (c) Aliasing now occurs for both $x_1(t)$ and $x_2(t)$
   (d) Aliasing always occurs for $x_1(t)$; aliasing only occurs for $x_2(t)$ if the sampling
        frequency is higher than $6\omega_0$

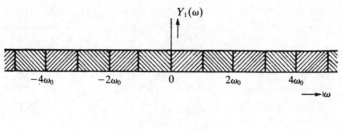

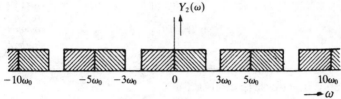

FIGURE IV.4 Answer to Exercise 3.3b.

IV.3 ANSWERS TO CHAPTER 4

4.1 The function is periodic for $\theta = 2\pi K/N$, with K and N arbitrary integers; the period is always $N_0 = 2\pi K_{min}/\theta$ where K_{min} is the smallest possible value of K. (At $\theta = 0.23\pi$ we have $K_{min} = 23$ and $N_0 = 200$; at $\theta = 0.24\pi$ we have $K_{min} = 3$ and $N_0 = 25$.)

4.2

n	a and b	c	d	e	f
-5	1	-1	$-0.5\sqrt{2}$	-1	0
-4	1	1	$0.5\sqrt{2}$	0	0
-3	1	-1	$0.5\sqrt{2}$	1	0
-2	1	1	$-0.5\sqrt{2}$	0	0
-1	1	-1	$-0.5\sqrt{2}$	-1	0
0	1	1	$0.5\sqrt{2}$	0	1
1	1	-1	$0.5\sqrt{2}$	1	0.8
2	1	1	$-0.5\sqrt{2}$	0	0.64
3	1	-1	$-0.5\sqrt{2}$	-1	0.512
4	1	1	$0.5\sqrt{2}$	0	0.4096
5	1	-1	$0.5\sqrt{2}$	1	0.32768

4.3

$x[n]$	n										
	-5	-4	-3	-2	-1	0	1	2	3	4	5
$\mathrm{Re}\{x[n]\}$	0	1	0	-1	0	1	0	-1	0	1	0
$\mathrm{Im}\{x[n]\}$	-1	0	1	0	-1	0	1	0	-1	0	1

4.4

n	$e^{j\pi n/4}$	e^{jn}
−5	−0.71 + 0.71 j	0.28 + 0.96 j
−4	−1	−0.65 + 0.76 j
−3	−0.71 − 0.71 j	−0.99 + 0.14 j
−2	−j	−0.42 − 0.91 j
−1	0.71 − 0.71 j	0.54 − 0.84 j
0	1	1
1	0.71 + 0.71 j	0.54 + 0.84 j
2	j	−0.42 + 0.91 j
3	−0.71 + 0.71 j	−0.99 + 0.14 j
4	−1	−0.65 − 0.76 j
5	−0.71 − 0.71 j	0.28 − 0.96 j

Only $x[n] = e^{j\pi n/4}$ is periodic (period = 8).

4.5 (a) $X(e^{j\theta}) = e^{-j3\theta}$

(b) $X(e^{j\theta}) = \dfrac{\sin\{(2N+1)\theta/2\}}{\sin(\theta/2)}$

(c) $x[n] = \begin{cases} \theta_0/\pi & \text{for } n = 0 \\ \sin(n\theta_0)/n\pi & \text{for } n \neq 0 \end{cases}$

4.6 (a) $I_1 = 2\pi(2N+1)$
(b) $I_2 = 2\pi/(1 - a^2)$

4.7 $S_1 = \theta_0/\pi$

4.9 (a)

n	$y_1[n]$	$y_2[n]$	$y_3[n]$	$y_4[n]$
−1	1	1	1	1
0	−a	1 − a	2 − a	−a
1	a^2	−a(1 − a)	−a(2 − a)	$1 + a^2$
2	$−a^3$	$a^2(1 - a)$	$a^2(2 - a)$	$−a(1 + a^2)$
·	·	·	·	·
·	·	·	·	·
·	·	·	·	·
n	$(-a)^{n+1}$	$(-a)^n(1-a)$	$(-a)^n(2-a)$	$(-a)^{n-1}(1+a^2)$

(b) The system is nonlinear and time-varying.
(c) This is because $y[-1] \neq 0$ ('the initial conditions are not zero'). If we had had $y[-1] = 0$, then the system would have been linear and time-invariant.

4.10

		$y[n]$		
n		a	b	c
<0		0	0	0
0		1	1	1
1		1.5	2	0
2		0.5	2	0
3		0	2	0
4		0	1	-1
≥ 5		0	0	0

4.11 (a)

$$h[n] = \begin{cases} 0 & \text{for } n<0 \\ 1 & \text{for } n = 0 \\ 3(1/2)^n & \text{for } n>0 \end{cases}$$

4.13 $N_4 = N_0 + N_2$ and $N_5 = N_1 + N_3$

4.14 $H(e^{j\theta}) = \dfrac{b_0 + b_1 e^{-j\theta} + b_2 e^{-2j\theta}}{1 - a_1 e^{-j\theta} - a_2 e^{-2j\theta}}$

4.15

	Stable	Causal	Linear	Time-invariant
(a)	Yes	Yes	Yes	No
(b)	No	Yes	Yes	No
(c)	Yes	No	Yes	Yes
(d)	Yes	Yes	Yes	Yes
(e)	Yes	Yes	No	Yes
(f)	Yes	Yes	No	Yes

4.16 (a) $X_a(z) = 1 + z^{-1} + z^{-2}$
   (b) $X_b(z) = 1 + \alpha z^{-1} + \alpha^2 z^{-2}$

   (c) $X_c(z) = \dfrac{1 - z^{-3}}{1 - \alpha z^{-1}}$

   (d) $X_d(z) = 1 + 2z^{-1} + 3z^{-2} + 2z^{-3} + z^{-4}$

4.17 (a) $x_e[n] = 2\delta[n] - (1/4)^n u[n]$
   (b) $x_f[n] = (-1/2)^n u[n]$
   (c) $x_g[n] = -2(1/2)^n u[n] + 3(3/4)^n u[n]$
   (d) $x_h[n] = 2(1/2)^{n-1}(n-1)u[n-1]$

4.18

	Zeros	Poles
4.16(a)	$z = -0.5 \pm 0.5\,j\sqrt{3}$	$2 \times$ at $z = 0$
4.16(b)	$z = -\alpha/2 \pm \alpha/2\,j\sqrt{3}$	$2 \times$ at $z = 0$
4.16(c)	$z = 1$	$2 \times$ at $z = 0$
	and $z = 0.5 \pm 0.5\,j\sqrt{3}$	and at $z = a$
4.16(d)	$2 \times$ at	$4 \times$ at $z = 0$
	$z = -0.5 \pm 0.5\,j\sqrt{3}$	
4.17(a)	$z = 1/2$	$z = 1/4$
4.17(b)	$z = 0$	$z = -1/2$
4.17(c)	$2 \times$ at $z = 0$	$z = 1/2$ and $z = 3/4$
4.17(d)	No zeros	$2 \times$ at $z = 1/2$

4.19 (a) $H(z) = \dfrac{z^2 - 1.5z + 2.25}{z + 0.9}$; $H(e^{j\theta}) = \dfrac{e^{2j\theta} - 1.5e^{ej\theta} + 2.25}{e^{j\theta} + 0.9}$

$$h[n] = \delta[n+1] + 2.5\delta[n] - 4.9(-0.9)^n u[n]$$

(b) The system is not causal.

(c) The system is stable.

(d) It is a non-minimum-phase system.

(e) A pole at $z = -0.9$ and zeros at $z = (2/3)e^{\pm j\pi/3}$.

(f) $H(z) = \dfrac{z^2 - 2z/3 + 4/9}{z + 0.9}$; $H(e^{j\theta}) = \dfrac{e^{2j\theta} - 2e^{j\theta}/3 + 4/9}{e^{j\theta} + 0.9}$

$$h[n] = \delta[n+1] + \frac{40}{81}\delta[n] - \frac{1669}{810}(-0.9)^n u[n]$$

(g) The system is not causal.

4.20 (a) $H(z) = \dfrac{b_0 z^2 + b_1 z + b_2}{z^2 - a_1 z - a_2}$

(b) The system has two zeros at $z = -1$ and poles at $z = 0.7 \pm 0.1\,j$. It is a stable minimum-phase system.

(c) The system has zeros at $z = -2$ and $z = -0.5$ and poles at $z = 0.5 \pm 0.5\,j\sqrt{7}$. It is an unstable non-minimum-phase system.

(d) $H(z) = \dfrac{z^2}{z^2 - z + 0.99}$

(e) $y[n] = 100A \cdot \cos(n\pi/3 - \pi/3)$

(f) $H(z) = \dfrac{z^2 - 0.35z}{z^2 - 0.7z + 0.49}$ and $h[n] = (0.7)^n \cos(n\pi/3)u[n]$

(g)

n	$h[n]$
≤ -1	0
0	1
1	0.35
2	−0.245
3	−0.343
4	−0.12005
5	0.08403
6	0.11765
7	0.04118
8	−0.02882
9	−0.04035

4.21 (a) The system of Fig. 4.35(b).
(b) The system of Fig. 4.35(b).

4.22 (a) $h[0] = 1; h[1] = \sqrt{2}; h[2] = 0; h[3] = -4\sqrt{2}; h[4] = -16; h[5] = -16\sqrt{2}.$

(b) $H(z) = \dfrac{z(z - \sqrt{2})}{z^2 - 2z\sqrt{2} + 4}$

(c) Zeros: $z = 0$ and $z = \sqrt{2}$; Poles: $z = \sqrt{2} \pm j\sqrt{2}$
(d) Unstable.

4.23 (a) $y_1[n] = Ay_1[n-1] - By_2[n-1] + x[n]; y_2[n] = By_1[n-1] + Ay_2[n-1]$

(c) $\dfrac{Y_1(z)}{X(z)} = \dfrac{z^2 - Az}{z^2 - 2Az + A^2 + B^2}; \dfrac{Y_2(z)}{X(z)} = \dfrac{Bz}{z^2 - 2Az + A^2 + B^2}$

(d)

	$H_1(z)$	$H_2(z)$
Zeros	$z = 0; z = \frac{1}{2}\sqrt{2}$	$z = 0$
Poles	$z = \frac{1}{2}\sqrt{2}(1 \pm j)$	$z = \frac{1}{2}\sqrt{2}(1 \pm j)$

(e) $h_1[n] = \cos(n\pi/4)u[n]$
$h_2[n] = \sin(n\pi/4)u[n]$

IV.4 ANSWERS TO CHAPTER 5

5.1 (a) $X_1[0] = 4; X_1[1] = -2; X_1[2] = 0$ and $X_1[3] = -2$
$X_2[0] = 6; X_2[1] = -1 - j; X_2[2] = 0$ and $X_2[3] = -1 + j$
(b) $\tilde{X}_1(e^{j\theta}) = e^{-j\theta} + 2e^{-2j\theta} + e^{-3j\theta}$
$\tilde{X}_2(e^{j\theta}) = 1 + 2e^{-j\theta} + 2e^{-2j\theta} + e^{-3j\theta}$

5.2 $x_3[n] = \frac{1}{4}\cos(\pi n/8) + \frac{1}{8}\cos(3\pi n/8)$

5.3 (a) $X[0] = 3; X[1] = 1 - 2j; X[2] = -1$ and $X[3] = 1 + 2j$

5.4 (a) $X_p[0] = 4; X_p[1] = 0; X_p[2] = 0$ and $X_p[3] = 0$

5.8 $N = 4$: $X_1[0] = 6; X_1[1] = -1 - j; X_1[2] = 0$ and $X_1[3] = -1 + j$

$N = 6$: $X_2[0] = 6; X_2[1] = -2j\sqrt{3}; X_2[2] = 0; X_2[3] = 0; X_2[4] = 0$ and
$\qquad X_2[5] = 2j\sqrt{3}$

$N = 16$:

k	$X_3[k]$	k	$X_3[k]$	k	$X_3[k]$	k	$X_3[k]$
0	6	1	$4.64 - 3.10j$	2	$1.70 - 4.12j$	3	$-0.56 - 2.88j$
4	$-1 - j$	5	$-0.26 - 0.06j$	6	$0.30 - 0.12j$	7	$0.18 - 0.28j$
8	0	9	$0.18 + 0.28j$	10	$0.30 + 0.12j$	11	$-0.26 + 0.06j$
12	$-1 + j$	13	$-0.56 + 2.88j$	14	$1.70 + 4.12j$	15	$4.64 + 3.10j$

5.9 (b)

	$N = 4$	$N = 6$	$N = 8$
$y_p[0]$	11/8	8/8	8/8
$y_p[1]$	13/8	12/8	12/8
$y_p[2]$	14/8	14/8	14/8
$y_p[3]$	7/8	7/8	7/8
$y_p[4]$	—	3/8	3/8
$y_p[5]$	—	1/8	1/8
$y_p[6]$	—	—	0
$y_p[7]$	—	—	0

5.14 (a) $X[0] = 10; X[1] = -2 + 2j; X[2] = -2$ and $X[3] = -2 - 2j$

IV.5 ANSWERS TO CHAPTER 6

6.2 $X_c(p) = \dfrac{2A}{T} \cdot \dfrac{(e^{pT/4} - e^{-pT/4})^2}{p^2}$

$$X_d(z) = \frac{A}{4}(z^3 + 2z^2 + 3z + 4 + 3z^{-1} + 2z^{-2} + z^{-3})$$

IV.6 ANSWERS TO CHAPTER 7

7.1 (a) First solution: $b_4 = 4, b_5 = 3, b_6 = 2$ and $b_7 = 1$
$\qquad$ Second solution: $b_4 = -4, b_5 = -3, b_6 = -2$ and $b_7 = -1$
(b) $A_1(e^{j\theta}) = |2\cos(7\theta/2) + 4\cos(5\theta/2) + 6\cos(3\theta/2) + 8\cos(\theta/2)|$
$\qquad A_2(e^{j\theta}) = |2\sin(7\theta/2) + 4\sin(5\theta/2) + 6\sin(3\theta/2) + 8\sin(\theta/2)|$

7.2 (a) There is one solution: $b_4 = 3$, $b_5 = 2$ and $b_6 = 1$

(b) $A(e^{j\theta}) = |2\cos(3\theta) + 4\cos(2\theta) + 6\cos(\theta) + 4|$

7.3 $h[0] = 0$; $h[1] = a + b$

7.4 $h[0] = bd + c$; $h[1] = ad + be$ and $h[2] = ae$

7.5 (a) $a_0 = 1$; $a_1 = 2 + j\sqrt{3}$ and $b_1 = 2 - j\sqrt{3}$

(c) $H(e^{j\theta}) = 5 - 4e^{-j\theta} + 2e^{-2j\theta}$

(d) $h[0] = 5$; $h[1] = -4$ and $h[2] = 2$

7.6 (a) Rewrite $H(z)$:

$$H(z) = (1 - z^{-5})\frac{1.2 - 1.2z^{-1}\cos(2\pi/5)}{1 - 2z^{-1}\cos(2\pi/5) + z^{-2}} = (1 - z^{-5})\frac{1.2 - 0.371z^{-1}}{1 - 0.618z^{-1} + z^{-2}}$$

This is the cascade connection of Fig. 7.16(a) with $N = 5$ and Fig. 7.10 with $b_0 = 1.2$, $b_1 = -0.371$, $a_1 = 0.618$ and $a_2 = -1$.

(b) A brief switching transient occurs at the output of the comb filter; the recursive part therefore continues to 'oscillate'.

7.7 $H(e^{j\theta}) = \dfrac{ab \cdot e^{-2j\theta}}{1 + b \cdot e^{-j\theta} - ab \cdot e^{-2j\theta}}$

7.8 $H(e^{j\theta}) = \dfrac{1}{1 + b(1 + a)e^{-j\theta} + a \cdot e^{-2j\theta}}$

IV.7 ANSWERS TO CHAPTER 8

8.1 (a)

$$h[n] = \begin{cases} 1/3 & \text{for } n = 0 \\[2mm] \dfrac{\sin(n\pi/3)}{n\pi} & \text{for } n \neq 0 \end{cases}$$

(b)/(c)

	Impulse response	
n	Answer (b)	Answer (c)
0 and 8	−0.069	0
1 and 7	0	0
2 and 6	0.138	0.069
3 and 5	0.276	0.236
4	0.333	0.333

8.2 (a) $h[n] = \delta[n] + \delta[n-1]$; $H(z) = 1 + z^{-1}$; $H(e^{j\theta}) = 2e^{-j\theta/2}\cos(\theta/2)$

(b) $h[n] = \delta[n] - \delta[n-1]$; $H(z) = 1 - z^{-1}$; $H(e^{j\theta}) = 2e^{j(\pi/2 - \theta/2)}\sin(\theta/2)$

(c) $h[n] = \delta[n] + \delta[n-1] + \delta[n-2]$; $H(z) = 1 + z^{-1} + z^{-2}$;
$H(e^{j\theta}) = e^{-j\theta}\{1 + 2\cos(\theta)\}$

(d) $h[n] = \delta[n] - \delta[n-2]$; $H(z) = 1 - z^{-2}$; $H(e^{j\theta}) = 2e^{j(\pi/2 - \theta)}\sin(\theta)$

(e) $h[n] = \delta[n] - \delta[n-4]$; $H(z) = 1 - z^{-4}$; $H(e^{j\theta}) = 2e^{j(\pi/2 - 2\theta)}\sin(2\theta)$

8.3 (a)/(b) $H(z) = \dfrac{1}{1 - e^{-0.25}z^{-1}} - \dfrac{2}{1 - e^{-0.5}z^{-1}} + \dfrac{1}{1 - e^{-0.75}z^{-1}}$

8.4 (a) $H_a(p) = \dfrac{2}{(p+1)(p+2)}$

The system has two poles: $p_1 = -1$ and $p_2 = -2$; the system is therefore stable.

(b) $h_a(t) = \begin{cases} 0 & \text{for } t < 0 \\ 2(e^{-t} - e^{-2t}) & \text{for } t \geq 0 \end{cases}$

$h_a(nT) = h_a(n/2) = \begin{cases} 0 & \text{for } n < 0 \\ 2(e^{-n/2} - e^{-n}) & \text{for } n \geq 0 \end{cases}$

(c) $H_1(z) = \dfrac{2}{1 - e^{-0.5}z^{-1}} - \dfrac{2}{1 - e^{-1}z^{-1}}$

(d) There is a zero at $z_1 = 0$ and there are two poles at $z_2 = e^{-0.5} = 0.6065$ and $z_3 = e^{-1} = 0.3679$. The system is therefore stable.

(e) $H_2(z) = \dfrac{z^2}{(3z - 2)(2z - 1)}$

(f) There are two zeros at $z_1 = z_2 = 0$ and two poles at $z_3 = \frac{2}{3}$ and $z_4 = \frac{1}{2}$. The system is therefore stable.

(g) $h_2[n] = \left[\left(\dfrac{2}{3}\right)^{n+1} - \left(\dfrac{1}{2}\right)^{n+1} \right] u[n]$

(h) $H_3(z) = \dfrac{1}{z(2z - 1)}$

There are two poles at $z_1 = 0$ and $z_2 = \frac{1}{2}$ (system is stable).

$h_3[n] = \left(\dfrac{1}{2}\right)^{n-1} u[n-1] - \delta[n-1]$

(i) For $0 < T < 1$

(j) $H_4(z) = \dfrac{(z+1)^2}{(5z - 3)(3z - 1)}$

There are two zeros: at $z_1 = z_2 = -1$ and two poles at $z_3 = 0.6$ and $z_4 = \frac{1}{3}$ (stable system).

$$h_4[n] = \left[\frac{16}{15}\left(\frac{3}{5}\right)^n - \frac{4}{3}\left(\frac{1}{3}\right)^n \right] u[n] + \frac{1}{3}\delta[n]$$

8.5 (a) The system has 5 poles: at $p_1 = -1$; $p_{2,3} = -\frac{1}{2} \pm \frac{1}{2}j\sqrt{3}$ and $p_{4,5} = -\frac{1}{2}\sqrt{3} \pm \frac{1}{2}j$

   (b) There are again 5 poles; now at $z_1 = 3/5$; $z_{2,3} = 0.714 \pm 0.330j$ and $z_{4,5} = 0.627 \pm 0.167j$

8.6 (a) $H_d(z) = \dfrac{(1 + z^{-1})(8.435 - 11.06z^{-1} + 8.435z^{-2})}{(5.032 - 2.968z^{-1})(19.79 - 29.19z^{-1} + 15.02z^{-2})}$

   (b)

i	θ_i (rad)	f_i (Hz)
1	0.28	0.09
2	0.44	0.14
3	0.49	0.16
4	0.77	0.24
5	0.85	0.27
6	1.29	0.41

IV.8 ANSWERS TO CHAPTER 9

9.1 The number of multiplications per unit time has been reduced by a factor of 7/10 (see Fig. IV.5).

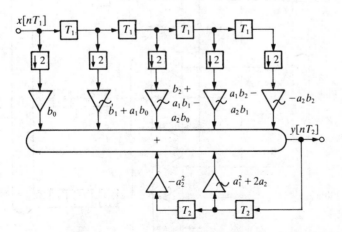

FIGURE IV.5 Answer to Exercise 9.1.

9.2 The number of multiplications per unit time has been reduced by a factor of 5/9 (see Fig. IV.6).

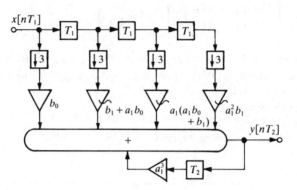

FIGURE IV.6 Answer to Exercise 9.2.

9.3 $c_0 = 1$; $c_1 = a_1$; $c_2 = a_1^2$; $c_3 = -b_1/a_1$ and $c_4 = (b_0 + b_1/a_1)$.   The   number   of multiplications per unit time has been reduced by a factor of 2/3.

9.4 The number of multiplications per unit time has been reduced by a factor of 1/3 (see Fig. IV.7).

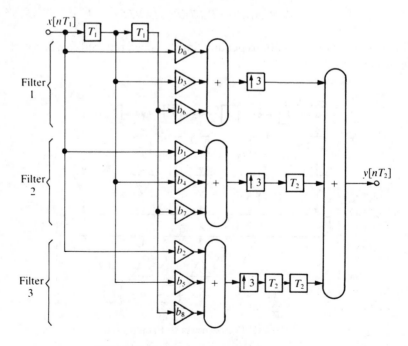

FIGURE IV.7 Answer to Exercise 9.4.

9.5 (a) We obtain the signal $\hat{x}[nT_2]$ from $x[nT_1]$ by inserting a zero between every two signal samples. The signal $y[nT_2]$ is the linearly interpolated version of $x[nT_1]$.

(b)
$$h[nT_2] = \begin{cases} 1 & \text{for } n = 0 \\ \frac{1}{2} & \text{for } n = \pm 1 \\ 0 & \text{elsewhere} \end{cases}$$

(c) $H(e^{j\omega T_2}) = 1 + \cos(\omega T_2)$
(d) See Fig. IV.8(a).
(e) See Fig. IV.8(b).
(f) The input spectrum of question 9.5(d).

(g) $y[nT_2] = \begin{cases} x[(n-1)T_1/2] & \text{for } n = \pm 1, \pm 3, \pm 5, \ldots \\ \frac{1}{2}\{x[(n-2)T_1/2] + x[nT_1/2]\} & \text{for } n = 0, \pm 2, \pm 4, \ldots \end{cases}$

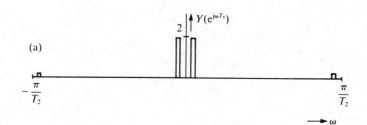

(a)

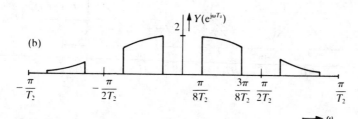

(b)

FIGURE IV.8 Answer to Exercise 9.5.

9.6 (a) See Fig. IV.9.
(b) See Fig. IV.10.
(c) See Fig. IV.11.

9.8 See Fig. IV.12.

9.9 (a) See Fig. IV.13.
(b) See Fig. IV.14.
(c) From Fig. IV.15 we have: $\tilde{X}(e^{j\omega T_1}) = 1/15\,X(e^{j\omega T_1})$.

9.10 See Fig. IV.16.

9.11 See Fig. IV.17.

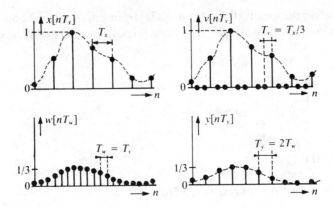

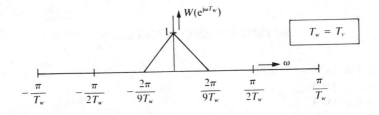

FIGURE IV.9 Answer to Exercise 9.6(a).

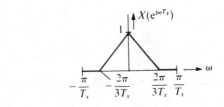

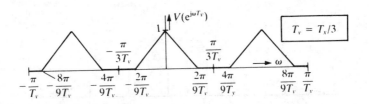

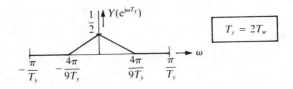

FIGURE IV.10 Answer to Exercise 9.6(b).

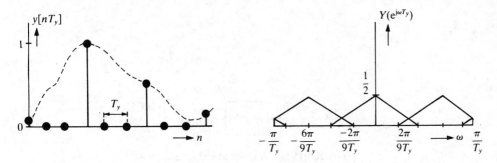

FIGURE IV.11 Answer to Exercise 9.6(c).

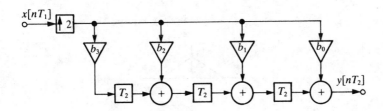

FIGURE IV.12 Answer to Exercise 9.8.

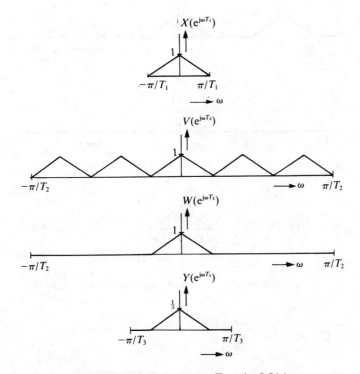

FIGURE IV.13 Answer to Exercise 9.9(a).

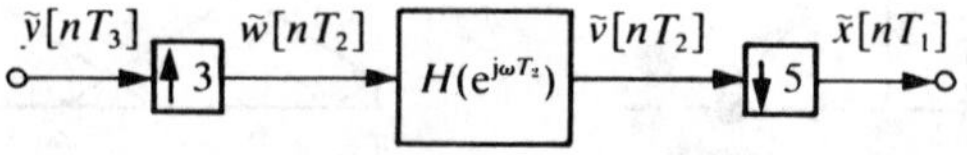

$$\bar{y}[nT_3] \xrightarrow{\quad} \boxed{\uparrow 3} \xrightarrow{\tilde{w}[nT_2]} \boxed{H(e^{j\omega T_2})} \xrightarrow{\tilde{v}[nT_2]} \boxed{\downarrow 5} \xrightarrow{\bar{x}[nT_1]}$$

FIGURE IV.14 Answer to Exercise 9.9(b).

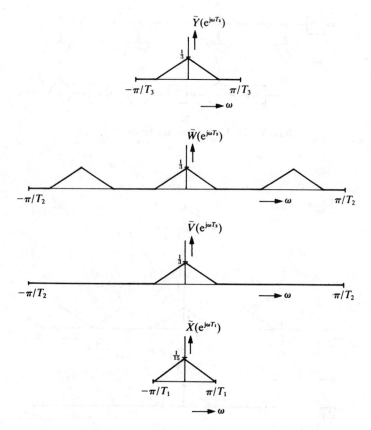

FIGURE IV.15 Answer to Exercise 9.9(c).

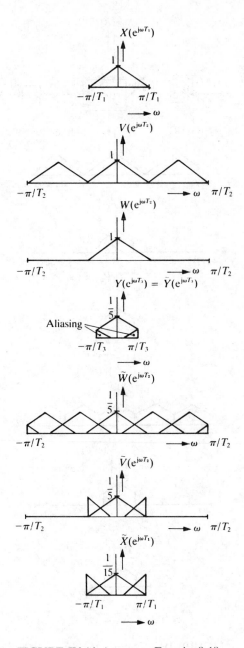

FIGURE IV.16 Answer to Exercise 9.10.

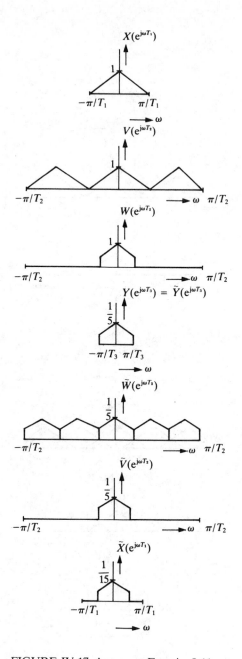

FIGURE IV.17 Answer to Exercise 9.11.

10.1 (a)/(c) Sign-and-magnitude representation:

Before quantization		After quantization (magnitude truncation)	
Decimal value x	Binary representation	Binary representation	Decimal value x_Q
+1.75	01.11		
+1.50	01.10		
+1.25	01.01	01	+1
+1.00	01.00		
+0.75	00.11		
+0.50	00.10		
+0.25	00.01	00	+0
+0.00	00.00		
−0.00	10.00		
−0.25	10.01		
−0.50	10.10	10	−0
−0.75	10.11		
−1.00	11.00		
−1.25	11.01		
−1.50	11.10	11	−1
−1.75	11.11		

10.1 (b)/(c) Two's complement representation:

Before quantization		After quantization (value truncation)	
Decimal value x	Binary representation	Binary representation	Decimal value x_Q
+1.75	01.11		
+1.50	01.10		
+1.25	01.01	01	+1
+1.00	01.00		
+0.75	00.11		
+0.50	00.10		
+0.25	00.01	00	+0
+0.00	00.00		
−0.25	11.11		
−0.50	11.10		
−0.75	11.01	11	−1
−1.00	11.00		
−1.25	10.11		
−1.50	10.10		
−1.75	10.01	10	−2
−2.00	10.00		

10.2 (a)/(b)

Sign and magnitude	Rounding	Value truncation	Magnitude truncation
$2.375 \longleftrightarrow 010.011$	$2.5 \longleftrightarrow 010.1$	$2.0 \longleftrightarrow 010.0$	$2.0 \longleftrightarrow 010.0$
$-2.375 \longleftrightarrow 110.011$	$-2.5 \longleftrightarrow 110.1$	$-2.5 \longleftrightarrow 110.1$	$-2.0 \longleftrightarrow 110.0$

10.2 (c)

Two's complement	Rounding	Value truncation	Magnitude truncation
$2.375 \longleftrightarrow 010.011$	$2.5 \longleftrightarrow 010.1$	$2.0 \longleftrightarrow 010.0$	$2.0 \longleftrightarrow 010.0$
$-2.375 \longleftrightarrow 110.101$	$-2.5 \longleftrightarrow 101.1$	$-2.5 \longleftrightarrow 101.1$	$-2.0 \longleftrightarrow 110.0$

10.3 See Fig. IV.18.

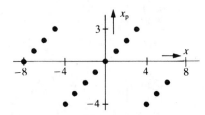

FIGURE IV.18 Answer to Exercise 10.3.

10.4 (a) $b_1 = \pm 1$
(c) No, the values $+1$ and -1 can always be represented exactly.

10.5 (a)

Zero z_1	Zero z_2	b_1	b_2
$+1$	$+1$	-2	1
$+1$	-1	0	-1
-1	-1	2	1
$e^{j\varphi}$	$e^{-j\varphi}$	$-2\cos(\varphi)$	1

10.5 (b) Quantization of the coefficients only has an effect in the case where $z_{1,2} = e^{\pm j\varphi}$; the zeros still lie on the unit circle then, but they may be slightly shifted.

10.6

n	$y[n]$
-2	$-111q$
-1	$-29q$
0	$+56q$
1	0
2	$-56q$
3	$-106q$
4	0
5	$+106q$
6	0
7	$-106q$
8	0
9	$+106q$
10	0

The period of the overflow oscillation is 4.

10.7

			$y[n]$		
n	(a)	(b)	(c)	(d)	(e)
0	$-9q$	$-5q$	$+9q$	$-9q$	$-5q$
1	$+8q$	$+5q$	$+8q$	$+8q$	$+4q$
2	$-7q$	$-5q$	$+7q$	$-7q$	$-4q$
3	$+6q$	$+5q$	$+6q$	$+6q$	$+3q$
4	$-5q$	$-5q$	$+5q$	$-5q$	$-3q$
5	$+5q$	$+5q$	$+5q$	$+4q$	$+2q$
6	$-5q$	$-5q$	$+5q$	$-3q$	$-2q$
7	$+5q$	$+5q$	$+5q$	$+2q$	$+q$
8	$-5q$	$-5q$	$+5q$	$-q$	$-q$
9	$+5q$	$+5q$	$+5q$	0	0
10	$-5q$	$-5q$	$+5q$	0	0

10.8

	$y[n]$	
n	(a)	(b)
0	$+7q$	$+6q$
1	0	$-q$
2	$-7q$	$-7q$
3	$-11q$	$-10q$
4	$-11q$	$-9q$
5	$-7q$	$-4q$
6	0	$+2q$
7	$+7q$	$+7q$
8	$+11q$	$+9q$
9	$+11q$	$+7q$
10	$+7q$	$+2q$
11	0	$-3q$

(a) We obtain a limit cycle of period 10.
(b) From $n = 27$ we have $y[n] = 0$.

References

[1] Dollekamp, H., Esser, L. J. M., De Jong, H., 1982. P^2CCD in 60 MHz oscilloscope with digital image storage. *Philips Tech. Rev.*, **40**: 55–68.

[2] Van Roermund, A. H. M., Coppelmans, P. M. C., 1983/84. An integrated switched-capacitor filter for viewdata. *Philips Tech. Rev.*, **41**: 105–23. (Reprinted as Chapter 11 of this book.)

[3] De Vrijer, F. W., 1976. Modulation. *Philips Tech. Rev.*, **36**: 305–62.

[4] Jerri, A. J., 1977. The Shannon sampling theorem – its various extensions and applications: a tutorial review. *Proc. IEEE*, **65**: 1565–96.

[5] Wiener, N., 1949. *Extrapolation, interpolation and smoothing of stationary time series*. Wiley, New York, N.Y.

[6] Robinson, E. A., 1982. A historical perspective of spectrum estimation. *Proc. IEEE*, **70**: 885–907.

[7] Schuster, A., 1898. On the investigation of hidden periodicities with application to a supposed 26-day period of meteorological phenomena. *Terr. Magn.*, **3**: 13–41.

[8] Rabiner, L. R., Gold, B., 1975. *Theory and application of digital signal processing*. Prentice-Hall, Englewood Cliffs, N.J.

[9] Oppenheim, A. V., Schafer, R. W., 1975. *Digital signal processing*. Prentice-Hall, Englewood Cliffs, N.J.

[10] Papoulis, A., 1980. *Circuits and systems: a modern approach*. Holt, Rinehart & Winston, New York, N.Y.

[11] Oppenheim, A. V., Willsky, A. S., Young, I. T., 1983. *Signals and systems*. Prentice-Hall, Englewood Cliffs, N.J.

[12] Persoon, E. H. J., Vandenbulcke, C. J. B., 1986. Digital audio: examples of the application of the ASP integrated signal processor. pp. 201–16 in 'Digital signal processing II, applications'. *Philips Tech. Rev.*, **42**: 181–238. (Reprinted as Chapter 13 of this book.)

[13] Welten, F. P. J. M., Delaruelle, A., Van Wijk, F. J., Van Meerbergen, J. L., Schmid, J., Rinner, K., Van Eerdewijk, K. J. E., Wittek, J. H., 1985. A 2-μm CMOS 10-MHz microprogrammable signal processing core with an on-chip multiport memory bank. *IEEE J.*, **SC-20**: 754–60.

[14] Special issue, 1982. 'Compact Disc Digital Audio.' *Philips Tech. Rev.*, **40**: 149–80.

[15] Van der Kam, J. J., 1986. A digital 'decimating' filter for analog-to-digital conversion of hi-fi audio signals. pp. 230–8 in 'Digital signal processing II, applications'. *Philips Tech. Rev.*, **42**: 181–238. (Reprinted as Chapter 14 of this book.)

[16] Goedhart, D., Van de Plassche, R. J., Stikvoort, E. F., 1982. Digital-to-analog conversion in playing a Compact Disc. *Philips Tech. Rev.*, **40**: 174–9.

[17] Peek, J. B. H., 1968. The measurement of correlation functions in correlators using 'shift-invariant independent' functions. (Thesis, Eindhoven 1967) *Philips Res. Rep. Suppl.*, **No. 1**: 1–76.

[18] Van Doorn, R. A., Verhoeckx, N. A. M., 1977. An I^2L digital modulation stage for data transmission. *Philips Tech. Rev.*, **37**: 291–301.

[19] Van Gerwen, P. J., Snijders, W. A. M., Verhoeckx, N. A. M., 1980. An integrated echo canceller for baseband data transmission. *Philips Tech. Rev.*, **39**: 102–17.

[20] Sluyter, R. J., 1983/84. Digitization of speech. *Philips Tech. Rev.*, **41**: 201–23.

[21] Claasen, T. A. C. M., Mecklenbräuker, W. F. G., 1978. A generalized scheme for an all-digital time-division multiplex to frequency-division multiplex translator. *IEEE Trans.*, **CAS-25**: 252–9.

[22] Locher, P. R., 1983/84. Proton NMR tomography. *Philips Tech. Rev.*, **41**: 73–88.

[23] Castleman, K. R., 1979. Digital image processing. Prentice-Hall, Englewood Cliffs, N.J.

[24] Peters, J. H., Kanters, J. T., 1986. CAROT: a digital method of increasing the robustness of an analog colour television signal. pp. 217–29 in 'Digital signal processing II, applications'. *Philips Tech. Rev.*, **42**: 181–238.

[25] Annegarn, M. J. J. C., Nillesen, A. H. H. J., Raven, J. G., 1986. Digital signal processing in television receivers. pp. 183–200 in 'Digital signal processing II, applications'. *Philips Tech. Rev.*, **42**: 181–238. (Reprinted as Chapter 12 of this book.)

[26] Brigham, E. O., 1974. *The fast Fourier transform*. Prentice-Hall, Englewood Cliffs, N.J.

[27] Papoulis, A., 1962. *The Fourier integral and its applications*. McGraw-Hill, New York, N.Y.

[28] Meade, M. L., Dillon, C. R., 1986. *Signals and systems – models and behaviour*. Van Nostrand Reinhold, Wokingham, Berkshire, England.

[29] Jury, E. I., 1964. *Theory and application of the z-transform method*. Wiley, New York, N.Y.

[30] Gray, A. H., Markel, J. D., 1973. Digital lattice and ladder filter synthesis. *IEEE Trans.*, **AU-21**: 491–500.

[31] Claasen, T. A. C. M., Mecklenbräuker, W. F. G., 1978. On the transposition of linear time-varying discrete-time networks and its applications to multirate digital systems. *Philips J. Res.*, **33**: 78–102.

[32] Widrow, B., *et al.*, 1975. Adaptive noise cancelling: principles and applications. *Proc. IEEE*, **63**: 1692–716.

[33] Herrmann, O., Rabiner, L. R., Chan, D. S. K., 1973. Practical design rules for optimum filter response low-pass digital filters. *Bell Syst. Tech. J.*, **52**: 769–99.

[34] Rabiner, L. R., Kaiser, J. F., Herrmann, O., Dolan, M. T., 1974. Some comparisons between FIR and IIR digital filters. *Bell Syst. Tech. J.*, **53**: 305–31.

[35] Kuo, B. C., 1980. *Digital control systems*. Holt-Saunders International Editions, New York.

[36] Bruton, L. T., 1975. Low-sensitivity digital ladder filters. *IEEE Trans.*, **CAS-22**: 168–76.

[37] Fettweiss, A., 1971. Digital filter structures related to classical filter networks. *Arch. Elec. Übertr.* **Bd. 25**, 78–89.

[38] Jacobs, G. M., Allstot, D. J., Broderson, R. W., Gray, P. R., 1978. Design techniques for MOS switched capacitor ladder filters. *IEEE Trans.*, **CAS-25**: 1014–21.

[39] Zverev, A. I., 1967. *Handbook of filter synthesis*. Wiley, New York, N.Y.

[40] Schafer, R. W., Rabiner, L. R., 1973. A digital signal processing approach to interpolation. *Proc. IEEE*, **61**: 692–702.

[41] Bellanger, M. G., Daguet, J. L., Lepagnol, G. P., 1974. Interpolation, extrapolation and reduction of computation speed in digital filters. *IEEE Trans.*, **ASSP-22**: 231–5.

[42] Crochiere, R. E., Rabiner, L. R., 1975. Optimum FIR digital filter implementations for decimation, interpolation and narrow-band filtering. *IEEE Trans.*, **ASSP-23**: 444–56.

[43] Rabiner, L. R., Crochiere, R. E., 1975. A novel implementation for narrow-band FIR digital filters. *IEEE Trans.*, **ASSP-23**: 457–64.

[44] Crochiere, R. E., Rabiner, L. R., 1981. Interpolation and decimation of digital signals – a tutorial review. *Proc. IEEE*, **69**: 300–31.

[45] Crochiere, R. E., Rabiner, L. R., 1983. *Multirate digital signal processing*. Prentice-Hall, Englewood Cliffs, N.J.

[46] Papoulis, A., 1965. *Probability, random variables and stochastic processes*. (McGraw-Hill series in systems science.) McGraw-Hill, New York, N.Y.

[47] Lawrence, V. B., Salazar, A. C., 1980. Finite precision design of linear-phase FIR filters. *Bell Syst. Tech. J.*, **59**: 1575–98.

[48] Claasen, T. A. C. M., Mecklenbräuker, W. F. G., Peek, J. B. H., 1976. Effects of quantization and overflow in recursive digital filters. *IEEE Trans.*, **ASSP-24**: 517–29.

[49] Jackson, L. B., 1970. On the interaction of roundoff noise and dynamic range in digital filters. *Bell Syst. Tech. J.*, **49**: 159–84.

[50] Jackson, L. B., 1970. Roundoff-noise analysis for fixed-point digital filters realized in cascade or parallel form. *IEEE Trans.*, **AU-18**: 107–22.

[51] Peek, J. B. H., 1985. Digital signal processing – growth of a technology. pp. 103–9 in 'Digital signal processing I, background'. *Philips Tech. Rev.*, **42**: 101–48.

[52] Van den Enden, A. W. M., Verhoeckx, N. A. M., 1985. Digital signal processing: theoretical background. pp. 110–44 in 'Digital signal processing I, background'. *Philips Tech. Rev.*, **42**: 101–48.

[53] Van den Enden, A. W. M., Leenknegt, G. A. L., 1986. Design of optimal IIR filters with arbitrary amplitude and phase requirements. *Proc. EUSIPCO-86 Conf.*, The Hague: 183–6.

[54] Rabiner, L. R. *et al.*, 1972. Terminology in digital signal processing. *IEEE Trans.*, **AU-20**: 322–37.

[55] Harris, F. J., 1978. On the use of windows for harmonic analysis with the discrete Fourier transform. *Proc. IEEE*, **66**: 51–83.

[56] Claasen, T. A. C. M., Mecklenbräuker, W. F. G., Peek, J. B. H., 1975. Quantization noise analysis for fixed-point digital filters using magnitude truncation for quantization. *IEEE Trans.*, **CAS-22**: 887–95.

[57] Butterweck, H. J., 1979. On the quantization of noise contributions in digital filters which are uncorrelated with the output signal. *IEEE Trans.*, **CAS-26**: 901–10.

[58] Van Gerwen, P. J., Mecklenbräuker, W. F. G., Verhoeckx, N. A. M., Snijders, W. A. M., Van Essen, H. A., 1975. A new type of digital filter for data transmission. *IEEE Trans.*, **COM-23**: 222–34.

[59] Van Meerbergen, J. L., 1988. Developments in integrated digital signal processors, and the PCB 5010. *Philips Tech. Rev.*, **44**: 1–14. (Reprinted as Chapter 15 of this book.)

[60] Lacroix, A., Witte, K.-H., 1986. Design tables for discrete-time normalized low-pass filters. Artech House, Inc., Dedham, MA.

[61] Blom, D., Hanneman, H. W., Voorman, J. O., 1973. Some inductorless filters. *Philips Tech. Rev.*, **33**: 294–308.

[62] Voorman, J. O., Brüls, W. H. A., Barth, P. J., 1982. Integration of analog filters in a bipolar process. *IEEE J.*, **SC-17**: 713–22.

[63] Ghausi, M. S., Laker, K. R., 1981. *Modern filter design: active RC and switched capacitor.* Prentice-Hall, Englewood Cliffs, N.J.

[64] Voorman, J. O., Snijder, P. J., Vromans, J. S., Barth, P. J., 1982. An automatic equalizer for echo reduction in teletext on a single chip. *Philips Tech. Rev.*, **40**: 319–28.

[65] Sharpe, R., 1983. LUCY-LUCINDA – A fully integrated solution to videotex terminal interfaces. *IEEE Trans.*, **CE-29**: 492–7.

[66] Special issue on MOS transistors and MOS circuits. *Philips Tech. Rev.*, **31**: 205–95, 1970.

[67] Brandt, B. B. M., Steinmaier, W., Strachan, A. J., 1974. LOCMOS, a new technology for complementary MOS circuits. *Philips Tech. Rev.*, **34**: 19–23.

[68] Martin, K., Sedra, A. S., 1979. Strays-insensitive switched-capacitor filters based on bilinear z-transform. *Electron. Lett.*, **15**: 365–6.

[69] Lee, M. S., Chang, C., 1981. Switched-capacitor filters using the LDI and bilinear transformations. *IEEE Trans.*, **CAS-28**: 265–70.

[70] Kuo, Y.-L., Liou, M. L., Kasinskas, J. W., 1979. An equivalent circuit approach to the computer-aided analysis of switched capacitor circuits. *IEEE Trans.*, **CAS-26**: 708–14.

[71] Lücker, R., 1982. Frequency domain analysis of switched-capacitor circuits using z-domain transfer function evaluation. *Arch. Elektron. & Übertragungstech.*, **36**, 383–92.

[72] Huber, B., Kunze, J., Lücker, R., van Roermund, A. H. M., Coppelmans, P. M. C., 1982. Design of a switched-capacitor filter for Viewdata modems. *Arch. Elektron. & Übertragungstech.*, **36**: 141–7.

[73] Brodersen, R. W., Gray, P. R., Hodges, D. A., 1979. MOS switched-capacitor filters. *Proc. IEEE*, **67**: 61–75.

[74] Lutz, P., 1981. Real terminations of SC-filters using LDI-Integrators. *Frequenz*, **35**: 93–5.

[75] Scanlan, S. O., 1981. Analysis and synthesis of switched-capacitor state-variable filters. *IEEE Trans.*, **CAS-28**: 85–93.

[76] Black Jr., W. C., Allstot, D. J., Reed, R. A., 1980. A high performance low power CMOS channel filter. *IEEE J.*, **SC-15**: 929–38.

[77] Radio progress during 1936, part IV: Report by the technical committee on television and facsimile. *Proc. IRE*, **25**: 199–210, 1937.

[78] Lerner, E. J., 1983. Digital TV: makers bet on VLSI. *IEEE Spectrum*, **20**, No. 2 (February): 39–43.

[79] Jackson, R. N., Annegarn, M. J. J. C., 1983. Compatible systems for high-quality television. *SMPTE J.*, **92**: 719–23.

[80] Carnt, P. S., Townsend, G. B., 1969. *Colour television, Vol. 2, PAL, SECAM and other systems*. Iliffe, London.

[81] Pritchard, D. H., Gibson, J. J., 1980. Worldwide color television standards – similarities and differences. *SMPTE J.*, **89**: 111–20.

[82] Drewery, J. O., 1976. The filtering of luminance and chrominance signals to avoid cross-colour in a PAL colour system. *BBC Eng. No. 104* (September), 8–39.

[83] Richter, H.-P., 1983. Verfahren zur digitalen Decodierung des PAL-signals. *Fernseh- & Kino-Tech.*, **37**: 511–19.

[84] Weltersbach, W., Jacobsen, M., 1981. Digitale Videosignalverarbeitung im Farbfernsehempfänger, 1. Teil: PAL-Farbdecoder. *Fernseh- & Kino-Tech.*, **35**: 317–23, and 2. Teil: Maßnahmen zur Verbesserung der Bildqualität, *ibid.*, 371–9.

[85] Veendrick, H. J. M., 1981. An NMOS dual-mode digital low-pass filter for color TV. *IEEE J.*, **SC-16**: 179–182.

[86] Veendrick, H. J. M., Pfennings, L. C., 1982. A 40 MHz multi-applicable digital signal processing chip. *IEEE J.*, **SC-17**: 40–3.

[87] Pelgrom, M. J. M. *et al.*, 1983. A digital field memory for television receivers. *IEEE Trans.*, **CE-29**: 242–50.

[88] Berkhoff, E. J., Kraus, U. E., Raven, J. G., 1983. Applications of picture memories in television receivers. *IEEE Trans.*, **CE-29**: 251–8.

[89] Conrads, W., 1983. Integrated feature TV concept with serial I²C-bus-control and field memory. *IEEE Trans.*, **CE-29**: 469–74.

[90] Schönfelder, H., Jacobsen, M., 1983. Qualitätsverbesserung einer PAL-Farbfernsehübertragung durch digitale Filtertechnik. *Frequenz*, **37**: 324–33.

[91] Schönfelder, H., 1983. Möglichkeiten der Qualitätsverbesserung beim heutigen Fernsehsystem. *Fernseh- & Kino-Tech.*, **37**: 187–96.

[92] Schmitz, H. J. R., Annegarn, M. J. J. C., Fekkes, W. F., 1984. A CCD memory controller for instant access to teletext. *IEEE Trans.*, **CE-30**: 442–6.

[93] Mertens, H., Wood, D., 1983. The C-MAC/packet system for direct satellite television. *EBU Rev. Tech.*, No. 200: 172–85.

[94] Kitzen, W. J. W., Boers, P. M., 1984. Applications of a digital audio-signal processor in T.V. sets. *Philips J. Res.*, **39**: 94–102.

[95] Schroeder, M. R., 1962. Natural sounding artificial reverberation. *J. Audio Eng. Soc.*, **10**, 219–23.

[96] Franssen, N. V. Artificial reverberation apparatus for audio frequency signals, U.S. Patent No. 4 352 954 (5th October 1982).

[97] Franssen, N. V., Immink, K. A., Dijkmans, E. C., Geelen, M. H. Arrangement for the transmission of audio signals, U.S. Patent No. 4 375 623 (1st March 1983).

[98] Dijkmans, E. C., Franssen, N. V. Artificial reverberation apparatus, U.S. Patent No. 4 366 346 (28th December 1982).

[99] Eggermont, L. D. J., Berkhout, P. J., 1983/84. Digital audio circuits: computer simulations and listening tests. *Philips Tech. Rev.*, **41**: 99–103.

[100] Eggermont, L. D. J., Höfelt, M. H. H., Salters, R. H. W., 1977. A delta-modulation to PCM converter. *Philips Tech. Rev.*, **37**: 313–29.

[101] Janssen, D. J. G., van de Meeberg, L., 1980. PCM codec with on-chip digital filters. *Electron. Components & Appl.*, **2**: 242–50.

[102] van der Kam, J. J., 1982. A telephony codec using sigma-delta modulation and digital filtering. Proc. Conf. on Communications equipment and systems (IEE Conf. Publ. No. 209), Birmingham: pp. 49–53.

[103] Agrawal, B. P., Shenoi, K., 1983. Design methodology for $\Sigma\Delta$M. *IEEE Trans.*, **COM-31**: 360–9.

[104] McClellan, J. H., Parks, T. W., Rabiner, L. R., 1973. A computer program for designing optimum FIR linear phase digital filters. *IEEE Trans.*, **AU-21**: 506–26.

[105] Van Meerbergen, J. L., 1986. Architectures and characteristics of commercially available general-purpose signal processors; paper presented at 'Workshop wave digital filters' IMEC, Louvain, Belgium.

[106] Van Meerbergen, J. L. *et al.*, 1986. A 2-μm CMOS 8-MIPS digital signal processor with parallel processing capability, Int. Solid State Circ. Conf. (ISSCC), Digest of technical papers, Anaheim, Cal.

[107] van Wijk, F. J. *et al.*, 1986. A 2 μm CMOS 8-MIPS digital signal processor with parallel processing capability. *IEEE J.*, **SC-21**, 750–65.

[108] Introducing the PCB 5010/PCB 5011 programmable DSPs, Philips Electronic Components and Materials Division, Eindhoven 1986 (16 pages).

[109] Rubinfield, L. P., 1975. A proof of the modified Booth's algorithm for multiplication. *IEEE Trans.*, **C-24**: 1014–15.

[110] Hellwig, K., Rinner, K., Schmid, J., Vary, P., 1986. Digitaler Signalprozessor für den Sprach- und Audiofrequenzbereich. *PKI Tech. Mitt.*, No. 1: 57–64. Philips Kommunikations Industrie AG, Nuremberg, Germany.

Index